A Guide to North America's Bees

A Guide to North America's Bees

The BEES In Your Backyard

Joseph S. Wilson & Olivia Messinger Carril

Princeton University Press
Princeton and Oxford

JOSEPH

I'd like to dedicate this book to my family. To my parents, Bruce and Marnae, for the encouragement and suggestions they provided. Also, to my kids, Ari, Isaac, and Viviann, who enthusiastically point out a bee every time they see one and who "helped" on every collecting trip we took. But most of all I dedicate this book to my amazing wife, Lindsey, who had to spend most of her recent summers both entertaining kids out in the desert while I photographed bees, and patiently enduring all the late nights I spent writing.

OLIVIA

As I wrote this book, I struggled through my first years as a mother, and I felt a connection to the female bees about whom I was writing, whose adult lives revolve completely around their offspring. For this reason I dedicate this book to all the mothers. And I dedicate this book to my husband and daughters, who opened the door to a true understanding of and appreciation for queen bees.

Published by Princeton University Press, 41 William Street, Princeton, New Jersey 08540

In the United Kingdom: Princeton University Press, 6 Oxford Street, Woodstock, Oxfordshire OX20 1TW

nathist.princeton.edu

ISBN 978-0-691-16077-1

Library of Congress Cataloging-in-Publication Data

Wilson, Joseph S., 1980–
The bees in your backyard : a guide to North America's bees / Joseph S. Wilson and Olivia Messinger Carril.
pages cm
Includes index.
ISBN 978-0-691-16077-1 (pbk. : alk. paper) 1. Bees—North America. 2. Bees—North America—Identification. 3. Bee culture—North America. 4. Honeybee—North America. 5. Bumblebees—North America. I. Messinger Carril, Olivia, 1976– II. Title. III. Title: Guide to North America's bees.

QL567.1.U6W55 2016
595.79'9097—dc23
2015007075

British Library Cataloging-in-Publication Data is available

This book has been composed in Arial MT and Avant Garde Goth PS

Printed on acid-free paper. ∞

Designed by D & N Publishing, Baydon, Wiltshire, UK

Printed in China

10 9 8 7 6 5 4 3 2 1

CONTENTS

1

INTRODUCTION

Did you know that more than 4000 species of bees live in the United States and Canada? To put that in perspective, there are 4 times more species of bees in these two countries than all the bird species north of Mexico, 6 times more kinds of bees than butterflies, and about 10 times as many bee species as mammal species. Despite their diversity, few people know anything about bees, even the ones in their own backyards. For example, everyone knows that robins nest in trees, that bears hibernate, and that butterflies start out as caterpillars, but most people don't know where bees live, how they spend the winter, or what they eat. This book is designed to introduce you to the bees of the United States and Canada, including their lifestyles and habitat preferences, and what you can do to attract them to your neighborhood. Understanding bees is beneficial not only to the bees, but also to your gardens.

An *Andrena* species visiting a prickly poppy (*Argemone*).

Gotham's bee

A new species of sweat bee was recognized in New York City in 2010 (with the scientific name *Lasioglossum gothami*). While it has probably always lived in New York City, it was until recently completely overlooked by scientists. There are likely many similar cases around the world.

Over 20,000 species of bees have been identified around the world. New species are being found every year, even in places like New York City. Because new species are continually discovered, scientists estimate that up to 30,000 species might exist worldwide. Bees can be found on every continent (except Antarctica), on small islands, on treeless mountaintops, in jungles and deserts, and on top of high-rises in Chicago. They are most abundant in dry and hot environments, like Mediterranean Europe, and the southwestern United States.

Though the drab reddish-brown honey bee is the default image conjured by most when they hear the word "bee," these creatures are in fact diverse and stunning beauties, and the menagerie includes blue and green jewels like *Osmia* and *Agapostemon*, fire-engine red *Nomada*, jet-black fuzz-balls like *Anthophora*, and zebra-striped *Anthidium*. Some of the smallest bees in the world are found in North America. *Perdita,* found in the southwest United States, measure only 0.1 inch, smaller than George Washington's nose on a quarter. At the other extreme, North America is home to giant bumbling carpenter bees

The short and the long of it

The smallest bee in the world measures only 0.08 inch and is found in South America (*Trigona minima*). The largest bee in the world lives in Malaysia (*Megachile pluto*); it is 1.5 inches long.

The largest and smallest kinds of bees found in North America, a *Perdita* (left), and a *Xylocopa* (right).

(*Xylocopa*). At more than an inch long, they sound like miniature helicopters as they hover near flowers.

Bees are thought to increase seed set in 70% of all flowering plants, including many of the fruits and vegetables we enjoy. The special relationship that exists between bees and the flowers they visit is not only economically (and gastronomically) important; it is also unique from a biological perspective. Although there are other organisms that are capable of pollination (and are, in fact, good at it), bees are the only ones to actively gather pollen from the flowers they visit, creating an evolutionary dynamic seen nowhere else in the animal kingdom. Despite the particular talents and unquestionable importance of bees, scientists have reason to believe that some bee species may be experiencing widespread population declines. While the specifics are still being assessed, some things are certain: bees are all around us, they enhance the quality of our lives, and they benefit from our improved understanding of them and their needs.

Our hope is that this book will turn amateur naturalists, gardeners, entomologists, and curious souls on to the amazing lives of the bees that not only reside in untamed wild areas, but also flourish in our very neighborhoods. With understanding comes appreciation; in addition to describing the life stories associated with the many bee species of the United States and Canada, we provide examples of ways to encourage these wonderful pollinators on your own plot of land.

1.1 IS THIS A BEE?

Even though bees are common in most neighborhoods, frequently seen on hikes, and ubiquitous residents of city parks, it is hard to tell whether an insect buzzing nearby is a bee or something else. It's no wonder people get confused. Because bees sting, resembling one is a successful strategy for vulnerable insects, and many a bug has evolved the appearance of a buzzing bee; however, a keen eye and a little practice are all you need to see past the ruse.

Bees and wasps are the most similar in appearance, and they are the most easily confused. It is not uncommon to hear complaints about the "bee" that landed on somebody's hamburger at a recent family picnic. Stories of the pesky nest dangling from a branch in the backyard abound. Hikers complain about the horrible buzzing creatures that swarmed from a log they used as a backrest halfway up the trail. And every summer, someone is attacked by "ground bees" while mowing the lawn. In all cases, the annoying insect was probably not a bee but a wasp. Wasps (including hornets and yellow jackets) and bees are close relatives, sharing in common a grandmother 100 million "greats" ago. In some instances the two are so similar that even trained scientists have difficulty distinguishing them. The bee called *Neolarra* (see section 9.1), for example, was thought to be a wasp by the first researchers to see it. It didn't help that the bee was dead and stuck to a pin, because the most telling

A *Megachile*, resting on a cactus flower (*Echinocereus*).

From highest to lowest, a fly, a wasp, and a bee visiting a "watering hole." On the fly, note the short antennae just visible on the face. On the bee, note the yellow masses of pollen on the legs, which the wasp is lacking. Photo by B. Seth Topham.

Meat-eating bees and pollen-eating wasps

There are exceptions to nearly every rule about the differences between bees and wasps. In South America, for example, a group of bees (*Trigona*) feeds its young with dead animal flesh, and in North America a group of wasps (*Pseudomasaris*) feeds its young with pollen.

This image shows some common flies (left), bees (middle), and wasps (right). You can see that the three groups commonly look a lot alike and that it takes an experienced eye to see the differences. Notice that all the flies have triangular heads (when viewed from above) with short little antennae and just one wing on each side of the body. Wasps and bees look even more similar; look for rough integument (skin) on wasps, with many tiny pits, as well as antennae that commonly are very close together on the face, and spindly legs.

differences between bees and wasps are their mannerisms and day-to-day behaviors.

Most important among these behavioral differences is that bees are pollen eaters. Wasps, in contrast, are meat eaters. While both visit flowers for nectar (the "energy drink" of the insect world), bees also visit flowers in order to collect pollen for their young. On the contrary, wasps pursue other insects and drag them back to the nest for their offspring to devour. This one dietary difference has resulted in very different bearings. To aid in the gathering of pollen, bees are usually hairy (pollen sticks to hair), and many species look like cotton candy with wings. Rooting around in flowers is messy business, and a few minutes rummaging among floral parts leaves a bee coated in hundreds of tiny grains of pollen. Using her many legs, the bee grooms herself, wiping all the pollen to the back of her body, where she stuffs it into the spaces between special stiff bristles on her legs or belly. These tufts or masses of special hairs are called scopa. Quite the opposite of the furry bee, wasps look like Olympic swimmers, devoid of all hair, skinny-waisted, and with long spindly legs.

A *Melissodes* bee foraging on a sunflower (*Helianthus*) with a large pollen load stuffed into the pollen-collecting hairs (scopa) on the back legs.

It's so *fluffy*!

Bees generally have lots of hair; wasps don't. In fact, one of the few consistent differences between the two is that, somewhere on their little bodies, bees have branched hairs similar to tiny feathers.

A wasp. Note the silvery hairs on the face, and the long spiny legs. The spines help hold onto insect prey the same way cat claws do.

There are exceptions to every rule, of course. Some bees have scant hair on their bodies and are wasp-thin. In these cases, look for silvery or golden hairs on the face; wasps tend to have glistening mugs, while bee hairs don't shimmer from any angle. Behavior, as mentioned above, can be telling, too. Bees spend more time on flowers than wasps do; wasps in contrast are more likely to raid your backyard barbeque in search of animal proteins accidentally left on a plate.

Since bees and wasps are difficult to distinguish, many stung victims often blame the hapless bee for crimes not committed. The culprit in these cases is likely a paper wasp, a hornet, or a yellow jacket. All live socially in hives. Ever the opportunists, these wasps take advantage of the many resources found in urban environments, often building their homes along fences, under eaves and decks, attached to windowsills, or in various holes or cavities. All three will collect fibers from dead wood and plants and then use their saliva to make a papier-mâché house of sorts. These nests often bear a strong resemblance to the honey bee hives depicted in Winnie the Pooh books, and it is thus not surprising that many people think these wasps are bees. These kinds of wasps also enjoy taking a bite of your grilled chicken back to the nest to feed their offspring, or stopping

A paper wasp ripping apart a katydid to transport back to the nest (wasps are voracious carnivores). Note the long narrow wings, folded in half and draped like thin ribbons down each side of the back.

The hive of a paper wasp at the height of activity. Most bees do not live in hives, and their homes can more accurately be called nests. Nests are typically in the ground, and each nest usually houses one individual, instead of a whole colony.

A yellow jacket's nest is built from a paper-like substance and is frequently mistaken for a beehive. While many people think these wasps' nests are beehives, bees do not make exposed paper-like hives. PHOTO BY LINDSEY E. WILSON.

Bee vs. wasp (physical differences)

Bee	Wasp
usually thick-bodied	skinny body with narrow waist
no silver hair on face	often with silver hair on face
often very hairy	generally hairless
pollen-collecting hair on legs or belly of females	no pollen-collecting hairs
stout legs with relatively few spines	long thin legs with spines

on the lip of your glass of root beer for a sugary sip. The gangly, thin-waisted, and hairless body gives them away as wasps and not bees, however. In addition, the wings of these wasps in particular are folded in a distinctive way. Rather than lying flat across their back (thorax) so they overlap over the abdomen, their wings run as parallel dark strips on either side of the thorax.

Though not close relatives of bees the way that wasps are, many flies mimic the bee look. For a fly, the advantages to playing copycat are huge. Bees have spent millennia evolving stings and every creature on land has learned that they are not to be messed with. For a fly, looking so painful can save them from becoming the lunch option of hungry birds, reptiles, and other potential predators. For a predator, of course, being able to tell the difference between a bee and a fly increases the number of options at the insect buffet. Over time, the discerning eye of the predator has therefore weeded out the not-so-good fly look-alikes, leaving behind flies that at first glance seem identical to bees—down to "pretend" pollen-collecting hairs on the legs!

Flies have several important characteristics that can help separate them from bees. First, flies have only two

Bee vs. fly (physical differences)

Bee	Fly
long slender antennae	short antennae
four wings	two wings
distinctly separated thorax and abdomen	"thick waist" where thorax connects to abdomen
pollen-collecting hair on legs or belly	no pollen-collecting hairs
eyes on sides of head	eyes large, often forward facing, sometimes touching on top of head

Five insects. Only one is a bee, though the other four are commonly mistaken for bees. From top to bottom: a mud dauber wasp, a paper wasp, a yellow jacket, a hover fly, and an actual bee (*Svastra*). Note the pollen-collecting hairs on the legs of the *Svastra*, the overall hairier body, and the stocky legs.

A fly that mimics a bumble bee. Though it is hairy like a bumble bee, notice that the eyes are very close together, almost touching, and that the antennae are hardly noticeable. It also has only two wings (one on each side).

A bumble bee (*Bombus*). Note the long antennae, the two wings on each side of the body (one bigger, and one smaller one behind), and the distinct and widely spaced eyes.

wings, while bees have four (a fore and a hind wing on each side). Second, flies usually have two short, blunt antennae that emerge from nearly the same place on their faces; bee antennae are longer (often *much* longer) and more widely spaced. Third, fly eyes are usually bigger and closer together than typical bee eyes, often almost touching at the top. As they do not carry pollen, flies have no dense tufts of stiff hairs on their bellies or legs (though a few species mimic this look with bright spots on their abdomens near the back legs). And finally, if you've actually captured a specimen for your collection, flies are much squishier, and piercing them with a pin is like piercing Jell-O. Bee bodies are much more resistant to the insect pin.

Even when a bee is properly identified as such, there are many common misconceptions about how it lives. Because of the importance and abundance of honey bees, we are most familiar with their life cycle. It is often assumed that all bees follow a lifestyle similar to that of the honey bee, when in fact honey bees are the exception rather than the rule for the habits of bees as a whole. Though extraordinary creatures, they are poor representatives of their fellow bee kin. First, honey bees live in hives, but 70% of all bees live in the ground. Second, honey bees are social and work together to build their hive nest; in contrast most other kinds of bees work alone. Third, honey bee mothers meet their offspring; the majority of bee mothers never encounter their young. And finally, honey bees make and store honey to eat in the winter, which few other bees do. We delve into each of these topics in more detail in the following sections.

What's the point?

You may have heard that bees can sting only once before they die. In reality, almost all bees and wasps can sting multiple times. It is only honey bees that die after their first sting. Not to worry, though: bees sting only to defend themselves.

1.2 BEE NAMES

If we are to talk about bees, we need to be able to distinguish between those that *are* bees, and those that aren't. This is trickier than you might think, and scientists have classified and reclassified bees and their relatives many times over the last 300 years.

Traditionally, scientists have used a system of classification known as taxonomy to group organisms together according to the way they look, and they use the same conventions across all living organisms. At the most inclusive and "highest" level of classification is domain, followed by kingdom, phylum, class, order, family, genus, and species. For example, all insects fall into the same class (Insecta) because they have a hard exoskeleton (rather than a soft skin like mammals), three distinct body parts (a head, a thorax, and an abdomen), six legs, compound eyes, and antennae. Within that large class, all butterflies are grouped together in the same order Lepidoptera, all flies are in the order Diptera, all bees are in the order Hymenoptera, and so on, according to certain characteristics that are shared in common by all members of the group. Each grouping in the taxonomic hierarchy is more exclusive than the one before it, until all the organisms in a group are considered the same species. For most purposes, these levels of organization suffice; however, bees require some additional divisions. Tribes and subfamilies are both smaller than the level of family, but larger than the level of genus. We use tribe and subfamily divisions frequently in this book. Subspecies are also discussed in this book; they are distinctive, often regional, variations of a species.

It's all in the sting

Interestingly, bees are really just a special kind of wasp. All wasps are divided into two main groups: those in which the ovipositor has evolved to function as a sting, and those in which it has not. Among those with stings are ants and some kinds of wasps; this group is called the Aculeata. Within the aculeates are two categories, one containing wasps and ants, and one containing the bees—this bee group is known as the Anthophila, which means "flower lover."

An example of how one bee (*Andrena sphaeralceae*) fits into the taxonomic hierarchy to which all living organisms belong. *Andrena sphaeralceae* belongs in the insect class (Insecta), along with wasps, ants, butterflies, and beetles. It is in the order Hymenoptera, which comprises bees, wasps, and ants. It belongs with the subset of wasps known as bees, and within these, it belongs to the bee family Andrenidae, which it shares with other bees like *Perdita (shown)*. It is in the genus *Andrena*, shared with *Andrena cerasifolii* (among others). But it is its own species, a designation it shares with no other types of bees, or any other organism for that matter.

Together, the combined genus and species name of an organism is entirely unique, and no other organism that has ever lived shares that combination of names. These two names are usually underlined or italicized (as we have done in this book). The genus name is always capitalized, and the species name is always in lower case. Quite often the scientific names of organisms are strange-sounding, as they are based on Latin or Greek roots that indicate something distinctive about the group. They are frequently unfamiliar words, and learning to pronounce them may seem overwhelming. Take comfort in knowing that you likely recognize, and are probably undaunted by, many scientific names. The common names of many flowers are the same as their scientific names: *Chrysanthemum*, *Hibiscus*, *Petunia*, and *Echinacea* are all scientific and common names. You probably also aren't intimidated by names like *Rhinoceros* or *Hippopotamus* even though they are scientific names. Dinosaur names, too, are scientific, but every six-year-old can rattle off his five favorites without a stutter. With practice, bee names are just as easy.

A note on scientific pronunciation. Scientific names are based on Latin or Greek and often use unusual combinations of letters. They also involve an abundance of syllables, and it seems that the emphasis is not always where a native English speaker might be inclined to place it. Nonscientists are often daunted by the strange-looking words and avoid saying them at all costs. Here is a secret: scientists are often daunted by the same words. What's more, even among scientists who study these bees (or plants, or butterflies), there are many different pronunciations, all approximating the same word. When Latin was a living language, there may have been an appropriate pronunciation, but today there is no one right way. *Hylaeus* may be pronounced "hi-*lee*-us", or "hi-*lay*-us" depending on who is saying the word; neither is more correct. With that in mind, take your best guess. If you are saying the word out loud, it is likely in conversation and you and your audience will together come to a common understanding with no harm done. We have provided one possible, popular pronunciation below each bee's summary information box in the following chapters.

The taxonomic hierarchy of the bee *Anthidium rodecki*

Kingdom: Animalia
- Phylum: Arthropoda
  - Class: Insecta
    - Order: Hymenoptera
      - Family: Megachilidae
        - Subfamily: Megachilinae
          - Tribe: Anthidiini
            - Genus: *Anthidium*
              - Species: *Anthidium rodecki*

A family affair

Very recently, scientists have been able to add a level of complexity to the classification of organisms. Researchers now use genetic or molecular techniques to complement taxonomic techniques. Specifically, they look at organisms' DNA in order to classify them according to their degree of relatedness. This capability is important because animals frequently seem to belong in the same category based on a similar appearance when, in fact, these distant relatives have converged on a similar appearance because it became advantageous for one reason or another. As an example, many flies and moths look like bumble bees (likely because the sting of a bumble bee packs a punch that may make predators think twice). Though they look superficially like bumble bees, genetic and taxonomic techniques confirm that they are as related as dogs are to cats. Genetic studies can also help with the problem of having only dead specimens to study; wasps and bees can look extraordinarily similar to each other when their behavior can't be observed. Molecular studies, which can be conducted using dead specimens, can confirm that they are truly distant cousins.

We recommend reading the scientific name first, then reading the pronunciation slowly, one syllable at a time, and then piecing them together. For clarification on our pronunciation guide, please see the appendix.

Why don't we just use common names, the way others do for most birds and mammals? First, the common names that exist for bees are not unique to different groups. For example, the name "long-horned bee" is a common name frequently used for more than 200 bee species in North America. These individual species have different life histories and geographic ranges and are not all closely related to each other. It is impossible to know which particular long-horned bee is meant without its scientific name. Second, even without the overlap in common names, there are more species of bees than there are birds (or mammals), and common names don't exist for most of them; also troublesome, because there is no standard set of rules, a few species have several common names (e.g., *Colletes* are known as cellophane bees, plasterer bees, or polyester bees). And finally, bee names based in Latin and Greek do not always translate well to common names.

Taxonomic shorthand

Throughout this book you may see the genus name abbreviated to the first initial. Once a bee genus has been introduced or mentioned, it is common for it to be shortened in this way rather than written out repeatedly.

Because all bees belong to the same order, bee classification takes place at the family level and below. Seven bee families have been identified in the world, six of which occur in the United States and Canada. Each chapter in this book contains all of the bees in one family (with the exception of the parasitic bees; they occur within three of the other families, but we have pulled them together into one chapter). Within each chapter, we have divided the bees further by subfamily, tribe, or genus depending on the size of the group; small genera (plural of the word genus) are put together in one section with other rare genera of the same subfamily or tribe. Genera with many species and those that are commonly seen are given their own sections.

You can tell whether someone is talking about a family, subfamily, tribe, or genus in one of several ways. The family name always ends in -idae, often abbreviated in conversation to -id, as in, "This plant has a lot of megachilids visiting it." A subfamily name always ends in -inae, and a tribe with -ini. Both tribe and subfamily names are abbreviated in conversation to "-ines," as in "You have a lot of beautiful anthidiines in your bee collection." This can be a minor source of confusion, though the rank can usually be understood by the context. Unlike the higher levels of classification, the genus name has no definitive suffix. In print, it will begin with a capital letter and be in italics.

1.3 THE BEE LIFECYCLE

One might assume that the life of a bee is spent zipping around, docking at flower after flower for a sip of nectar or a dollop of pollen. After all, that's what we witness in our gardens all summer long. In fact, the majority of a bee's lifecycle is not even seen by us, occurring out of our sight in a carefully built nest. Though most bees live a year (some even longer), they are "adults" hurrying from flower to flower for only three to six weeks of that year. The rest of the time is passed in relative darkness, inside the nest, undergoing a variety of developmental changes. While human offspring pass from baby to toddler to teenager to adult, a bee moves from egg to larva to pupa to adult. It may surprise you to learn that these are the very same phases through which a butterfly passes, and, in fact, many bees even spin cocoons just as butterfly caterpillars do. Because the end product looks so very different from the grub-form that characterized its youth, the growth of a bee (and butterfly) is known as complete metamorphosis (the scientific way of saying "really big change").

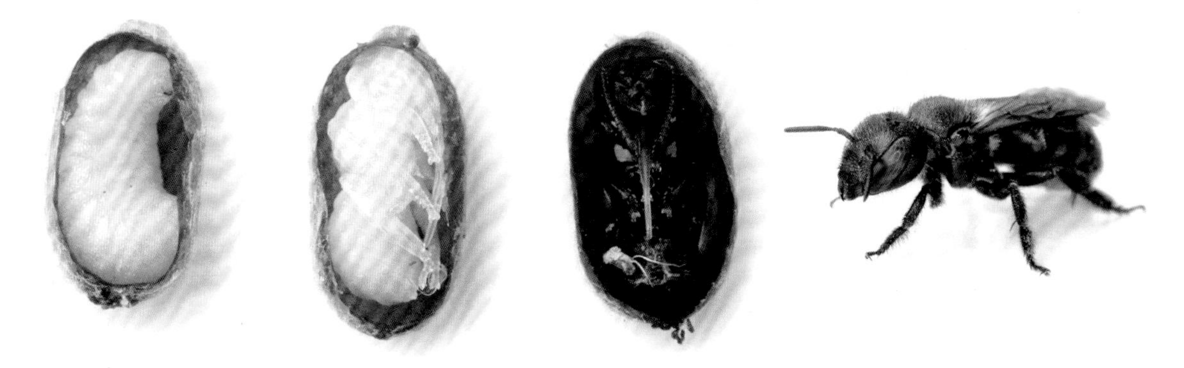

Four images showing stages through which a bee passes in its life (here, *Osmia*). The egg stage is not shown. From left to right: larva, prepupa, pupa, and adult. PHOTO COURTESY OF USDA, W. P. KEMP, AND J.S. WILSON.

The story of a bee's life and the genesis of its offspring could be told from any starting point. Because most of us are already familiar with them, we'll start with the adults.

ADULT. A bee emerges from its nest for the first time anytime between February and September, depending on species, and length of the flowering season. With few exceptions, male bees emerge before the females, often up to a week earlier. They wait expectantly near the nest sites from which they emerged, or on flowers close by, for females to make their appearance. When females finally do emerge, males eagerly pounce, hoping to be the first and perhaps only bee to mate with them. Newly emerged females don't have mature ovaries, and many species spend the first few days drinking nectar and eating pollen; the pollen is thought to help speed along the maturation of their reproductive parts. Even before their ovaries have matured, however, female bees can mate and store sperm in their spermatheca, an organ in their abdomens for just this purpose. This stored sperm is kept alive and viable for the remainder of the female's life, and she can use it whenever she is ready to fertilize an egg. Mating just once is enough to provide a lifetime supply of sperm for a female bee. Interestingly, males often appear able to recognize a female that has been mated and will avoid her; likely it is a difference in the odors she emits that gives her away.

A mating pair of *Perdita*. The female is easily identifiable by the blobs of pollen on her hind legs (she is the bee to the left). The male is on the right, and the tip of his abdomen is curled underneath hers.

Bee mothers (as well as some wasp and ant mothers) have the amazing ability to determine whether each egg they lay will develop into a male or a female bee. When an egg is laid, the female decides whether sperm will be released from her spermatheca as the egg passes by it. Fertilized eggs (an egg plus sperm) develop into female bees, and unfertilized eggs (no sperm) develop into males. This system is called haplodiploidy, referring to the fact that male bees are haploid, containing only genetic material from their mothers, and females are diploid, containing genetic material from both their mothers and their fathers.

Eggs are placed in individual nest cells (akin to nurseries) inside nests, and there are often several nest cells inside a nest. Each egg is provided with a mass of pollen mixed with some nectar, which the larva will ingest upon hatching. Most mother bees never

A pollen loaf in a *Nomia* bee nest cell, carefully prepared and shaped by the mother, using both pollen and nectar. PHOTO BY L. SAUL-GERSHENZ © 2014

meet their young, dying after they have done all they need to do in order to provide for the offspring that will develop from the eggs they lay.

EGG. Eggs have a soft outer shell (called a chorion). They are usually white, and oval-shaped. Bee species in North America lay eggs that range in size from extremely small (the size of a pinhead) to almost two-thirds of an inch long (a little less than the diameter of a dime). The egg stage is the shortest-lasting stage of a bee's life, with most eggs "hatching" just a few days after having been laid. Hatching isn't the same as in birds, though. The eggshell is so soft that it usually dissolves or is shed rather than cracking open.

An *Osmia* bee egg. The outer shell is relatively soft and is referred to as the chorion. It splits or dissolves, revealing the young bee larva (called a first instar) inside. PHOTO BY JAMES H. CANE.

LARVA. Once the egg hatches, a bee enters its larval stage. A larva is soft and white with a mouth, tiny nubs where the antennae will someday be, and a caterpillar-like body. These "grubs" do nothing but eat the pollen and nectar their mothers have left them. As they grow, they periodically shed their outer skin, molting into larger and larger grubs. These molts happen five times, and each represents a different larval stage (called an instar). Interestingly, even though larval bees eat constantly and grow continually, they produce very little poop (called frass) until the last larval stage.

The larva of a *Nomia* bee. There are no distinct legs or eyes (who needs them in the dark?), only a giant mouth with which to devour the pollen, and little leg-nubbins for moving around the pollen loaf. PHOTO BY L. SAUL-GERSHENZ

The last stage of larval development for a bee is known as the prepupa. Bees may remain in the prepupal stage for long periods of time; for example, for an egg laid in the spring, the larval bee might take all summer to develop to its prepupal stage, and then may stay in this stage for all of fall and winter. Some bees may even wait several years in this state before finishing their development and emerging as adults (see bet-hedging, opposite).

Just before transitioning to the next stage of their lives, bees defecate, getting rid of all the waste matter from the huge loaf of pollen they've ingested. Then, just like butterflies and moths, some bees spin a silky cocoon around their bodies—a mouth-made sweater to keep them protected as they transform to pupae.

PUPA. The bee pupa looks very much like the adult, with legs, antennae, eyes, and three body sections (but just wing buds—no actual wings). Though the form is similar to an adult, the color is a light gray to milky white, lacking any of the bright colors and fuzzy fur associated with adult bees. The pupal stage in a bee usually doesn't last long. In a short period the wings grow and stretch out, fur sprouts from the body, and the bee becomes a fully formed adult. It then chews and digs its way out of its nest and commences life as an adult.

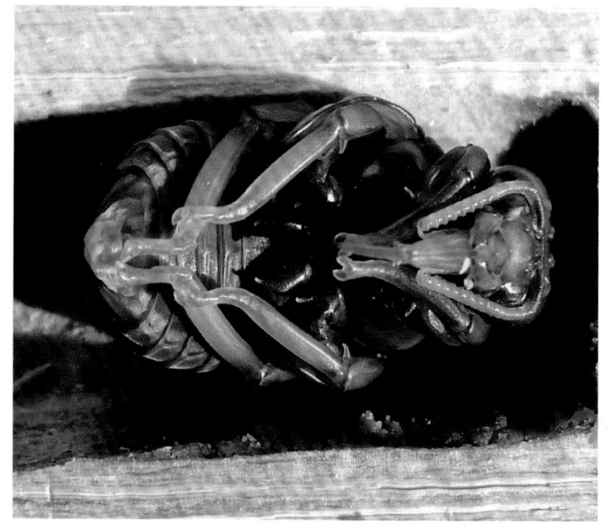

A pupa of a *Xylocopa*. This is the last stage before adulthood. As adults, *Xylocopa* are pure black bees. Note that this one is still getting its dark adult coloration. The legs are not completely differentiated. On the upper side of the image, the beginning of a wing is just noticeable wrapping around the body. PHOTO BY JAMES H. CANE.

Bet-hedging in bees

When only some of the bees that hatched and developed in a particular year actually finish development (pupate) and emerge the following year, it is called "bet-hedging." The rest wait in their nests for an additional year (or two, or three) before emerging. It is thought that this behavior is advantageous to bees that live in the desert where rainfall and flower bloom can be unpredictable. If only half the bees emerge in any given year, then only half potentially perish if there is little floral bloom.

Of course, every bee species has evolved a unique spin on the general bee lifecycle, and their individual differences are part of what makes studying bees so interesting, and rearing them for crop pollination so challenging.

1.4 WHERE DO BEES LIVE?

Though a mother solitary bee seldom meets her offspring, she spends a great deal of time preparing them for their year-long life, choosing the perfect nest location and constructing the nest so that it protects her precious young from weather elements and harmful organisms. Finally, she makes sure they have enough food to fuel their development into the next generation of adults.

If you believe everything you see on TV, you might think that bees live in hives that dangle from tree branches, tempting hungry bears. In fact, 70% of bees nest in holes in the ground, dug gopher-style by a single female bee.

Those bees that don't nest in the ground choose pre-existing holes (including hollowed-out twigs, pock-marks in rocks, abandoned beetle burrows, tiny holes in bricks, and even empty snail shells). A few kinds build nests on the outside of structures, using rocks and mud they collect and cobble together. Even hive bees prefer not to nest in something so exposed as a blob hanging from a tree branch. Honey bees and bumble bees, the two hive bees in North America, typically choose crevices in rocks, hollow tree stumps, or mouse and rabbit burrows for their colonies.

A bee nest contains anywhere from one to several dozen nest cells. Each nest cell serves as a chamber, analogous to a nursery, for a single developing bee.

For bees that nest in the ground, the nest is carefully excavated and prepared before the female lays any eggs in it. Generally the nest site is a sunny south- or east-facing patch of exposed earth. At the spot she deems to be just right, a female bee wets the ground, either with water she has gathered, or with her own saliva. Once it is softened, she uses her legs and mandibles (jaw) to worry the surface. As it gives, she shovels the dirt away, burrowing deeper and deeper into the earth and leaving a mound of dirt behind her. Inside, the nest is either one long tunnel with nest cells at the very bottom, or a tunnel with several side branches and offshoots, each of which may contain one or a few nest cells. The female uses the tip of her abdomen (specifically a structure called the pygidial plate) to tamp the walls of her excavated nest until they are smooth, maybe applying a little mud, extra soil, or some other substance to get it just so. Some bees paint the walls with a waxy or cellophane-

An *Anthophora* digging the most common kind of bee nest: a tunnel in the ground.

A *Diadasia* beginning the nest-building process in hard-packed earth. Note the slightly moistened ground around her head, which is pressed firmly to the ground. She will spin furiously in circles, chewing at the ground with her tough mandibles (jaw) until she has created a tunnel.

Let me out!

When a bee has completed its preadult life stages, it bullies its way out of the nest, charging through anything in its way. Ground-nesting bees will often dig straight up, forgoing the tunnel prepared by their mother for a more direct route to the surface. Twig nesters go back the way their mothers came, pushing through nest cells in front of them, and often destroying any of their siblings still in the process of growing. It is probably a good thing that the mothers lay the slower-developing females at the back of the nest so their brothers don't destroy them as they bore their way out!

like material produced with a special gland on the female's abdomen (called the Dufour's gland) to keep out moisture and any bacteria in the soil. Other bees line their nests with pieces of leaves, the hairs that often coat plant leaves, or even flower petals. When the nest meets her satisfaction, the bee gathers pollen and nectar from flowers and carries them back with her scopa (pollen-collecting hairs) and in her crop (nectar stomach), respectively. The bee then mixes or layers the pollen and nectar together in the nest cell and, finally, lays an egg somewhere near this "loaf."

Holes in a cactus flower probably made by an *Ashmeadiella* who took home a little "wallpaper" to line her nest cells. PHOTO BY LAURELIN EVANHOE AND MATTHEW HAUG.

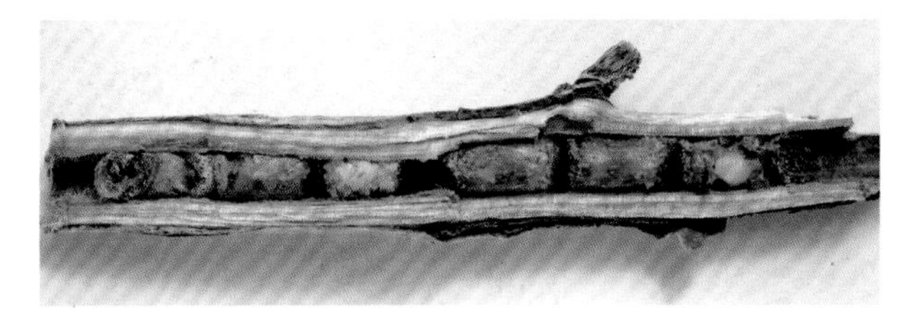

ABOVE: The twig nest of a *Hoplitis sambuci*. Note that the nest cells, each of which contains one bee, are all in a row. PHOTO BY JAMES H. CANE.

RIGHT: The external nest of a *Dianthidium*, cobbled together from mud and rocks, each of which is carted in by the female bee. Inside this lump can be found one to a few bee nest cells, each containing one growing bee.

Bees that nest in preexisting cavities similarly prepare for their offspring. They may modify the cavity as needed, either packing in mud or rocks to shrink the diameter of the tunnel, or chewing at the insides to widen it to the right volume. Many cavity-nesting bees construct nest cells all in a row. In such cases, each egg, and its pollen loaf, is sealed off from its siblings by a wall or partition built by the mother between the cells. The walls can be made of packed-together soil, pebbles, resin, bits of plant foliage, or sawdust left from digging into the wood around the nest. Finally, the tunnel, either with a series of cells, or with just one cell, is back-filled with debris or soil and the entrance capped, often with mud or plant materials.

1.5 BEE SOCIALITY

Even as well known as they are, honey bees and bumble bees make up less than 1% of all bee species, and their life history is hardly illustrative of bees at large. Honey bees and bumble bees are highly social, meaning they live in hives with a queen and workers that divide up the various duties necessary to keep the hive thriving. In contrast, most North American bees are solitary, with each female bee building her own nest, gathering the food resources to feed her offspring, and laying all the eggs. There are also bees that, though solitary, share nest constructing and guarding duties to some extent, or live close together but maintain the solitary lifestyle. In fact there are many cooperative strategies, most of which don't involve a queen bee, and it appears that different types of sociality have arisen more than once in different bee lineages.

SOLITARY. The vast majority of the bees native to North America are solitary. Rather than dividing household chores, as honey bees do, each individual female builds her own nest. Like all individualists, she selects her nest site based not on what other bees are doing, but on habitat characteristics that are important to her (e.g., soil type, cavity size, surrounding vegetation). As a solitary bee, she also gathers all the resources she needs to provide for her offspring (like a worker honey bee), and lays all the eggs (like a queen honey bee)—she is the full package. A mother solitary bee never meets her young, dying soon after she has built her nest, loaded it with pollen, nectar, and eggs, and sealed it shut.

A solitary *Halictus* in her nest entrance.

AGGREGATIONS. Some solitary bees nest in aggregations; though each nest is built and maintained by a single bee, many nests can be found close together, often with several hundred, or even several thousand, nests in the same area. The largest aggregations are found among bees that nest in the ground (woody material in one area is seldom sufficient for thousands of wood-nesting bees to gather together). Sometimes these aggregations may persist in one area for several years or even several decades. Why bees nest in aggregations is not entirely known. Nesting aggregations might form because a particular area has desirable characteristics that are hard to find elsewhere. There is also evidence that nesting in aggregations may lessen the chances of any one bee being picked off by a predator, or attacked by a parasite. With so many bees flying in the same area, a neighbor is always nearby to chase away ne'er-do-wells.

A hard-packed dirt road with hundreds of *Diadasia* nests. This particular aggregation has been seen in the same spot for at least ten years.

COMMUNAL. Other species of solitary bees nest communally, with several bees sharing the same nest entrance. Think of it as the hymenopteran equivalent of city dwellers in a high-rise apartment complex. Though all inhabitants share the same foyer and revolving doors, they live separate lives once inside. Forty or more bees may share the entrance to a nest (most often in the ground), but within the nest each bee has her own nest cell or cells that she prepares for her own eggs.

Some bees, like these *Andrena*, nest communally. They share a nest entrance, but each bee has her own nest cells inside, not unlike humans living in an apartment building. Photo by Africa Gómez (http://abugblog.blogspot.com).

SEMISOCIAL. Semisocial bees use the same nest and also cooperatively provide for the offspring in it. Working together to prepare one nest cell at a time, they fill it with pollen and nectar, lay an egg, and close the nest cell off before moving to the next. This work is divided up, so that some bees gather the pollen and build the nest cells, and some bees just lay eggs. The size of the bee appears to be a major factor in deciding who stays home to lay eggs and who leaves the nest, with smaller individuals doing the foraging. Generally the bees are all the same age, possibly even sisters. Many species of sweat bees (called *Lasioglossum*, see section 6.3) in the United States and Canada cooperate in this way. As with all social types described above, semisocial mother bees and their aunts never meet the offspring for which they provide, dying after the nest and its cells are completed. The designation

Quasisocial: maybe, or maybe not

A rare type of sociality, known as quasisociality, seen occasionally among bees, is thought to be more a matter of circumstance than a stable lifestyle. *Quasi* means "sort of, but not really," and these bees share characteristics with both social species and solitary species. These fence-sitters tend to be sisters and work together to build a nest and gather pollen and nectar supplies for the cells within. Rather than one, several of the bees in the nest are designated egg layers; they take turns placing eggs in completed cells. All the sisters die before their offspring emerge, however, and like solitary bees, they never meet their young. Whether quasisocial behavior exists among bees in the United States or Canada has not been verified, but it has been repeatedly documented among orchid bees of Central and South America.

semisocial means "halfway social," as these bees live much like social bees, but their colonies are not as complex and they last for just one generation.

EUSOCIAL. Eusocial bees are the pinnacle of sociality (*eu-* means "true" or "well done" in Greek, thus these bees are "perfectly social"). In addition to sharing a nest and dividing the duties of nest maintenance and offspring care, eusocial bees actually meet their young, and the mother bee works alongside her daughters. A eusocial colony of bees, therefore, has three unique qualities: all individuals share a nest, they split nest-making and reproductive responsibilities, and they comprise a mother and all her daughters (meaning that the mother meets her offspring).

There are several examples in the United States and Canada of eusocial bees, including honey bees, bumble bees, and some kinds of sweat bees. Honey bees are the only *highly* eusocial bees, though; bumble bees and sweat bees are thought of as *primitively* eusocial. There are several differences, in terms of both appearance and behavior. A queen honey bee (the egg layer) looks different from her workers (the housekeepers)—she is bigger, with a swollen abdomen and no pollen-collecting appendages. She cannot survive without her workers, and new colonies are started by several hundred bees working together. Behaviorally, honey bees are capable of advanced communication, and their hives are often gargantuan, with several thousand individuals living and working together.

In contrast, bumble bee and sweat bee queens differ from workers only in size. They are fully capable of existing on their own, and in fact they build the founding nest by themselves. In it, they lay a batch of eggs which they carefully nurture to adulthood. Once the daughters are mature, the bumble bee or sweat bee mother stays home, giving up the duties of nest care and taking on the sole responsibility of laying eggs. Her daughters, the worker bees, take over the duties of building additional nest cells and providing their younger sisters with pollen and nectar or honey. Neither bumble bees nor sweat bees are capable of advanced communication, and colonies tend to be small, with only 50 to a couple hundred individuals. Sweat bee nests may be even smaller than this, with fewer than a dozen members.

Honey bees live socially, with many females working together to maintain the nest and the growing larvae within. They are one of the few bees that live in such close proximity to each other. Photo by Mike Wells of Harvest Lane Honey.

It's all in the hips

It is not only the enormous size of their colonies or their complex social structure that make honey bees unique. Honey bees are also capable of advanced communication, achieved through an elaborate game of charades. When a worker bee finds an important resource, she performs a particular type of dance that indicates to her sisters both how far away and in what direction the treasure can be found. For example, when a good source of pollen is more than the length of football field away, she'll do a "waggle dance," the well-tuned steps of which resemble a figure eight. First, the bee dances in a straight line, waggling her abdomen back and forth as she walks. The direction of the resource from the nest is indicated by the angle of the walk relative to the top of the honeycomb—straight toward the top indicates that a bee should fly toward the sun. Any deviation from straight up indicates the angle away from the sun. She'll circle around to her starting position after her waggle and repeat the run again. How long she dances tells the other bees how far away the resource is. The workers whose job it is to collect nectar and pollen intently watch this dance, then leave the hive following the directions laid out by the dancing bee.

1.6 WHAT DO BEES EAT?

Almost entirely, the nutritional needs of a bee are met by flowers (for a detailed explanation of pollination, see section 2.1). Adult bees eat a little pollen and a lot of nectar, and they feed their larvae a lot of pollen, often mixed with a little nectar. Pollen provides bees with an excellent source of protein, as well as chemical compounds important in the process of egg development in females. When females collect pollen from flowers, they store it in special tufts of stiff hairs on their abdomen or on their legs—these tufts are called scopa. They use their four front legs to scrape pollen from the flowers, scooping it to their two back legs and packing it into these hairs. Any pollen that inadvertently sticks to another part of the body is generally scraped off during elaborate grooming sessions and shoved to the scopal hairs. Many bees carry the pollen dry, but others will mix it with nectar, making a pollen "dough" to take back to the nest. A few bees mix the pollen with floral oils before carrying it home. In addition to the bees that carry pollen on their scopa, one group of bees (known as *Hylaeus,* section 4.2) carries its pollen in a crop—a "stomach" of sorts—and then regurgitates it back at the nest.

Bees have many adaptations to aid them in harvesting pollen from flowers. The hairs on the body of a bee are electrostatically charged, so that pollen naturally sticks to it, a charge that builds up as the bee flies. Other adaptations include special hairs on other body parts. For example, bees that collect pollen from the flowers of cat's eye (*Cryptantha*) often have hooked or wavy hairs on their faces. When they stick their heads into the very narrow floral tubes on these flowers, the pollen sticks to the hooked hairs on their faces. Once they back out of the flower they can move the pollen from their faces to their scopa. Similarly, some bees that collect from these tiny flowers have pollen-collecting hairs on the underside of their faces. Bees that collect oil instead of nectar to mix with pollen have thick pads of hair for sopping it up.

Hooked and wavy hairs on the face of an *Anthidium* bee. The bee uses these hairs to pull pollen from flowers too narrow or short to fit a whole body into. PHOTO COURTESY OF USDA-ARS BEE BIOLOGY AND SYSTEMATICS LABORATORY.

Bees have also evolved behavioral adaptations to aid in gathering pollen. As an example, many plants in the potato family (Solanaceae), including tomatoes, potatoes, chili peppers, tomatillos, and eggplants, keep their pollen inside anthers (pollen-holding structures) that work like salt shakers. The pollen stays securely inside until the anthers are tipped upside down and shaken. Only some bees know how to shake the anthers; they will hang below the flower and buzz their flight muscles to vibrate the pollen loose. This process, known as buzz pollination,

An *Augochloropsis* furiously buzz pollinating a meadow beauty (*Rhexia*). PHOTO BY BOB PETERSON.

Busy as a bee

Bees are known as hard workers, and for good reason. In general, just one nest cell may be filled with three or four times a bees' body weight in pollen, which is many trips to flowers. The bee *Anthemurgus passiflorae* takes 35 trips to gather enough pollen for one egg. Another bee, *Andrena vaga*, takes nine trips, whereas *Colletes cunicularis* needs just seven trips. Each "trip" involves visiting tens to hundreds of flowers, so that one nest cell contains pollen from a huge floral bouquet. Interestingly, the tiniest bees are able to carry the heaviest pollen loads, and some have been found carrying more than their body weight in pollen and nectar on their delicate hind legs!

A *Diadasia* with a full pollen load on her hind legs. The stiff bristles making up the scopa (pollen-collecting hairs) on these bees are widely spaced in order to accommodate the large pollen grains of the cactus flowers on which they specialize.

Don't bees eat honey?

People often assume that most bees eat honey; in fact, in the United States and Canada, only honey bees make and eat honey. Honey bees gather large amounts of nectar, bring it back to the hive, regurgitate it into special honeycombs, and evaporate out some of the water to thicken it, creating honey. Honey bees then eat this stored food during winter or during other times of the year when nectar is scarce. Bumble bees store nectar, but it is not true honey and is used as nourishment for the stuck-at-home queen rather than as a stockpile for the whole colony to survive the winter. All other North American bees eat nectar and pollen as larvae and adults, but not honey.

is an essential step in the pollination of many plants. Interestingly, honey bees do not know how to buzz pollinate and are, therefore, not very effective pollinators of some of our tastiest foods.

Nectar provides bees with important carbohydrates in the form of sugars, as well as amino acids. Bees use their tongues to either suck or lap up nectar. Some bees have very long tongues, giving them the ability to probe deep into flowers with long necks, like beardtongues (*Penstemon*). Other bees have very short tongues, which they use to sop

LEFT: An *Anthophora* visiting a Russian sage flower (*Perovskia*) for nectar, with its long tongue fully extended like a straw to stick into the narrow flower tube.

BELOW: A *Calliopsis* visiting a spring composite for nectar, with its short tongue out to probe the shallow flowers.

The picky eater is not a picky drinker

A specialist bee may be seen on many plants other than its "host", as it will stop by many kinds of flowering plants for a quick sip of nectar. This can make it difficult to determine whether a bee is a specialist based solely on the flower it is visiting.

up nectar from shallow flowers like those in the sunflower and carrot families (Asteraceae and Apiaceae, respectively).

Some bees are very particular about the pollen they collect for their offspring, limiting themselves to just a few plant species, even when many plant species may be available. These picky bees are known as specialists. No matter where they are found, they always specialize on the same flowers, and it appears to be an inherited trait—the female offspring of a specialist bee collects pollen from the same species of flowers her mother did, even though they never met.

1.7 A BEE'S ENEMIES

Enemies of bees fall into broad categories. Predators are those that attack and kill adult bees. A parasitoid (which means "parasite-like") feeds on the body of a living host—either from the inside or while attached to the outside, eventually killing the host. Parasitoids are not the same as parasites; a true parasite does not kill its host, while parasitoids do. Cleptoparasites are a special class of parasite. The word "host" takes on a different meaning here; rather than gleaning nutrients from the body of their host, they mooch off their host's labor. Specifically, they lay an egg in another bee's nest. When the cleptoparasite egg hatches into a larva, it eats the pollen and nectar provisions the host bee had gathered for its own offspring. Finally, scavengers are those organisms that root around bee nests. While they don't directly destroy bees, they often decimate nests and nest cells in the process of finding debris on which to feed.

Descriptions of the organisms that populate these categories could be a book all its own. They include viruses, bacteria, fungi, mites, birds, mammals, other insects, and even other bees. The life histories of many of these predators and prey sound like fodder for horror films, complete with monstrous beings eating their hosts from the inside out, and psychotic beasts controlling muscle function with an injection of poison. Below, we highlight those enemies that you are most likely to see while looking for bees.

PREDATORS

ROBBER FLIES (ASILIDAE). Robber flies are some of the best aerial predators in the insect world. With gigantic eyes for seeing in all directions and an amazing ability to maneuver in the air, they can nab an insect midflight, grasping it tightly with their spiky legs. The prey is paralyzed immediately with a dose of toxic saliva, injected when the robber fly pierces the exoskeleton with its strawlike beak. The poison then dissolves the tissues inside, turning the bee's organs into a thick broth which the robber fly sucks up.

A robber fly and its prey. PHOTO BY JILLIAN H. COWLES.

BEE-WOLVES (*PHILANTHUS*). With such a name, is there any guess as to their prey? A female bee-wolf is an unyielding predator in her attempt to provide for her offspring. Stalking hapless bees as they forage on flowers, she stings them on the bottom of the thorax (the "chest"), in the weakened crease between the front legs. Her toxic venom paralyzes her victim so that it can't crawl, fly, sting, or gnaw at her. With her victim alive but immobile, she then drags it back to her nest and begins a pile of bee bodies. When she has stockpiled five or so, she lays an egg near them and seals the nest entrance. The larva emerges and gobbles up the living, paralyzed bees. Bee-wolves will attack any bee they can catch, typically medium to large-sized ones.

A bee-wolf (*Philanthus*) drinking nectar from a buckwheat flower. PHOTO BY ALICE ABELA.

ASSASSIN BUGS (REDUVIIDAE). Assassin bugs are vicious predators that hide on foliage or flowers, waiting for unsuspecting prey to stop for a bite to eat or a rest. Bees, alas, are one such prey. The assassin bug nabs the bee with its long spiny forelegs and, hanging on tightly, pierces it with its rostrum (long tubular mouthparts). It injects its prey with toxic saliva that dissolves the insides of the bee. The assassin bug then feasts on its homemade "bee juice".

An assassin bug with a doomed bee.

SPIDERS (ARANEAE). Spiders hunt bees in one of two ways, either ambushing them on flowers, or building webs and catching them in flight. Those that ambush bees on flowers are often spectacularly camouflaged, with beautiful hues to match the colors of flower petals. Spiders gorge themselves on whatever species they catch.

MAMMALS, LIZARDS, AND BIRDS. Bees that nest in aggregations are particularly easy prey for foraging mammals, lizards, and birds. Mice, skunks, weasels, badgers, voles, and other scavengers will happily dig up bee larvae in the ground for a quick snack. Even armadillos have been seen rooting for bee nests in areas of the southeastern United States. Birds and lizards don't seem

A well-camouflaged spider and its most recent victim.

A spiny lizard (*Sceloporus*) in search of insects, including bees, for its next meal.

to mind the potential sting of an adult bee and will hunt near nesting aggregations, picking off bees that take too long to crawl into their nesting tunnel. The birds known as flycatchers and shrikes will nab them out of the air. And woodpeckers will sometimes drill into plant stalks containing nests and eat the larvae they find inside.

PARASITOIDS

TWISTED-WING INSECTS (STREPSIPTERA). The life cycle of a twisted-wing insect is jaw-droppingly different from that of "typical" insects. Females never leave their hosts' bodies, living burrowed into the exoskeleton with just a small piece of head exposed. They have no legs, no wings, and no eyes. At least males look insect-like for the few short hours of their adult lives. During those hours, they seek out a bee (or other insect) that is disfigured by the head of a female twisted-wing. He mates with her by piercing the exposed part of her exoskeleton (just behind her head) and filling her full of semen. The semen diffuses throughout the body, and some of it eventually arrives at the ovaries, fertilizing an egg. The resulting offspring develop by consuming their mother's body from the inside out, eventually emerging from the same opening created by their fathers during mating. The tiny grubs then actively seek new bees to infect. Once inside, they induce their hosts to create a layer of tissue around them, protecting them from attack by the host's immune system. Thus secured, they molt into adults. Males leave the body, but females continue to grow inside and can occupy up to 90% of the host's abdominal region. Inevitably, female bees thus infected are unable to reproduce.

A male twisted-wing insect (*Strepsiptera*) lands on a bee's abdomen to mate with the female living in the bee. PHOTO BY JACO VISSER.

WEDGE-WINGED BEETLES (RIPIPHORIDAE). These beetles sport elaborate and showy antennae, giving them the appearance of miniature elk or deer grazing on flower heads. It is on flower heads (usually buds) that the females lay eggs. By the time the flowers bloom, the small larvae (called triangulins) are ready to go, and will hitch a ride on female bees that stop to forage. They are then carried back to a bee's nest. Inside the nest, they wait in the dark for a bee egg to hatch and then burrow *into* the larva, where they grow along with their unlucky host. Once the larval bee pupates, the beetle larva tunnels back out of the bee and proceeds to eat it from the outside in. Ripiphoridae are thus internal parasites in the early larval stages, and external parasites during their later larval stages.

A wedge-winged beetle (*Ripiphorus*) with ornate antennae. PHOTO BY JILLIAN H. COWLES.

THICK-HEADED FLIES (CONOPIDAE). With their bulging eyes and wide heads, the name of these flies is a good fit. They often sport yellow or white markings on their black bodies, so they superficially resemble bees and wasps. Though they drink nectar and are often seen at flowers, thick-headed flies are vicious. Their abdomens have various spines and spears they use to great effect. Thick-headed flies catch foraging adult bees, often in midair, then, grasping them tightly, use appendages on their own

A thick-headed fly on a milkweed (*Asclepias*).

abdomen to pry open two segments of the bee's abdomen. They insert an egg, then let the poor bee go. The bee is apparently unfazed by the growing larva in its abdomen. However, when the bee dies (either from natural causes, or because the larva is too big), the thick-headed fly completes its development and wriggles back out between two segments of the dead bee's abdomen, as an adult.

BEE FLIES (BOMBYLIIDAE). Bee flies are fuzzy, resembling many of the bumbling bees on which they rely. They are common visitors of flowers, where adults forage for nectar and pollen. Some bee flies are generalists in that they will parasitize a diverse range of insect hosts, but others are very specific; for example, several bee flies in the genus *Anthrax* will parasitize only *Xylocopa*. Bee flies have an amazing ability to "throw" their eggs into a bee nest. The female first rubs the end of her abdomen on some nearby debris or a piece of wood. Then, as she whips the tip of her abdomen forward quickly, she hurls an egg deep inside the tunnel of a bee nest (it can be in the ground or in a plant stem). It is thought that the debris from the end of the abdomen sticks to the egg's surface and perhaps helps to camouflage it inside the nest, where it bounces to a stop. Eventually the larva that hatches kills and eats the bee larva.

FUNGI (ASCOSPHERACEAE). In the dark and dank living quarters of bees, fungus seems inevitable. Many mother bees do their best to minimize fungus by laying down fungicides and water-repellent barriers to keep cells dry; unfortunately some fungal contamination is inevitable. Chalkbrood is the most common fungus to pester bees, and it is quite common in honey bees, bumble bees, and even many solitary bees. Chalkbroods (*Ascosphera apis* and *A. aggregata*) develop first in the gut of a growing larva after it has consumed infected pollen. The unlucky bee is then devoured from the inside out. Once dead, the mummified larva takes on a chalky appearance (thus the common name). People who cultivate these bees for crop pollination do much to try to limit the spread of fungal infection.

MITES (VARROIDAE). There are many species of mites that infect bees, and there are as many life history stories as there are species. The best known is the

A spindly legged bee fly sipping nectar from a spring mustard (Brassicaceae).

An *Osmia* infected with mites. PHOTO BY RICK AVIS.

The varroa mite

The varroa mite is probably native to Russia; honey bee species native to Russia and Japan appear to have some immunity to it. When the western or European honey bee (*Apis mellifera*) was introduced into Asia in the late 1800s, the varroa mite transferred from native Asian honey bees (*A. cerana*) to the western honey bee. Having not evolved with this pest, *A. mellifera* has no defenses against it. Infected colonies made their way west, back to Europe, and finally to the United States in the 1980s. The impact on honey bees in North America was immediate and devastating. Drastic measures have been taken to combat these (and other) mites. One approach has been to invest money in other native pollinators that are never infected by these mites. The initiative to manage the alfalfa leafcutter bee (section 7.7) and the blue orchard bee (section 7.3) can be attributed in part to this redirection. Another approach has been to combat the mite directly. Miticides can be used only minimally, since honey bees produce honey for human consumption. Recently scientists studying the Asian honey bee noticed that the pupae in infected brood cells were removed by the worker bees, thus limiting the spread of the mite in the colony. This behavioral trait is specifically being crossbred into European honey bees with some success. While it doesn't eliminate the mites entirely, it does keep them at a low enough level that the colony does not fail for lack of healthy bees.

varroa mite (*Varroa destructor*), considered one of the biggest decimators of western honey bee (*Apis mellifera*) populations around the world. This mite burrows through the exoskeleton of a honey bee, sucking out the hemolymph (the insect equivalent of blood). Just as mosquitoes and ticks can transfer diseases in humans, so can mites make honey bees sick by infecting them with viruses and bacteria. In addition, they cause open sores on the bees' bodies, making them susceptible to further infections. In honey bees, mites hitch a ride on a female back to a colony and then crawl into brood cells before they are sealed. Inside, they lay eggs on the growing bee larvae. Young mites attach to a bee larva, and when it emerges from its cell, they are carried throughout the nest. Interestingly, it appears that mites prefer to attack drones (male bees) over worker bees.

CLEPTOPARASITES

OTHER BEES. Many genera of bees are known as cleptoparasites, a name that comes from the Greek word for thief (*clepto*). In a manner similar to cuckoo birds, these bees sneak into a "normal" bee's nest and discreetly lay a

Good neighbors

One reason that bees may nest communally, with many bees sharing one nest entrance, is that it lessens the risk of attack by cleptoparasites. With at least one host female likely always present in the nest, sneaking in to lay an egg is harder for a cleptoparasite; however, the more bees nesting together, the easier the bees are to find, and social colonies with hundred of individuals are the jackpot for other types of predators. This may be why social bees (and also social wasps) have the most painful stings—they have to work the hardest to defend themselves!

tiny egg near the pollen and nectar loaf being made by the host bee for its own offspring. The egg of a cleptoparasite tends to be flat and inordinately small, and it is often inserted into the cell wall or put behind the pollen loaf where it is hard for the host mother bee to see. When the larva of the cleptoparasite hatches, it has giant pincers for mandibles. It uses these sharp pincers to pierce and kill the offspring of the host bee and then eats the pollen loaf that was prepared for the host young. See chapter 9 for more information on the cleptoparasites.

CUCKOO WASPS (CHRYSIDIDAE). These tiny, beautifully colored wasps lay an egg inside the nest of a bee as she is packing it full of pollen and nectar. The egg hatches, after the egg of the host bee has hatched, and the wasp larva kills and eats both the bee larva and whatever pollen may be left (some genera eat only the larva once it is grown and not the pollen). These little wasps have lost the ability to sting and have very little defense against bees that might catch them sneaking around their nests. Like an armadillo or hedgehog, though, cuckoo wasps are capable of folding themselves into a tiny ball so that only their extra-hardened exoskeletons are exposed. Defensive female bees bounce them from their nests, whence they unfold, unscathed, and scurry off to find another untended nest.

A cuckoo wasp, one of the many enemies of bees.

ROOT-MAGGOT FLIES AND FLESH FLIES (ANTHOMYIIDAE AND SARCOPHAGIDAE). Though many root-maggot and flesh flies do not bother bees, within both familes are a few genera that are cleptoparasites of bees. These flies wait at flowers for female bees to visit and then surreptitiously follow them back to their nests (thus the other common name "satellite flies"). Even as the bee is inside her nest, they blatantly back into the nest entrance and lay an egg. When the egg hatches, it eats the bee larva.

SCAVENGERS

DERMESTID BEETLES AND SPIDER BEETLES (DERMESTIDAE AND ANOBIIDAE). Tiny, seemingly harmless beetles, dermestids and spider beetles are known for eating anything that decays, including skin cells, molds, plant detritus, textile fibers, and, to the dismay of entomologists everywhere, pinned insect collections. Dermestids are also common inhabitants of bee nests. While they are not known specifically to harm or prey on living bees, they have been observed opening closed bee nest cells, in the process exposing and killing the fragile larvae inside. While the biology of spider beetles is little known, it appears that females lay eggs in bee nests. The emerging beetle larvae eat bee eggs, larvae, and dead bees. Adults, too, feed on detritus in bee nests; in the process of scrounging for scraps, they can be quite destructive. Scavengers are a common cause of disappointment among gardeners and landscapers who build artificial nests to attract bees only to find that the bees never establish well.

1.8 A BEE'S BODY

Like Inspector Gadget's many tools and appendages, each part of a bee's body seems to have a use. Millions of years of evolution have acted to make it the epitome of efficiency and functionality. From ridges and spines, to fine brushes and spatulas, every appendage has a purpose. The following section, by necessity, contains many technical terms that will be foreign to most readers. Our hope is that the included pictures will prove useful as you become comfortable with the names of a bee's many body parts, because knowing a little about them can make it much easier to identify bees.

A bee's body comprises three main sections: the head, the thorax, and the abdomen. Scientists who study bees and their relatives generally refer to the three body regions of a bee as the head, mesosoma, and metasoma, because the abdomen in bees technically includes a part of the back of the thorax. For simplicity, we will use the former, more general, terms.

head thorax abdomen

An *Anthophora* showing the three parts of a bee's body: the head, thorax (which includes the wings and legs), and the abdomen.

THE BODY

The overall look of the bee body can be very telling with bees—their "gestalt", if you will. Important features include the shape, which may be flattened or very thick. The abdomen may be rounded or more oval-shaped. The head may be small or extremely robust. Hair patterns and colors can also be important. Their placement on the abdomen and thorax are often diagnostic. The "skin" of a bee is called the integument and is marked with small indentations (punctations). The patterning of these small marks is often very important in species identification. Also important is the shininess of the integument between the punctations.

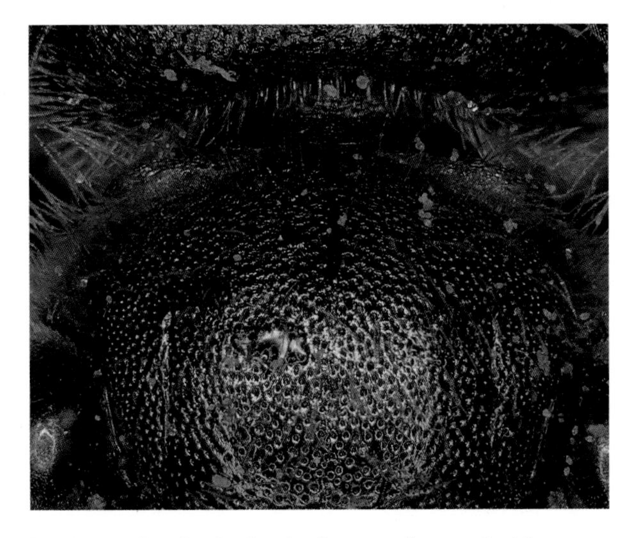

The thorax of an *Osmia*, showing the many thousands of tiny indentations that make up the surface of the integument (skin). The spacing between these, the size of the punctations, and the appearance of the spaces between them are important in bee identification. PHOTO COURTESY OF USDA-ARS BEE BIOLOGY AND SYSTEMATICS LABORATORY.

A female *Anthidiellum*, showing many of the remarkable kinds of hairs common to bee bodies. Note the thick scopal hairs on the abdomen, the spiny hairs on the legs, and the tuft of softer hairs closer to the thorax. Photo courtesy of USDA-ARS Bee Biology and Systematics Laboratory.

THE HEAD

Many important appendages are found on a bee's head, including the antennae, the mouthparts (tongue), the mandibles (jaw), and two sets of eyes.

ANTENNAE. In female bees, the antennae are made up of 12 segments, while the antennae of males generally have 13. The first segment—the one closest to the face—is called the scape. The second, which is often round and a little knobby, is called the pedicel, and the next 10 or 11 segments are called flagellomeres (all together called the flagellum). They are numbered beginning with the one closest to the pedicel. The antennae are attached to the bee's head in

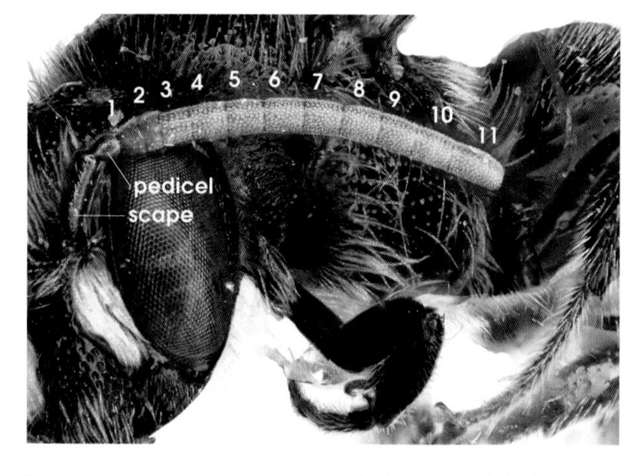

The antenna of a male *Afroheriades* (not found in North America), showing the many segments. Closest to the bee's head is the scape followed by the pedicel, a more or less round joint. Next are 10 or 11 flagellomeres. Typically, males have 11 flagellomeres and females have 10. Photo courtesy of USDA-ARS Bee Biology and Systematics Laboratory.

In a few groups of bees, males have 12 segments in the antennae instead of the usual 13. This is very rare, and generally found in cleptoparasites.

antennal sockets. If you look closely at the segments of the flagellum under a microscope, you can see they are covered with microscopic hairs and little pits. The hairs and pits are called sensilla, and bees use them to "smell" the world around them. The sensilla act as tiny sensors, capable of processing the many chemical compounds in the air, especially the fragrances produced by flowers.

MOUTHPARTS. Taken together, the mouthparts of a bee are called the proboscis, and they consist of many jointed pieces, each of which has an odd-sounding name. Bees fall into one of two categories based in large part on the features of their proboscis: they are either short-tongued, or long-tongued. In the short-tongued category are the bee families Andrenidae, Halictidae, Colletidae, and Melittidae. In the long-tongued category are the bee families Megachilidae and Apidae.

The tongue, or glossa, is not actually the feature most indicative of a long-tongued bee—it is instead four tiny segments that run next to the tongue called the labial palps. In long-tongued bees the first two segments (those closest to the bee's head) are much longer than the second two. In short-tongued bees all the segments are more or less the same length. In both cases the labial palps are sensory organs for "tasting" the world.

In the long-tongued bees, the tongue can be folded from base to tip into roughly three parts that collapse into the

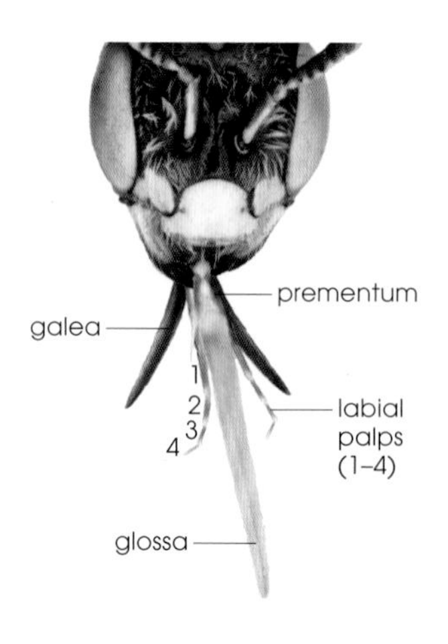

The tongue (glossa) and some of the other mouthparts of a *Perdita*, a short-tongued bee. While the tongue does not look particularly short, the first and second segments of the labial palps (one set is found on each side of the bee) are the same length as the second two (numbered 1–4 on the left side). Compare this with the long-tongued bee tongues in the following images. Also shown in this picture are the prementum, which is the beginning of the glossa, and the galea. The galea is two sheaths that fold around the glossa.

BELOW: The tongue and some of the mouthparts of a *Chelostoma*, a long-tongued bee. Notice that the first two segments of the labial palps are elongated and flattened—they are much longer than the second two segments, which are angled to the side just slightly (the segments are labeled on the right side of the bee). Also shown in this picture is the labrum, a long plate that folds up underneath the clypeus. When folded up it is hard to see unless you look underneath the bee (this is unfortunate because the shape of the labrum can aid in identification). PHOTO COURTESY OF USDA-ARS BEE BIOLOGY AND SYSTEMATICS LABORATORY.

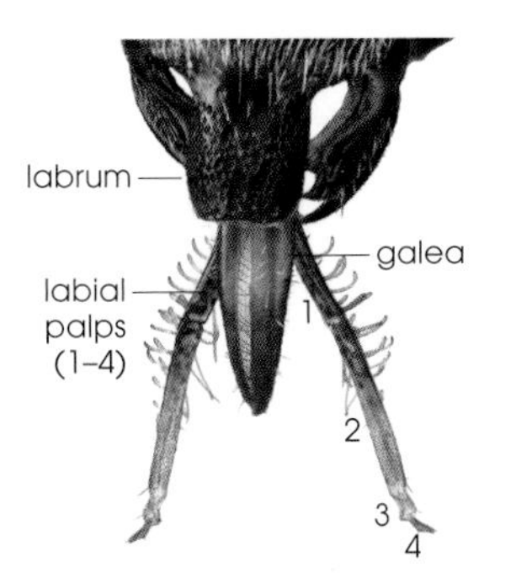

shape of a Z. Similar to the way a bird folds its legs at the elbow and ankle joints to press up against the body, long-tongued bees fold their tongues up under themselves when flying. Quite often the proboscis is so long that even when folded up, it extends the full length of the underbelly. When the proboscis of a long-tongued bee is extended, its other pieces can be folded together to form a straw for sipping. When the bee wiggles its tongue within this straw, it can suck nectar up toward the mouth. This kind of tongue is very porous and has special hairs that aid in sopping up minute amounts of nectar from inside flowers.

In short-tongued bees, the tongue is stout and sometimes has two lobes. The labial palps and galea do not fold around the tongue to create a long straw, and the straw formed by just the maxillary palps is much smaller. These bees may lap or dab at nectar more than suck it up.

In addition to slurping nectar, or the sweat on your skin, bees use their tongues for "painting" the lining inside their nests, for a little bit of grooming, and for communicating—many of the chemical receptors used by bees to understand their environment are on their tongues (the rest are on the antennae and the feet). Bees may also lick other bees or surfaces in order to sense pheromones or other important compounds.

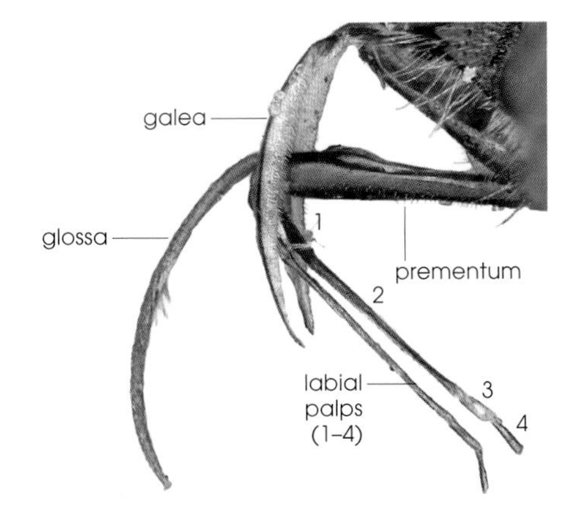

The tongue of a *Protosmia*, a long-tongued bee, showing how it folds up to be tucked under the body. The galea are obvious above the glossa and labial palps. Note that the first two segments of the labial palp are much longer than the second two segments. PHOTO COURTESY OF USDA-ARS BEE BIOLOGY AND SYSTEMATICS LABORATORY.

MANDIBLES. The bee equivalent of the opposable thumb is the mandible. As a pair, these "jaws" are like small pinchers and span the lower section of the face. The mandibles are very important to many bee activities, including sparring with enemies, building nests, filling a nest with leaves, carrying pebbles, and hanging onto a plant stem (in the case of males, which do not have nests to sleep in at night). At the tip of the mandibles (the end farthest from the connection to the face) are a number of teeth. Some bees have very simple mandibles with few teeth, whereas others have many teeth that may be pointed or rounded.

The mandibles of an *Osmia*. These particular mandibles are toothed, but many mandibles are much simpler, ending in narrowed points like a pair of tweezers. PHOTO COURTESY OF USDA-ARS BEE BIOLOGY AND SYSTEMATICS LABORATORY.

EYES. A bee has two types of eyes. Near the top of the head are the simple eyes (known as ocelli). The "simple" refers to the fact that each eye has just one lens. There are three ocelli, with two near the back of the head and one just forward of those. Bees use these eyes for orientation, triangulating on the position of the sun in the sky and using this information to guide them to and from their nests. Those bees that fly at dusk or dawn tend to have larger ocelli in order to sense light from the sun when it is low in the sky.

The three simple eyes at the top of the head of an *Osmia*. The distance between these eyes and the back of the head (called the preoccipital carina) can be important for identification. PHOTO COURTESY OF USDA-ARS BEE BIOLOGY AND SYSTEMATICS LABORATORY.

The compound eyes are more visible, with one on each side of the head. Each compound eye comprises thousands of individual lenses. Through these many lenses, bees see color, shapes, and light. Each lens sees the same view from a slightly different angle, and a bee must compile all these images in order to create a picture of the world around it. Because the lenses are so tiny and because none sees a very big slice of the world, bees are fairly near-sighted and cannot make out sharp images from great distances.

However, bees are very good at sensing movement, because they process the images they see so quickly. A bee eye processes images 15 times more quickly than our own eyes do; while something might move so quickly that it is a blur to us, bees can perceive every detail. As an example, whereas a fluorescent light may seem to give off an even glow to us, bees see the light as flickering many times per second. This speedy processing influences how bees perceive flowers. It makes rounded objects hard to see, whereas those with edges and points, which appear to flicker across a bee's field of view as the insect moves, are much more easily recognized. This is why bee-pollinated flowers often have many-pointed petals.

One of the compound eyes of a *Megachile*. This eye is made up of thousands of tiny lenses, each of which sees the world just a little differently. PHOTO COURTESY OF USDA-ARS BEE BIOLOGY AND SYSTEMATICS LABORATORY.

A bee couldn't see Dorothy's ruby slippers

Because bees perceive solid red colors poorly, those flowers that rely on bees for pollination are seldom shades of burgundy or crimson. Bee-pollinated flowers are more typically shades of purple, yellow, or orange.

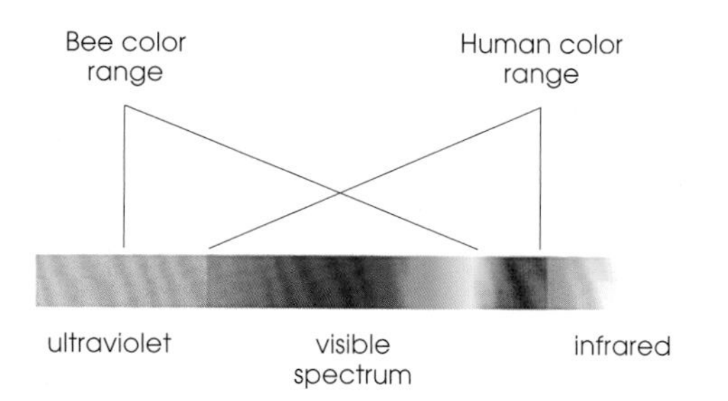

The light spectrum. Humans see only the visible spectrum, and not ultraviolet or infrared colors. Bees can recognize most of the colors seen by humans, but they do not distinguish shades of red well. However, they can see ultraviolet reflections that our own eyes cannot perceive. This subtle shift is enough to change bees' perception of the world significantly from our own.

Bees have three color receptors behind each lens, and each responds to a different wavelength of light (representing a different color). Bees' eyes are most in tune to green, blue, and ultraviolet light, while they are much less sensitive to red hues. As a result, bees regard the world much differently than we do, picking up on floral cues that are completely invisible to us. For example, flowers often have ultraviolet stripes or "guides" on their flowers indicating the most direct route to floral rewards like nectar and pollen. Our eyes, which do not perceive ultraviolet light, cannot see these guides.

OTHER FACIAL FEATURES. Several other important structures on a bee's head are worth mentioning because they aid in bee identification. Between the antennal sockets and above the mandibles is a plate known as the clypeus, located about where our nose and upper lip would be. While the clypeus has no primary function, it is often colored with yellow, red, or white markings. These markings may aid bees in recognition, territoriality, and other social interactions. It has recently been found among social wasps that such

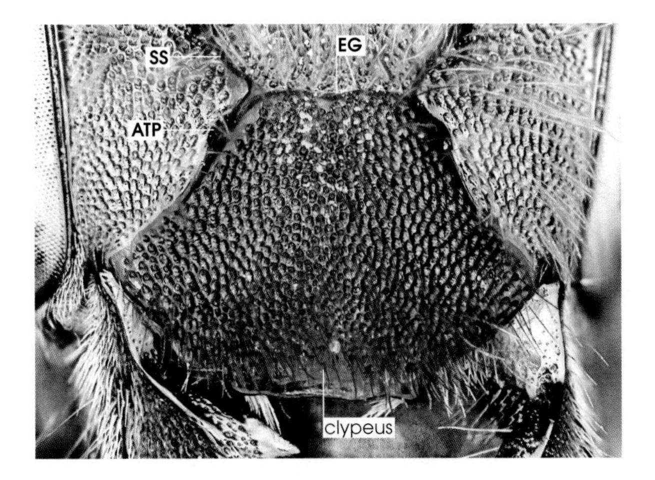

The lower portion of the face of an *Osmia*. Several important features can be seen when looking straight at a bee's face. The clypeus is the large, more or less six-sided plate occupying the majority of the face. Along the top edge of the clypeus are two holes (one on each side) known as the anterior tentorial pits (ATP). Running from the clypeus to the edge of the antennal socket are either one or two lines in the face known as the subantennal sutures (SS)—the ATPs may occasionally be along the SS. Finally, the top margin of the clypeus is a groove known as the epistomal groove (EG). Photo courtesy of USDA-ARS Bee Biology and Systematics Laboratory.

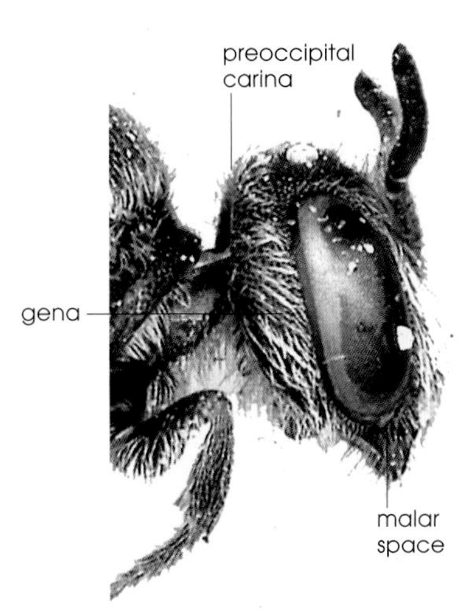

The side of the face of a *Perdita*, showing the preoccipital carina (the area between the upward facing and backward facing parts of the head), the gena (the area behind the compound eye), and the malar space (the area between the bottom of the compound eye and the start of the mandibles). Photo courtesy of USDA-ARS Bee Biology and Systematics Laboratory.

markings help them to distinguish between individuals. While many of the bees that have spots and stripes on their face are solitary and not social, they may nonetheless rely on these visual cues for some communication; males, for instance, flash their bright faces in the sun to let other bees know they have staked out a territory.

Underneath the clypeus is a smaller plate known as the labrum. It is hinged all along the edge it shares with the clypeus and is often not seen unless you look at the bee from below (this is unfortunate, since the shape of the labrum can be quite distinctive). Beside each compound eye, near the top of the head, are two indentations known as fovea—these depressions may have hair in them or may be bare. See accompanying images for other important facial features.

THE THORAX

The middle section of a bee's body, complete with wings and legs, is called the thorax.

WINGS. Bees have two wings on either side of their bodies, known as the forewing (the one toward the head of the bee) and the hind wing. The wings hook together with minute hooks (called hamuli) on the front edge of the hind wing. Wings are supported by a series of veins; the spaces between them are so distinctive on a bee's wing that they can be used to tell groups of bees apart. Both the forewing and the hind wing have identifying characters, described in the accompanying image.

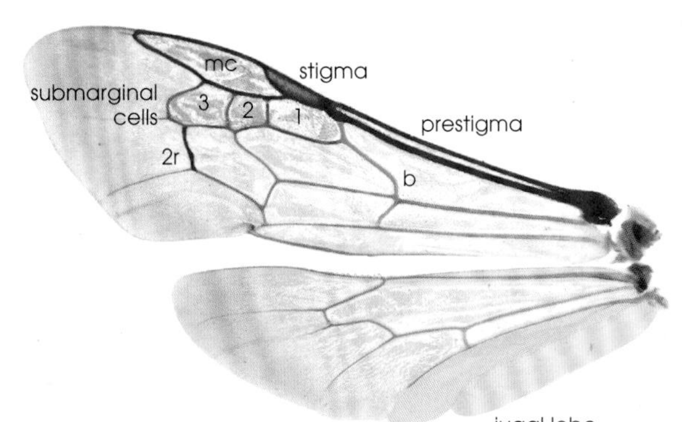

The fore and hind wing of a *Halictus*. The forewing is the top wing, and the hind wing is the bottom. Important veins and spaces are colored. In purple are three important cells. Outermost is the marginal cell (mc). The stigma is the dark area on the bee's wing just beside the marginal cell. The prestigma is the very long cell along the wing margin and closest to the bee's body. The submarginal cells (colored pink) are called the first, second, and third, named from the one closest to the body and working out. In green is the basal vein (b). In this bee it has a distinct arc in it; other bees have a straight basal vein. The red vein at the outside edge of the wing is called the second recurrent vein (2r). Its shape, and how much curve it has, are also important. At the front edge of the hind wing are tiny wedges or hooks (hamuli, barely visible here) that attach the hind wing to the forewing. Finally, two important lobes are found at the back edge of the hind wing: the jugal lobe, colored yellow (entirely missing in some bees), and the vannal lobe, colored light blue, just in front of or above it.

Depending on the bee, those membranous wings may carry the little creature (and its heavy pollen harvest) a great distance. Bumble bees have been known to travel up to 5 miles (but more typically travel a mile). Even medium-sized bees may travel up to a mile from their nest in search of plants (and then back again). The tiny bees known as *Perdita*, on the other hand, may travel only the length of a football field (or less) in their lifetimes.

LEGS. Bees have three pairs of legs (six total). In order from front to back, they are known as the foreleg, the midleg, and the hind leg. Each leg is made up of distinct parts (see accompanying image), and the joints and the number of sections making up the leg are similar for each of the pairs, but each pair has distinctive added features. Each foreleg, for example, has a notched area just the diameter of the antennae, used for keeping those important sensory organs clean and well functioning. The hind tibiae have hairs on their inner margins (called keirotrichia) that are used to clean the wings. And in many bees, the midlegs also often have brushes and combs on the tibia and basitarsus that are movable, and are used to transfer pollen to the hind legs.

Most bees possess electrostatically charged pollen-collecting hairs known as scopa on their hind legs. The location of the scopa differs between bees, as does the size and shape of the scopal hairs. For example, bees that carry large pollen grains tend to have widely spaced scopal hairs to incorporate the bigger diameters of the pollen they collect. Bees in the family Megachilidae have pollen-collecting hairs on the underside of the abdomen instead of on their legs. Some bees (including bumble bees and honey bees) have pollen baskets (known as corbiculae) on their legs instead of, or in addition to, scopa. In many bees this is a concave area on the tibia surrounded by curved hairs. It allows these bees to carry pollen moistened with nectar back to the nest, instead of

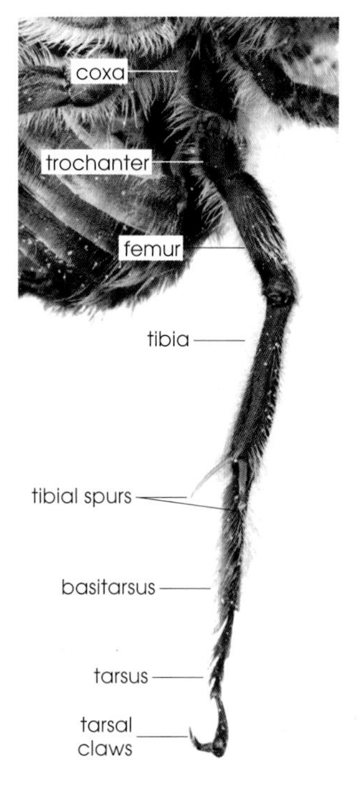

The hind leg of a *Eucera*. Moving from the end closest to the body (the base) to the tip, the distinct sections are the coxa, the trochanter, the femur, the tibia, and the tarsal segments (the segment closest to the tibia is often called the basitarsus and is often much longer than the others). There may be an additional little pad between the front tarsal claws known as the arolium. On the tibia, a basitibial plate is often found at the end closest to the femur, the insect equivalent to our knee cap. At the other end are tibial spurs that vary in length and sometimes number. Every hair, spine, and specially curved surface on each leg seems to serve a distinct purpose and can aid in identification.

A female *Eucera*, showing the pollen-collecting hairs (scopa) on the legs; on this bee they are filled with bright yellow pollen.

A female *Hoplitis* showing white pollen-collecting hairs (scopa) on the underside of the abdomen.

The corbicula of a bumble bee (*Bombus*). The corbicula is a large flattened tibial section on the back leg with hairs that curl around to its front. It serves a function similar to that of scopal hairs, except that it is better equipped to carry the wet mixture of nectar and pollen collected by some bees.

just dry pollen like bees with scopa do. Some bee species have pollen-carrying structures on their hind femurs, and even on the sides of their abdomen instead of, or in addition to, on their tibiae.

Bees that nest in the ground or in soft wood have spines or spurs on their tibiae. These protrusions are used to maneuver when inside their tight burrows. By pushing out with their legs, they can grip the tunnel walls. Bees that are cleptoparasites (chapter 9) also have extra spines on their legs, and it is thought that these help the bees brace themselves when they are attacked in their host's nest. There is a plate where the tibia meets the femur (known as the basitibial plate) that serves as a knee pad in most ground-nesting bees. Also on the hind legs, bees that nest in the ground often have a paint brush (called a penicillus) on their hind basitarsi that helps to apply the cell lining.

THORAX. The thorax is made of a series of plates fused together. Many of them serve no immediately obvious purpose. Some may be defensive, others may represent points of anchor for internal muscles, and the rest may be nothing more than carryovers from ancestors past. Regardless, many are very important for bee identification and are detailed in the acompanying images.

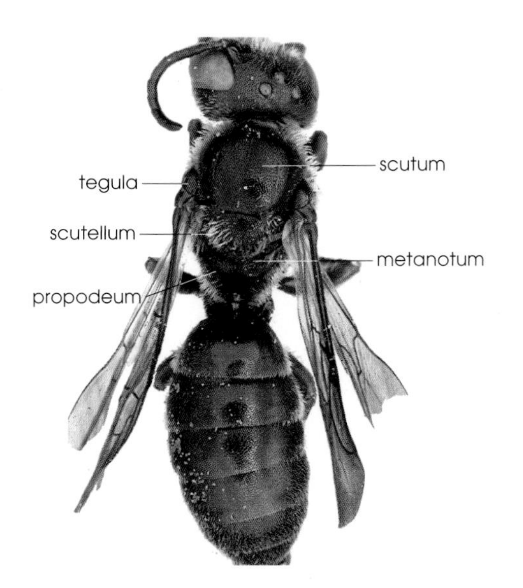

The thorax of a *Chelostoma* showing some of the major sections. The tegula are the little caps over the wing joints. The scutum is the largest section of the thorax, taking up most of its surface. The scutellum is the smaller plate just behind it. The metanotum is quite small and generally curves downward, meeting the propodeum, which slopes down to face the abdomen. Finally, on the scutum itself are two tiny indentations known as the parapsidal lines that can aid in identifying some bee genera (not labeled here, see Osmiini, chapter 7 identification tips). PHOTO COURTESY OF USDA-ARS BEE BIOLOGY AND SYSTEMATICS LABORATORY.

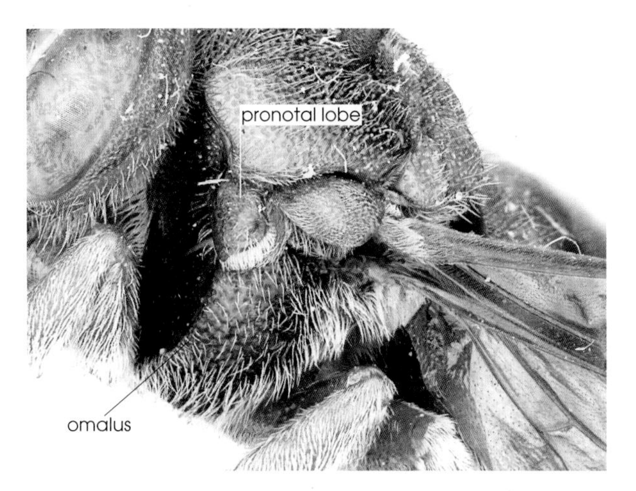

The thorax of a *Heterostelis* (not found in North America) showing two features that can be important in bee identification: the pronotal lobe, and the omalus. The omalus is not a plate or face, but a ridge separating the forward-facing and side-facing portions of the abdomen. PHOTO COURTESY OF USDA-ARS BEE BIOLOGY AND SYSTEMATICS LABORATORY.

THE ABDOMEN

The abdomen is covered in a series of telescoping plates. On the top (dorsal) surface, the plates are known as tergal segments (one segment alone is called a tergum or tergite, and all together they are referred to as the terga). Researchers number them from front (i.e., closest to the head) to back as T1 through T6 or T7; females have six segments and males have seven. The terga furthest from the head are often difficult to see because the bee can collapse each segment into the one before it like nesting Russian dolls.

A *Dianthidium*, with the tergal segments (sometimes called tergites), numbered. Because the segments telescope into each other, the last several can sometimes be hard to see. Many important distinguishing features are found on the first one, which has two parts—the front (anterior) side, which faces the thorax, and the top (dorsal) side, which faces straight up. The tergites are referred to as T1 through T7. PHOTO COURTESY OF USDA-ARS BEE BIOLOGY AND SYSTEMATICS LABORATORY.

A *Lasioglossum*, showing the sternal segments, numbered. The first is hardly visible in this image, and is not numbered. The sternites are difficult to see because they run underneath the edges of the tergal segments. The hind (apical) margins of the sternal segments sometimes have features on them that are important for distinguishing between species. Photo courtesy of USDA-ARS Bee Biology and Systematics Laboratory.

On the bottom (ventral) surface of the abdomen are sternal segments, numbered exactly like the tergal segments (S1 through S6 or S8). The sternal segments can have extra protrusions, flaps, or other projections, but these can be difficult to see. Just as with the tergal segments, females have fewer (six) than males (eight). Because of the additional segments, the abdomen of males sometimes appears longer and may be curled under.

At the tip of the abdomen there is a pygidial plate (not all bees have this, and its presence or absence can aid in identification), which is used in constructing and tamping down the earth in a ground-built nest. At either side of the pygidial plate is a brush of hair called the pygidial fimbria. A brush of hair may also be seen on the tergal segment above the pygidial plate. This is called the prepygidial fimbria.

Pygidial plate of a *Eucera*. Bees that nest in the ground use this for tamping down the dirt in their nest. The sting is just visible below the plate, at the very tip of the abdomen.

The sting of a bumble bee (*Bombus*) and a honey bee (*Apis*). Honey bee stings have barbs, and they are the only bees that do. When a honey bee stings her victim, the barbs catch in the skin. Then, as the bee flies away, the sting is pulled from her abdomen and left behind; thus, a honey bee stings only once. This is not true for other bees, which can sting multiple times.

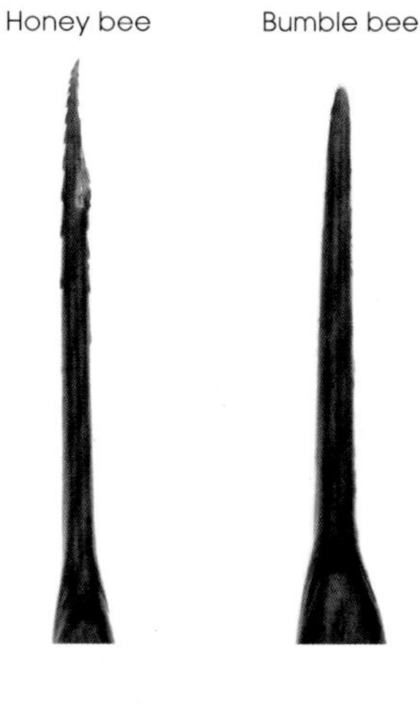

The sting of a bumble bee. The sting is the modified ovipositor or egg-laying tube of a bee. Because only females lay eggs, only females can sting.

STING. Finally, at the tip of the abdomen is the sting, a highly modified ovipositor (egg-laying tube). Over evolutionary time scales, the sting of a bee has become so modified for stinging that eggs are no longer laid *through* it, but from a separate opening near its base. Since the sting evolved as an apparatus for laying eggs, males don't have one, and therefore they cannot sting.

1.9 HOW TO STUDY BEES

Studying bees is a rewarding pastime, but it means learning to see the world in a way that may be new for you. Many bees are miniscule, and though you may see them often, it is only with practice that you will recognize them as bees. Spend more time looking closely at the flowers in your yard. You may see what you thought were little gnats hovering near a bloom. When you stick your nose close, however, the giant pollen masses on either side of the "gnat's" body reveal its true identity. Watch for hovering beasts guarding a patch of mint and vigorously attacking any insect that comes too close. Listen, too. You may hear the buzz of a bee hovering over the ground, checking out several potential nesting sites and trying to decide on the best location for her new home. Even the nonbees can tell you something. Do you see bee flies? A ground-nesting bee is likely somewhere nearby, and if you follow the flight path of the bee fly you may even find it. Is there a robber fly? What is it hunting? The presence of the cleptoparasitic bee *Nomada* can tell you that *Andrena* are likely in the vicinity. Leaves that look as though a hole-punch has been taken to them tell you that a *Megachile* is building a nest in the area.

Knowing the identity of a bee gains you access to the plethora of facts already gleaned by other researchers about that group and can guide your efforts to learn even more. A few tools can aid in your quest to learn more about bees:

- A dissecting microscope is ideal but pricey, especially for a good one.
- A hand lens will also suffice. Loupes, as used by botanists and geologists, work well if you have bright light. As you might expect, the more you pay for a hand lens or loupe, the better the image. High-quality magnifying glasses will also work. Believe it or not, entomologists at the turn of the nineteenth century quite often identified insects to species with just a magnifying glass.
- A camera with a macro lens or zoom capability can be as effective as a microscope for most tasks. Close-up photos of insects are especially useful if collecting insects isn't your style. Try to get pictures of the face, the abdomen, the wings, and the legs. With few exceptions, these shots will allow you to determine at least a genus, and occasionally the species.
- An insect net can be purchased from scientific supply stores, major online retailers, or entomological specialty companies; alternatively, you can build your own. A dowel, hose clamps, and a wire coat hanger are enough for a satisfactory net to use around your yard. If high-end construction is more your style, excellent nets can be made with golf clubs that have had the club removed. Use a welding agent to stick a female screw-end inside the hollow handle. Make a hoop from piano wire, with tiny loops at either end. Holding the hoop closed, stick a washer and bolt through the two loops, and insert the bolt into the golf club handle. You'll also need material for the net itself. We recommend wedding veil material because it flows well, isn't stiff, and allows you to see your "catch" inside the net. Use duck canvas cloth for the edge that threads through the hoop because it stands up well to being slapped against the ground or thrashed through vegetation.

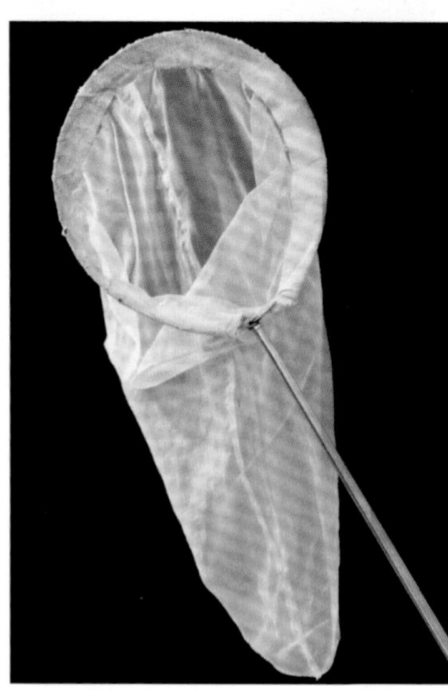

Example of a homemade net using canvas duck cloth, wedding veil material, a golf club, piano wire (or other thick gauge wire), and a female screw-end welded into the handle of the golf club, with a screw to secure the wire.

Collecting bees with a net is more challenging than you might first think. Bees are fast and accustomed to fleeing any slight movement (it might be a bird or robber fly). Keep your shadow off the bee. Move slowly (don't scratch your nose suddenly). Keep the net low, and hold the tip of the fabric part so that it can't blow in the breeze. Sweeping up from below a bee quickly (it's all in the wrist) can catch it by surprise and help you succeed in its capture. Swatting down from the top, trapping it between the ground and your net is also a good technique. With the net hoop firmly on the ground, hold the top of the net up. Bees always fly up when alarmed, and you can easily coax them to the tip of your net, and then lift it off the ground. If you sweep from the side, be ready to flip the end of the net quickly over on itself so that the bee can't fly back out.

If you've collected a bee in order to learn its identity you have two options: you can kill it, or you can identify it and let it go. If you have no desire to store mounted dead insects in your house, you can put the bee in a plastic container, and chill the bee in your refrigerator for a few hours. When it is still, use a hand lens or microscope to get a good look at crucial body parts before it comes to. If you have a plastic vial you can put the bee in this, so that you don't have to touch it (there is no harm in feeling its fuzzy body while it is sleeping, though as it starts to wake up, it can sting!). When it finally awakens, you can let it go. With most chilled bees, you'll have about 10 minutes to look it over before it is fully awake again (some wake up much faster than others; *Nomada* seem to bounce back almost immediately). Chilled bees stay alive for anywhere from one to three days, depending on the size of the bee. If it starts to come to before you are done looking it over, put it back in the fridge until it is still again.

A budding entomologist may choose to kill the bee and pin it for his or her collection. "Killing agents" are hard to come by for the amateur, but good methods include (1) putting the bee in the freezer, (2) submerging it in alcohol, or (3) exposing it to the fumes of ethyl acetate (dowse a cotton ball in ethyl acetate, and seal the bee and the cotton ball in a small container together). Each technique has its drawbacks. Freezing the bees obviously works only if you are close to your house and not out hiking. Submerging them in alcohol is fast, but it can also make them brittle if they are left in the alcohol for too long. And ethyl acetate sometimes takes a long time to kill a large insect, though it has the advantage that the bees often die with their tongues out. Since many useful features for identifying bees are on the mouthparts, this is especially important to beginners. Straight ethyl acetate can be purchased from biological supply companies; alternatively nail polish remover can be used. Dead bees can be mounted on pins (also available from the same types of places that sell nets). Generally, number 1 or number 2 pins are the best size for small and medium to large bees, respectively. Very small bees should be mounted to the side of a pin (we describe this technique below).

Medium-sized to large bees can be pinned directly. Hold the bee gently but firmly, pinching its body between your thumb and index finger so that the thorax (back) is facing out. With your other hand poke the pin through the right-hand side of the body. As you push the pin through, try to angle it so the bee's thorax will be parallel to the ground once it is fully mounted. You don't want the bee to be cockeyed on the pin, so go slow to make sure you'll be

An *Anthidium* specimen on a pin. Notice that the pin goes through the bee so that when the pin is straight up and down, the bee's thorax is exactly perpendicular to it. The same should be true when viewing the bee from the front, with the thorax completely level, not tipped. PHOTO COURTESY OF USDA-ARS BEE BIOLOGY AND SYSTEMATICS LABORATORY.

Isn't killing bees bad?

The answer is yes and no. Obviously, passing judgment on this activity is a matter of personal opinion; however, if you are on the fence and worried about bee health, here are a few points to consider. Determining whether or not bee populations are in decline means knowing the size of the populations in an area to begin with, and also the names of the species making up those populations. Most bees cannot be identified to species without a good look under a microscope, and this is also true for some genera; therefore, at least some bees must die at the hands of scientists in order to determine the baseline population levels for different species. For the amateur naturalist, however, identifying the general group of bees (i.e., the genus or the tribe) to which a bee belongs can be done *with* a little practice and *without* killing any bees. Learn the important features to look for, and leave the bees to go about their business. If you would like a bee collection, however, collecting a few specimens that you see is not generally harmful. Use your discretion. Nabbing one or two *Perdita* for your collection is not, at this point, going to make the difference between a bee's extinction and its survival. Nabbing a bumble bee, however, is not a good idea. In particular, when you collect a queen bumble bee (the giant ones that fly early in the year), you are in fact collecting an entire potential nest, as the queen is responsible for laying all the eggs that result in a colony of many individuals. Many bumble bees can be identified on the wing instead, by looking at their abdominal stripes. Should you decide to collect, your specimens will be most valuable if you choose to participate in a citizen science program and report what you've found to an interested group.

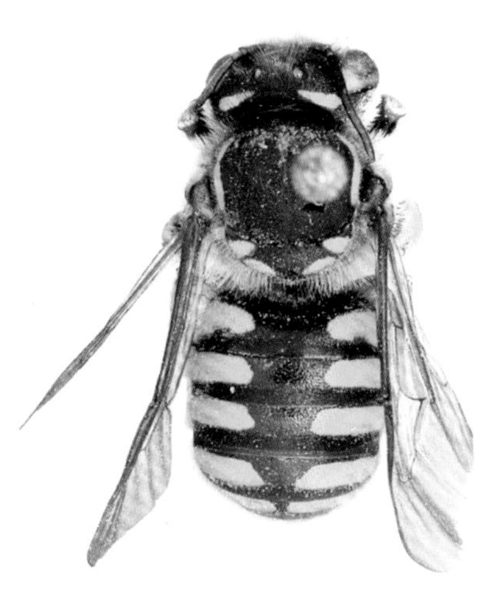

An *Anthidium* specimen on a pin. Note that the pin always goes through the *right* side of the body, leaving the left side completely unmarred for identification purposes. PHOTO COURTESY OF USDA-ARS BEE BIOLOGY AND SYSTEMATICS LABORATORY.

satisfied. Sometimes it is helpful, as you pin, to rest your hands on each other for maximum stability. As you slide the bee up, leave enough room at the top of the pin so that you can easily grasp it without touching the bee itself. If you spin the pin quickly in your hand once the bee is mounted, you can often get the abdomen and legs to stretch out some, making identification easier. Depending on your patience level and the eventual goals for your specimens, you can stick the pinned bee into a piece of Styrofoam so that its abdomen and legs are resting on the foam. This spreads the legs and abdomen as the bee dries, making for a more aesthetically pleasing final specimen. Use extra pins to brace unruly legs as the bee dries. Note that the final dried specimen will be very fragile.

After a bee is pinned, you can press the pin into a piece of Styrofoam so that it dries with its abdomen and legs resting on the foam.

A *Micralictoides* on a paper triangle mount. Note that the triangular mount is on the *right* side of the bee, and that it is smaller than the bee's body; i.e., less than 0.1 inch wide and 0.25 inch long.

The goal of pinning or mounting an insect is to leave the left side unmarred—free from the giant hole that a pin leaves. When a bee is very small, sticking a pin through it would obscure the entire surface of the thorax. For these small bees, you can cut an isosceles triangular point out of a piece of cardstock and use a dollop of clear nail polish or craft glue to affix the bee to its narrowest point. Stick the pin through the wide end of your paper triangle, and mount the bee so that its thorax rests on the narrow end. It is easy to make the points too large; try to keep them less than a quarter of an inch long, and a tenth of an inch wide.

Another option that is faster, but aesthetically less pleasing, is to adhere the bee directly to the pin with glue or nail polish (glue tends to be more viscous and dries faster—it is the better choice, but both work). Put a pinpoint of glue on the side of a pin, and then lay the pin on the *right side* of the bee's body, just behind the wings. Use the least amount of glue necessary to keep the bee affixed without gumming up too much of the insect. Make sure the glue wraps all the way around the pin, though, so the bee doesn't fall off. Leave the glued bee lying on your work surface, with the weight of the pin holding the glue to the body until it has dried (30 seconds to a minute should be enough).

A *Perdita* specimen glued to the side of a pin. Note that the glue is on the *right* side of the body, leaving the left unmarred.

Bees that have been mounted on pins can be stored on pieces of

Rehydrating dead bees

There are times when a bee specimen dies curled up, with its mouthparts hidden away and its feet tangled together and held tightly under the body. If it is imperative that you see a particular feature that isn't immediately visible, and the bee is too dry to manipulate without breaking, you can relax the bee. Stick the pinned insect into a piece of play dough, Styrofoam, or a cork. Put the bee on its stand in a small plastic storage container (the kind you would use for leftovers) along with a very wet sponge. Leave it sitting out at room temperature. Depending on where you live and the natural amount of moisture in the air, the bee may relax and become movable after just half a day, or it may take several days. Keep a close eye on your specimen, though, as mold will begin growing on the bee very quickly.

Styrofoam. A shoe box with a cutout piece of Styrofoam inside works just fine if you don't want to spend a lot of money on fancy insect drawers. You can also buy cardboard boxes specifically for storing insects from entomology supply stores. In many parts of the country, beetles and other insects that eat dead animals will happily devour pinned specimens, leaving you with a pile of debris below an empty pin. Freeze your box of bees from time to time for a full 24 hours to minimize beetle problems. And invest in some clove oil. A little bit of clove oil soaked on a cotton ball and kept in a small pill cup in the corner of your insect box may help keep other insects at bay (until it dries out).

Pinned specimens should be labeled so you can keep track of their identity, as well as the specifics of their habits when you collected them. Labels generally include the country, state, county, and location where the insect was collected. They also include the date, the name of the collector, and the name of the flower from which it was collected (if known). You can write labels by hand, or type them on your computer and print them out. A 4-point font makes a nice 1-inch label. Labels are also important if you decide someday to share your collection with others.

So little is known about bees. Distributions, seasonality, and floral preferences are poorly understood for many species. As the landscape on which they reside shifts, as habitats become fragments, and as climate patterns change, researchers fully expect that distributions, seasonality, and even floral preferences may change too. Any bit of information that can be shared by natural historians, armchair entomologists, and curious gardeners is extremely helpful. If you would like to contribute your own findings, find an online website that is actively seeking such information, or, if you have no desire to keep your collection of identified specimens, consider donating them to local insect collections (most universities have them), to the US National Pollinating Insects Laboratory, or to your local science museum.

An alternative to collecting and preserving your bee specimens is to make a photographic collection. Investing in a good-quality macro lens will improve your chances of getting high-quality photographs. Two main methods are used for getting good photos of bees: taking photos of them in the wild, or photographing them in captivity. Bees often come back to the same patch of flowers time and again as they collect resources for their nests, so, as a photographer, you can stalk a flower much like a hunter who waits near a water hole. Frequently it is easiest to set up a tripod, point the camera at a flower that is getting a lot of attention from bees, then wait for bees to come. Just as when you are collecting bees as specimens, you need to be fairly still while waiting for bees to visit. It is often a good idea to set your camera on rapid fire or an equivalent setting so you can take several pictures in a short amount of time, increasing the likelihood of getting a sharp image. Also, because many bees are shiny, and nearly all bees are most active when it is sunny, we suggest positioning yourself so the sun is at your back, ensuring that the side

An example of a label; this specific label identifies the country, county, location, date, and collectors' names.

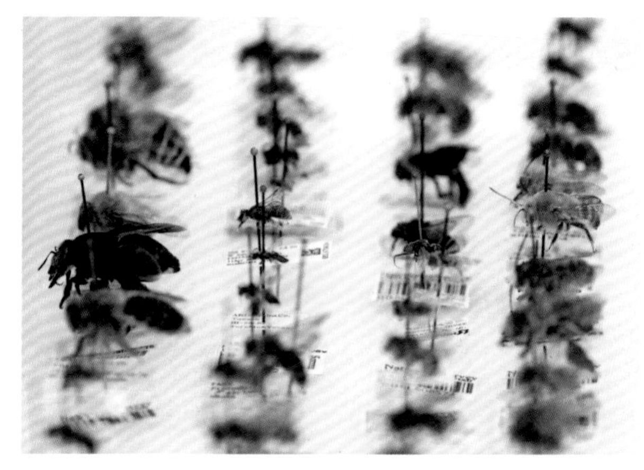

Pinned and labeled bees can be useful to scientists and can add to our understanding of bees around the world.

of the bee facing the camera is fully illuminated. You'll have to experiment with your exposure settings and depth of field to get a shot to look just the way you want it, but generally a larger depth of field and even a bit of overexposure are best. Balancing these two things with a fast enough shutter speed to keep the bee from appearing blurry is the challenge. Half the fun is in the hunt, however, and photographing bees in the field can be as rewarding as it is challenging. It also gives you a great opportunity to observe bees in their natural environment.

If taking pictures in the field proves too challenging, consider the option, used by many photographers, of taking pictures of a chilled bee. A bee left in the refrigerator for an hour, then placed back on a flower will soon wake up and slowly begin moving. It may forage for nectar or stop to sun itself and warm up. This gives photographers a few seconds of relative calm to snap a photo. Try to think about placing the bee so that, when it revives, it looks natural. Remember to think about the background behind your bee in these situations as well.

1.10 IDENTIFYING BEES

The stories that go with each genus and even species of bee are unique, and the most pleasure for a bee enthusiast can be gleaned from knowing these backstories. The majority of this book comprises brief accounts of each bee genus found in United States and Canada; however, in order to learn the inside story of a particular bee, you must know how to distinguish it from the other bees you see. Daunting though this may seem, most bees carry with them subtle hints as to their identity. Once you know what to look for, you will be amazed at how many kinds you can find in your flower garden, beside the trail as you hike, or even among decorative flower boxes at the zoo. Identifying bees may be based on years of science, but it is also an art; if you ask a bee expert to tell you how he or she knows that a bee is an *Osmia*, the likely reply will be, "Because it looks like an *Osmia*." There is a gestalt sense to basic bee identification that has little to do with microscopic details and more to do with the overall picture. What flower was the bee visiting? Where in the country did you see the bee? What time of year was it? Were there any distinctive features on the bee? Being aware of these factors can tell you a lot.

Nonetheless, there are more than 100 genera of bees in the United States and Canada. A few quick reference tricks can cue you in to the right chapter or section of this book. Below are some hints and tips, followed by a true bee key. If you can't identify the bee based on the tips, use the key. Note the following images are not to scale.

■ BEES THAT ARE YELLOW AND BLACK

Anthidium or Anthidiini (sections 7.5–7.6)

Nomada; some species (section 9.1)

Perdita (section 3.3)

■ BEES THAT ARE METALLIC GREEN OR BLUE

Osmia (section 7.2)

Agapostemon (section 6.1)

Augochlorini (section 6.4)

■ BEES THAT ARE METALLIC GREEN OR BLUE *CONTINUED*

Lasioglossum; some species (section 6.3)

Euglossa (section 8.12)

Andrena (section 3.1)

■ BEES WITH A BLACK THORAX (BACK) AND RED ABDOMEN (BELLY)

Lasioglossum; some species (section 6.3)

Sphecodes (section 9.2), and other parasitic bees (chapter 9)

Andrena (section 3.1)

Ashmeadiella (section 7.4)

Anthophorula; some species (section 8.3)

Perdita and *Macrotera* (section 3.3)

■ BEES THAT HAVE LONG ANTENNAE

Male Eucerini (sections 8.5–8.7)

■ BEES THAT ARE BLACK WITH BRIGHT WHITE STRIPES OF HAIR ON THE ABDOMEN

Halictus (section 6.2)

Lasioglossum (section 6.3)

Caupolicana (section 4.3)

■ BEES THAT ARE BLACK WITH BRIGHT WHITE STRIPES OF HAIR ON THE ABDOMEN *CONTINUED*

Eucerini; some species (sections 8.5–8.7)

Megachile; some species (section 7.7)

Osmiini, some species (section 7.4)

Many parasitic bees (chapter 9)

Anthophora; some species (section 8.8)

Colletes; some species (section 4.1)

Protandrenini and Panurgini (section 3.2)

Andrena (section 3.1)

Lithurgopsis (section 7.1)

Melitoma (section 8.4)

■ BEES THAT ARE COMPLETELY RED

Nomadinae; some species (section 9.1)

■ BEES WITH A TRIANGULAR OR POINTY ABDOMEN

Coelioxys and *Dioxys* (section 9.3)

BEES WITH YELLOW OR WHITE STRIPES ON THE ABDOMEN ITSELF (NOT HAIR BANDS)

Calliopsis (section 3.4)

Perdita and *Macrotera* (section 3.3)

Stelis (section 9.3)

Nomia (section 6.5)

BEES THAT ARE EXTREMELY LARGE (BIGGER THAN THE TOP OF THE THUMB)

Bombus (section 8.10)

Centris (section 8.9)

Xylocopa (section 8.1)

Caupolicana (section 4.3)

Protoxaea (section 3.5)

BEES THAT ARE GNAT-SIZED

Perdita (section 3.3)

Lasioglossum; some species (section 6.3)

Ceratina (section 8.2)

BEES WITH DARK WINGS

Xylocopa (section 8.1)

Dieunomia (section 6.5)

Bombus (section 8.10)

BEES THAT ARE REALLY FUZZY

Apidae (chapter 8, sections 8.3–8.9).

BEES THAT ARE SMALL, SHINY, AND BLACK

Ceratina (section 8.2)

Protandrenini and Panurgini (section 3.2)

Hylaeus (section 4.2)

Osmia; some species (section 7.2)

Dufourea (section 6.6)

Osmiini; some species (sections 7.3–7.4)

KEY TO THE CHAPTERS OF THIS BOOK

A *dichotomous key* is a taxonomic "if-then" tool that scientists use to determine the species, or genus, or even family to which an organism belongs. Think of dichotomous keys as Choose-Your-Own-Adventure novels for the naturalist. Each couplet in a key consists of two options that can be answered with yes or no; typically the couplets are written out as two long descriptive sentences, but we have simplified the style in our key below. The nature of your answer, whether yes or no, determines the next couplet on your path to discovering an organism's identity.

While a key may seem intimidating at first, if you follow a few simple rules, it is manageable even for a beginner. First, always read *all* of a statement. Second, keep in mind that a sentence comprising several pieces usually lists those pieces with the most important and noticeable features first. Third, if you can't decide which path to choose (yes or no), try following both to the next set of couplets and see if your uncertainty is cleared up. Finally, remember that none of these statements stands alone. The bees that belong with statement 5 have the characteristics mentioned in all statements that led to statement 5. Reading that a bee has a long tongue in one statement does not necessarily mean it is an Apidae, because Megachilidae also have long tongues but are split out early in the key.

This key requires a microscope, a hand lens, or several macrophotographs. A final note: some of these characters are on the tongue of the bee and may be hard to see. For a key to bee families, there is no way around this, unfortunately, except to learn the tribes, subfamilies, or genera of bees by sight, so that you can skip the family key entirely. Refer to section 1.8 for clarification on body parts.

1. Does the bee have pollen-collecting hairs on its abdomen or legs?

   YES (it is a female) (A and B)..................Go to statement 2

   NO (it is a female parasitic bee, a *Hylaeus*, or it is male)...Go to statement 10

(A) Scopal hairs on a female *Eucera* leg.

2. Are the pollen-collecting hairs on the underside of the abdomen?

   YES (it is in the family **Megachilidae**) (B). *You might also note that bees with pollen-collecting hairs on the underside of the abdomen have just two submarginal cells on their wings* ....**Go to chapter 7**

   NO. *These bees may have two or three submarginal cells on their wings*...............................Go to statement 3

(B) Scopal hairs on a female *Megachile* abdomen.

3. Are there two grooves (subantennal sutures) on the face, under each antennal socket, running all the way to the top of the clypeus (C)?

   YES (it is in the family **Andrenidae**) *Note that there are a few Andrenidae in the Panurginae subfamily that have only one subantennal suture; see chapter 3 identification tips for more information.* ... **Go to chapter 3**

   NO ....Go to statement 4

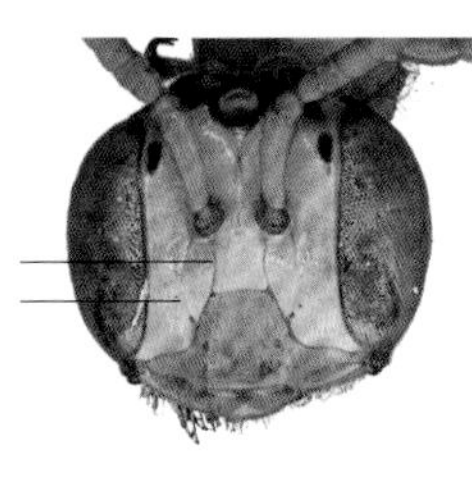

(C) A species of *Perdita*, showing the two subantennal sutures beneath each antennal socket. All Andrenidae have two subantennal sutures. PHOTO COURTESY OF USDA-ARS BEE BIOLOGY AND SYSTEMATICS LABORATORY.

4. Is the tongue long, with the first two sections of the labial palp elongated?

   YES (it is in the family **Apidae**) (D) .... **Go to chapter 8**

(D) A long tongue, showing two long labial palps (numbered 1 and 2), and two short palps (3 and 4). PHOTO COURTESY OF USDA-ARS BEE BIOLOGY AND SYSTEMATICS LABORATORY.

4. *continued*

   NO (it is a short-tongued bee) (E) ....Go to statement 5

(E) A short tongue, showing the four more or less equally long palpal segments. PHOTO COURTESY OF USDA-ARS BEE BIOLOGY AND SYSTEMATICS LABORATORY.

5. On the wing, is the second recurrent vein strongly curved, so that it looks like the letter S (F)?

   YES (it is in the family **Colletidae**, the subfamily Colletinae). *The tongue will also have two little lobes at the ends (G), or may occasionally be rounded, but it is never pointy* .... **Go to section 4.1**

   NO ....Go to statement 6

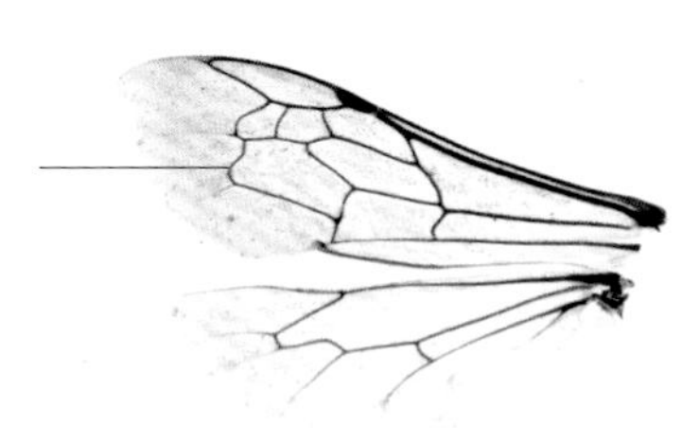

(F) The wing of a *Colletes*, showing the strong curve in the second recurrent vein (in red). Compare this with the *Halictus* wing below; the second recurrent vein in that bee is not curved.

(G) Colletidae all have short tongues—note that the labial palps are all the same length in this image. Also, the tip of the glossa (the middle portion), is split in two. In *Colletes* the two tips are often rounded. PHOTO COURTESY OF USGS BEE INVENTORY AND MONITORING LAB

6. On the wing, is the basal vein strongly curved, looking like the letter J rotated on its side (H)?

   YES (it is in the family **Halictidae**, subfamily Halictinae) .... **Go to chapter 6**

   NO ....Go to statement 7

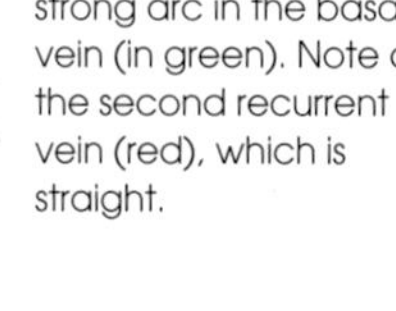

(H) The wing of a *Halictus*, showing the strong arc in the basal vein (in green). Note also the second recurrent vein (red), which is straight.

7. Are the antennae very low on the face, only just above the top edge of the clypeus, and clearly emerging well below the halfway point of the compound eyes?

   YES (it is in the family **Halictidae**, subfamily Rophitinae) (I) *The diameter of the antennal socket is typically greater than the distance from the bottom of the antennal socket to the top of the clypeus.* .... **Go to section 6.6**

   NO ....Go to statement 8

(I) The face of a *Protodufourea* showing the low placement of the antennae on the face (just above the clypeus).

8. Are the wings either completely black or tipped with black (J), OR does the abdomen have beautiful pearly bands running across each section of the abdomen (K)?

   YES (it is in the family **Halictidae**, subfamily Nomiinae) .... **Go to section 6.5**

   NO ....Go to statement 9

(J) A *Dieunomia*, showing the typically darkened wing tips; some species have wings that are entirely black.

(K) A *Nomia*, with pearly, almost iridescent bands on the abdomen.

continued overleaf

KEY TO THE CHAPTERS OF THIS BOOK *continued*

9. Is the bee large, more than half an inch, and extremely fuzzy?

   YES (it is in the family **Colletidae**, subfamily Diphaglossinae) (L) *You might also note that the tongue is forked, with each fork pointed, and that the stigma on the wing is tiny to almost invisible*.................................... **Go to section 4.3**

   NO (it is in the family **Melittidae**) (M) .... **Go to chapter 5**

(L) A large, hairy *Caupolicana*. All bees in the Diphaglossinae bee tribe are large like this, quite the opposite of the Melittidae in the other couplet of this section of the key.

(M) A *Hesperapis*. All bees in the Melittidae bee family are petite, as opposed to the large Diphaglossinae in the other couplet of this section of the key. Note also the flattened abdomen, which is typical of pinned specimens of *Hesperapis*.

10. Is the bee a male? *Count the number of segments making up one antenna; males will have 13 segments, females have 12. Look also for a sting protruding from the abdomen. It may or may not be visible, but if it is, the bee is clearly a female.*

    YES (it is a male)...................................Go to statement 12

    NO (it is female).....................................Go to statement 11

11. Does the bee have yellow markings on its face, running beside each compound eye?

    YES (it is in the family **Colletidae**, genus *Hylaeus*) (N).. **Go to section 4.2**

    NO (it is a cleptoparasitic bee) .... **Go to chapter 9**

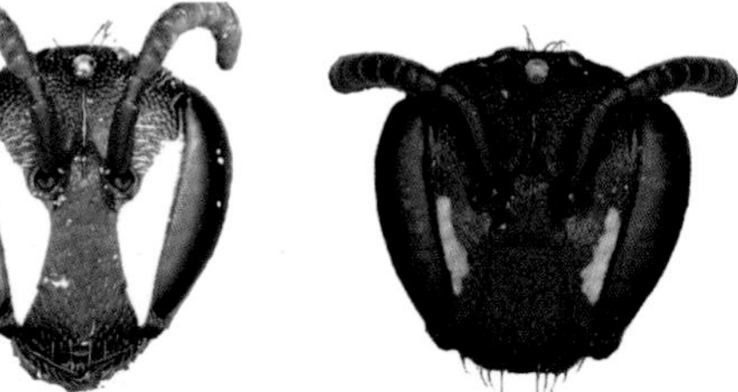

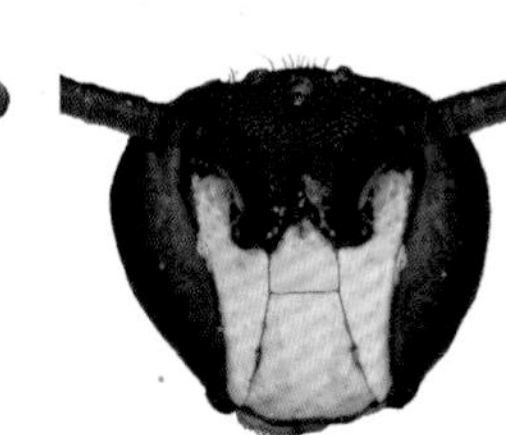

(N) Three *Hylaeus* faces, demonstrating the variability in color patterns. Despite the difference, the yellow or white consistently runs up beside the compound eyes. They share a similar overall look with cleptoparasitic bees, including the lack of pollen-collecting hairs on the females; however, only *Hylaeus* has facial markings like these. PHOTOS COURTESY OF USDA-ARS BEE BIOLOGY AND SYSTEMATICS LABORATORY.

12. Are there two grooves running from each antennal socket all the way to the top of the clypeus (called subantennal sutures) (O)?

    YES (it is in the family **Andrenidae**)........... **Go to chapter 3**

    NO ....Go to statement 13

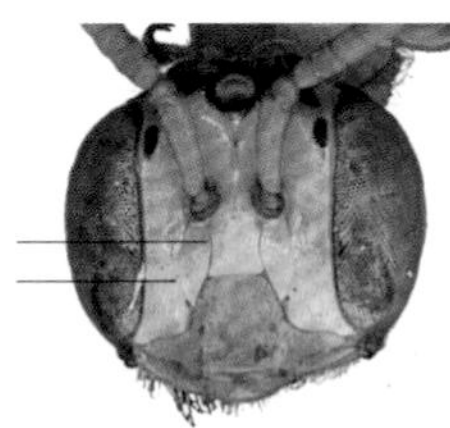

(O) A species of *Perdita*, showing the two subantennal sutures beneath each antennal socket. All Andrenidae have two subantennal sutures. PHOTO COURTESY OF USDA-ARS BEE BIOLOGY AND SYSTEMATICS LABORATORY.

13. Is the tongue long, with the two top segments of the inner portion of the tongue (the labial palp) being at least twice as long as the bottom two segments of the labial palp?

    YES (it is a long-tongued bee) (D)..........Go to statement 14

    NO (it is a short-tongued bee) (E) ....Go to statement 15

14. Is the labrum (the part beneath the clypeus that folds under the bee—look from below) attached to the clypeus for just a small portion of its width, because its basal edges, and/or the end of the clypeus, are rounded?

14. *continued*

YES (it is in the family **Apidae**) (P). *The labrum also tends to be square, or wider than long. These bees have either two or three submarginal cells and tend to be hairy (though not always).* .... **Go to chapter 8**

NO (it is in the family **Megachilidae**) (Q). *In these bees, the base of the labrum attaches to the clypeus for its entire width. It tends to have a straight base and is usually longer than wide. These bees only ever have two submarginal cells. The abdomen of these bees tends to be cylindrical, though not always.* .... **Go to chapter 7**

(P) The square labrum of a *Eucera* (Apidae). Its base (where it attaches to the clypeus) is curved so that it only attaches to the clypeus for part of its width.

(Q) The rectangular labrum of a *Chelostoma* (Megachilidae). The labrum is attached, or hinged, to the clypeus for its entire width, which is only just visible here. Photo courtesy of USDA-ARS Bee Biology and Systematics Laboratory.

15. Does the second recurrent vein have a strong curve in it, giving it a distinctive S-shape? (R)

YES (it is in the family **Colletidae**, subfamily Colletinae). *The tongue will also have two little lobes at the ends, or it may occasionally be rounded, but it is never pointy.* (S)........ **Go to section 4.1**

NO ....Go to statement 16

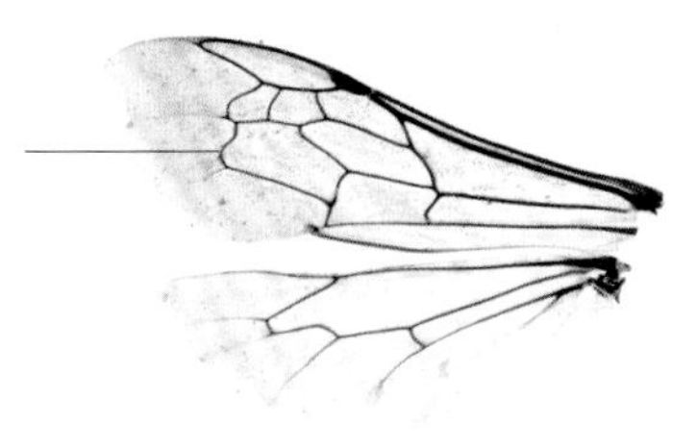

(R) The wing of a *Colletes*, showing the strong curve in the second recurrent vein (in red). Compare this with the *Halictus* wing below; the second recurrent vein in that bee is not curved.

(S) Colletidae all have short tongues—note that the labial palps are all the same length in this image. Also, the tip of the glossa (the middle portion), is split in two. In *Colletes* the two tips are often rounded. Photo courtesy of USGS Bee Inventory and Monitoring Lab

16. On the wing, is the basal vein strongly curved, looking like the letter J rotated on its side?

YES (it is in the family **Halictidae**, subfamily Halictinae) (T) .... **Go to chapter 6**

NO .... **Go to statement 17**

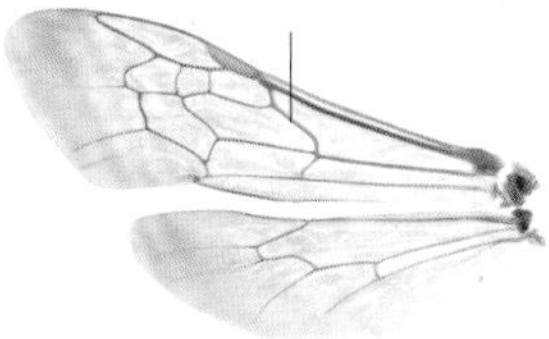

(T) The wing of a *Halictus*, showing the strong arc in the basal vein (in green), making it look like the letter J turned on its side. Note also the second recurrent vein (red), which is straight.

17. Are the antennae very low on the face, only just above the top edge of the clypeus, and clearly emerging well below the halfway point of the compound eyes?

YES (it is in the family **Halictidae**, subfamily Rophitinae) (U) *The diameter of the antennal socket is typically greater than the distance from the bottom of the antennal socket to the top of the clypeus.* **Go to section 6.6**

NO ....Go to statement 18

(U) The face of a *Protodufourea* showing the low placement of the antennae on the face (just above the clypeus).

continued overleaf

KEY TO THE CHAPTERS OF THIS BOOK *continued*

18. Are the wings either completely black or tipped with black (V), OR does the abdomen have beautiful pearly bands running across each section (W)?

    YES (it is in the family **Halictidae**, subfamily Nomiinae)..**Go to section 6.5**

    NO ....Go to statement 19

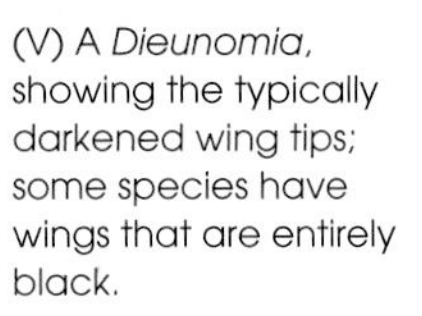

(V) A *Dieunomia*, showing the typically darkened wing tips; some species have wings that are entirely black.

19. Is the bee large, more than half an inch, and extremely fuzzy?

    YES (it is in the family **Colletidae**, subfamily Diphaglossinae). *You might also note that the tongue is forked, with each fork pointed, and that the stigma on the wing is tiny to almost invisible*................................**Go to section 4.3**

    NO (it is in the family **Melittidae**)..........**Go to chapter 5**

(W) A *Nomia*, showing the pearly, almost iridescent bands on the abdomen that characterize this genus.

1.11 APPRECIATING BEES

Each evening, millions of people sit in front of the television absorbing the latest twists and turns in the stories of their favorite fictional (or nonfictional) characters. We process new details in the relationships between friends and enemies, the historical reasons for ongoing romantic trysts, new battles and life-threatening problems, the lives of heroes, villains, and those in between. With enthusiasm we can recite the details of seven years' worth of story line to our coworkers and friends the next day, and plot lines, quips, and interesting facial expressions become memes recognizable by the majority. Our minds are capable of remembering incredibly complicated narratives, and we devote a considerable portion of our brain to keeping track of the plots unfolding in our lives, as well as the fictional ones from books and television, when the real ones aren't enough.

Amazingly, the real world—specifically the natural world—abounds with stories as complex and engaging as any fictional account, with the same twists and turns, interesting characters, and nail-biters. Right under our noses, plots unfold each year involving exquisite creatures upon which we usually unknowingly rely for both the aesthetic beauty of the flowering world and the gastronomic divinity embodied in the fruits and vegetables gracing our dinner plates each evening: bees.

Several pieces of evidence suggest that some bees may be experiencing population declines as a result of rapid environmental changes in the last 25–50 years. The use of ever-stronger pesticides, the expansion of monocultures of nonflowering crops, the fragmenting of habitats, and the introduction of deadly diseases are all likely contributors. Our habits have pushed bees from thriving among us to getting by on the sidelines. We appreciate best those things we most clearly comprehend; therefore, a first step in helping our tiny pollinators is understanding them. When we are armed with information, a good rallying cry, and the support of the many, bees stand every chance of being welcome, beneficial, and integral parts of our neighborhoods. Besides, helping bees is so easy; they are opportunists, as interested in their perseverance as we are.

A bee in the genus *Diadasia* resting on a cactus flower that has recently been showered by a late spring downpour.

Abundant, spectacular, and important to countless aspects of our daily lives, bees have a phenomenal story to tell. Biology is perceived as difficult and beyond the average citizen. The stigma surrounding science may dissuade many from learning about the amazing life histories of bees. As a result, the details of bees' lives have been replaced with generalities and myths: "all bees sting," "bees are dangerous," "bees live in hives and make honey." The truth is much more fascinating and completely accessible to anybody who loves a good story. Our intent with this book is to make the tales of the bees exciting and informative for those we believe could become their biggest advocates; those most likely to lend them a helping hand: you.

While most people recognize and appreciate the beauty and diversity of flowers found in North America, like these blooming during a rare wet spring in California, most people are not familiar with the similarly stunning array of bees that pollinate them.

2

PROMOTING BEES IN YOUR NEIGHBORHOOD

It is thought that more than 70% of flowering plants are pollinated by insects, and the majority of those pollinators are bees. Specialists, generalists, males, and females: all kinds of bees can be exceptional pollinators, far surpassing other flower-visiting invertebrates. Why are bees better pollinators than other insects? Mostly because the number of offspring they leave behind is directly correlated to the number of flowers they visit. They are thus highly motivated to visit as many flowering plants in a day as they can.

The colorful fruits and vegetables that line the shelves in the supermarket are by and large pollinated by bees, as are many of the spices and herbs that brighten our favorite dishes. Even several of the delectable beverages we enjoy are brought to us by pollen-moving bees. Moreover, bees indirectly aid in the production of the beef and dairy products that we enjoy by pollinating alfalfa, the major fodder of cattle. Finally, the natural sources of many of the vitamins important to our well-being are a result of bees pollinating flowers and providing us with an assortment of "power foods" rich in vitamins A, C, K, E, potassium, and folic acid.

Many flowers are visited by a wide variety of bees. Here a large carpenter bee (*Xylocopa*) forages on a wild rose (*Rosa*) while a *Ceratina* flies by.

No bees, no ice cream

Most people recognize the fact that bees help in the production of our fruits and vegetables. Because alfalfa is a major food source for dairy cows, and because alfalfa needs bees in order to produce seeds, bees are responsible for many of the dairy products we eat (including ice cream!). Even the vanilla in your ice cream is made possible by bees.

The fruits and vegetables (blue), nuts (brown), beverages (yellow), herbs and spices (purple), oils (pink), and forage crops (green) known to be pollinated by bees. The size of the font indicates the importance of bees to its successful pollination, with a bigger font meaning that the crop is more reliant on bees.

Flowers and the bees on which they rely are often more abundant near roads, particularly in dry areas like the Sonoran Desert (pictured here). Because the little rain that falls collects at the edges, roads provide above-average water for wildflowers. PHOTO BY LINDSEY E. WILSON.

There is no doubt that bees are important to the lifestyle we enjoy, which is why it is disconcerting that evidence has been found suggesting that at least some species of bees are in decline. Honey bee declines are the most well documented (we discuss potential reasons for these declines in section 8.11), but some bumble bee populations are also in peril, and the status of other native bees is unknown.

Documenting changes in bee populations is tricky because (1) few researchers are studying them now, (2) even fewer studied them 50 years ago, and therefore (3) establishing a baseline for accurate measurements is difficult. Regardless, we know enough about bees to know that certain modern trends are likely detrimental to bees; in other words, we can guess that bees are being negatively affected even if we can't say by how much. Bees require nesting habitat, which is becoming difficult for ground-nesting bees to come by in a heavily paved world (and the majority of bees are ground nesters). Habitat is becoming fragmented; large patches of flowers are reduced to overgrown remnants, surrounded by neighborhoods or shopping centers. Enormous chunks of land are planted with one type of crop—often corn or soybeans—with little room for the profusion of native wildflowers that used to support an abundance of bees. Often these monocultures are treated with pesticides to kill harmful insects, but beneficial insects in the area may become unfortunate collateral damage. Finally, non-native pathogens now found in the United States and Canada can be especially deadly to native bees that have not evolved with them.

On the other hand, paved roads have allowed the expansion of many native flowers; runoff from concrete or asphalt creates an unnatural moist patch next to the road. Flowering plants (and their bee visitors) have an easy avenue for expansion into new territories. And perhaps more importantly, citizen scientists, amateur naturalists,

Are roads good for bees?

Flowers are often more abundant along roadsides, making those areas a boon to bees. In the arid west this is because of the precious extra water that runs off the roads when it rains. In the Midwest and east, shade-producing trees are cleared from roadways, producing areas of copious sunlight for wildflowers otherwise unable to flourish. On the other hand, roads, depending on the speed limit, can also be bad for bees, which end up smashed on windshields, as they try to make their way from one side of the road to the other.

A bumble bee (*Bombus*) visiting cat's eye (*Cryptantha*) flowers. PHOTO BY B. SETH TOPHAM.

As an *Agapostemon* (left) and a bumble bee (*Bombus*, right) forage for pollen and nectar, they transfer pollen from one plant to another, pollinating them.

Bees are not only responsible for pollinating many of the foods we eat, they are also the main pollinators of many wildflowers. PHOTO BY LINDSEY E. WILSON.

and even entire communities of inquisitive urban dwellers are showing an interest in their native pollinators. Through the planting of native bee-loving wildflowers and the careful placement of bee nesting materials, the loss of habitat and resources that may be affecting our native bees can be mitigated. Most bees are resilient and opportunistic, more than willing to adjust to changing landscapes in order to survive. Look at the carpenter bees (*Xylocopa;* see section 8.1) that have incorporated decks into their list of acceptable structures for nesting. Or the sweat bee (*Lasioglossum;* see section 6.3) species that live in Central Park, New York City, surrounded by, and apparently oblivious to, walls of concrete. Look at the blue orchard bees (*Osmia;* see section 7.2), which happily visit apples, cherries, and almonds all spring long, filling their nests with this pollen that their ancestors 200 years ago never knew.

2.1 BEES AS POLLINATORS

Bees are essential pollinators of many flowers, both wild and cultivated. As stationary beings, flowers require a matchmaker to play go-between in their relationships. Bees are excellent intermediaries, transporting pollen from one flower to another and assuring the production of plant offspring in their quest to provide for their own offspring.

Pollen is a tiny vessel containing the male reproductive portion of a flower, which makes it the plant equivalent of sperm. This pollen is found inside packages called anthers, so sperm is double packaged—inside pollen that is then packaged in the anthers. The anthers are presented at the ends of long filaments in flowers. Among other things, the female portion of a flower includes a platform designed specifically for catching pollen, called the stigma. It is

Some bees like this *Ceratina* have very little hair and are therefore less effective pollinators than hairier bees.

Hairy bees like this *Melissodes* often transfer a lot of pollen between plants because pollen sticks to their hair as they move from one plant to another. Note the yellow dusting on its thorax (back).

Anthers hold all the pollen for a plant, and within each grain of pollen is sperm, which must be delivered to another flower in order for fertilization to occur. Bees are one of the primary transporters of pollen between flowers. Note also the gray balls among the anthers. These are stigmas, the female parts of a flower that are receptive to grains of pollen.

slightly sticky and variously shaped depending on the flower species. Pollen that lands on its surface takes root. A tiny tendril extends from the pollen grain, boring a hole down into the flesh of the stigma, eventually making its way to the ovaries. The sperm rolls down the tunnel thus created and fertilizes the plant's eggs.

Most flowering plants have both anthers and a stigma and could conceivably self-fertilize, depositing their own pollen on their own stigma; however, just as inbreeding is less than ideal in animals, the best seeds form when flowers are fertilized by pollen from different flowers of the same species. Wind can pollinate some flowering plants, but animals, especially insects, and most especially bees, are estimated to be essential to the pollination of the majority of the world's flowering species.

Flowers generally provide nectar below or behind the reproductive parts of the plant. This *Andrena* has to pass by anthers full of pollen, and a stigma (hidden among the anthers in this photo), before reaching sweet nectar. In this way, flowers ensure that visiting bees pollinate them.

Just as businesses advertise with showy signs and freebies, flowers have showy petals and elaborate scents to lure in bees (and other pollinators). Once enticed to stop, bees are rewarded with free sweet nectar, usually found at the base of a flower. To enjoy a sip, therefore, a bee must pass by pollen-loaded anthers and also the stigma. Generally they pass the stigma first, inadvertently loading it with pollen from flowers previously visited. As they maneuver lower, they harvest some new pollen and also wet their whistles. Bees are not conscious of their importance, but their self-serving efforts inadvertently promote the production of seeds and the next generation of flowers.

Bees have evolved to be efficient. Unlike other insects, the number of their young that survive to adulthood (their "reproductive success") is determined in large part by the amount of pollen they can collect. Bees, in other words, have a huge incentive to visit hundreds of flowers in a day. From a plant's point of view, this busy buzzing is beneficial not just because it moves pollen, but also because it can move pollen a great distance. Rather than "mating" with their next-door neighbors, bees enable plants to breed with individuals in entirely separate patches. This mixing of pollen between populations keeps the plant's offspring healthy, vigorous, and more resilient.

Winter rains enable the growth of large fields of wildflowers in California. Many bee species can be found in these fields, both on the flowers and nesting in the ground beneath.

Truckloads of bees

The California almond is a crop that relies on honey bees for seed set. The flowers cannot self-fertilize, and in the absence of bees, there is almost no almond yield. Almonds cover over 800,000 acres of California and are a large part of the state's agricultural industry. To ensure a harvest, more than 1.5 million colonies of honey bees (and each colony has several thousand bees) are placed in almond orchards each spring!

In addition to being important pollinators of many wildflowers, bees are a necessary component in the production of many valuable crops. Bee crop pollination is in fact a multi-billion dollar industry in the United States and Canada, as well as in Mexico, where many bees are now reared. So important are bees that we ship them from crop to crop, moving boxes of honey bees (and, increasingly, other bees as well) all across the country so that, when orchards bloom, the bees are there to increase fruit-set. Many crops benefit not only from the bees that we actively manage, but also from the native pollinators that naturally occur in the area around the crop. Those fields and orchards that are near wild areas (where bees are, unsurprisingly, more abundant) often produce more fruit than fields and orchards that are surrounded by monocultures or urban areas. While some crops would produce a harvest without bees, either wild or managed, in many cases the harvest would be significantly less.

Though bees can be helpful and, in some cases, essential pollinators, not all bee visits benefit the flower. For example, tiny bees like *Perdita* often visit just the anthers of a flower, collecting pollen without ever making contact with the stigma. As a result, they "steal" pollen from the flower, lowering the potential number of seeds it could make. Other bees are nectar robbers. Because their tongues are not long enough to reach all the way down the throat of a long-necked flower, they create a back entrance, chewing a hole at the base of a flower's petals so they can access nectar inside. Since they don't run the anther and stigma gauntlet, they don't pollinate. Bees that aren't very hairy, like *Hylaeus* and *Ceratina*, probably carry less pollen and are less effective pollinators than hairy species.

Bees have evolved for eons along with their floral counterparts. Fossils reveal that the first bees appeared during the Cretaceous time period, roughly 125 to 100 million years ago. The climate at that time would have been warm and wet, like a humid summer day in the Midwest. Pine trees and ferns shaded an abundance of dinosaurs, with many small mammals and birds underfoot. Sea levels were higher, and shallow estuaries, swamps, and marine bays made for long coastlines, where early flowering plants rooted in the mud and bloomed just above the water's surface. A few flowering plant species occurred on land, probably looking similar to today's magnolias. The ancestors of grasshoppers, aphids, beetles, and butterflies were likely visitors, where they were preyed on by another common group: wasps, including the wasp that is the ancestor to all modern-day bees.

Using genetic techniques, we know that the closest relatives to bees are meat-eating wasps, particularly those in the family Sphecidae. Interestingly, pollen and insect flesh are not that different in their nutritional value. Both are high in

A honey bee (*Apis mellifera*) foraging among the petals of a Rocky Mountain bee plant (*Cleome*) while a tiny *Perdita* (on the flower stalk to the left) harvests pollen from the anthers at the end of long filaments. Many small bees are poor pollinators of flowers because they collect pollen but never come in contact with the stigma. PHOTO BY B. SETH TOPHAM.

These manzanita (*Arctostaphylos*) flowers have been robbed by a thirsty bee. The thief chewed holes near the base of many of the flowers to access the nectar. Now this *Halictus* is taking full advantage of the easy access. These flowers won't be as attractive to other bees that might enter in the "appropriate" manner, and fruit set will likely diminish as a result.

Bees are most closely related to wasps in the family Sphecidae like this digger wasp.

nitrogen and contain all the other building blocks necessary for growth and function (proteins, amino acids, starches, and even vitamins). Plant pollens also produce sterols, the basic building blocks of many hormones, which bees cannot synthesize on their own. Without these hormones, bees cannot develop from egg to adult. The exact scenario in which the first bee chose to forage on flowers instead of hunt for live prey can only be imagined, but considering that even present-day wasps visit flowers for nectar, the switch is not too hard to envision. Early "proto-bees" were probably carnivorous, specializing on flower-visiting insects (as Sphecidae still do today). Over time they may have developed a preference for the pollen that dusted the bodies of their prey.

The first flower to be visited and pollinated by early bees was probably disk- or bowl-shaped, with easily accessible pollen. Scientists don't know which plant groups were first visited by bees, but they do know that by 55 to 34 million years ago, bees frequently visited flowers in the sunflower family (Asteraceae). This family is still the main host for many bees at the base of the bee family tree.

Because few other insects were using pollen as a source of food, there was likely little competition for this wonderful and underexploited resource. With a wealth of pollen all to themselves, early bees spread and diversified. Though the first bee probably originated in what is now Africa, by the end of the Cretaceous (60 million years ago) bees ranged at least as far as present-day New Jersey. It is thought that bees and flowering plants aided each other in expanding their respective ranges across the globe, each providing an important service to the other; sustenance in exchange for pollen-delivery services.

2.2 PROVIDING HABITAT

The needs of native bees are simple: food and a home. Secondarily, accents for the home can aid in nesting success. Of course, depending on the bee, the requirements for a home differ. For example, ground-nesting bees need different materials than twig-nesting bees. Below, we list several simple ideas for attracting all kinds of bees to your yard and gardens.

PROVIDE AREAS FOR GROUND-NESTING BEES

- The "ground" in which ground-nesting bees choose to nest varies between species. Some nest in urban lawns, while others prefer bare areas, ranging from hard-packed earth to loose and sandy soil. Thus, preparing habitat for ground-nesting bees is easy—leave an area, any area, undisturbed so the bees can nest in peace.
- Garden pathways of packed dirt are often excellent ground-nesting bee habitat because flowering plants are close by.
- Unpaved driveways or small dirt patches are also good potential habitat.
- Mounds of soil, left undisturbed, can provide not only material for twig-nesters to use in modifying their nest cavities, but also nesting habitat for ground-dwelling bees.
- Build a sand pit. If you live in an area with well-drained soil, dig a hole about 2 feet deep and 2–3 feet in diameter. Fill it with fine-grained sand, or a sandy loam. Use a tamp to compact the sand grains slightly.

Even small areas of bare ground are beneficial for bees, especially if they are near clumps of flowers. Females and males both like to rest on warmed earth while they sun themselves. And males use bare areas as staging grounds for aerial combat as they defend their territory from other males.

Many ground-nesting bees prefer to nest in areas with bare patches of soil. Instead of mulch, try leaving some areas between flowers empty, providing areas for bees to nest. PHOTO BY LINDSEY E. WILSON.

- If your soil is not well drained, a sand pit will not work well because it will fill with water and destroy the bee nests. In these cases, a pile of sand, a raised bed with sand in it, or planter boxes filled with sand can be just as effective.

PROVIDE NATURAL HABITAT FOR TWIG-NESTING BEES

- Plant perennial plants, and leave the dried stems at the end of the flowering season for bee nests the following year. *Yucca* stems, in the southwest, are good for *Xylocopa* bees. Elderberry (*Sambucus*) are good for many other bees in latitudes farther north. Many other plant stalks can also be good homes.
- Leave logs and other woody debris for bees. Stumps can be used for aesthetically pleasing landscaping, while at the same time providing habitat for bees.

Whether you intend to provide habitat or not, many bees are opportunistic and will wriggle their way into your eaves, or other small spaces in buildings and decks. Here, a *Xylocopa* looks for a new nesting site near the roofline.

What if I *don't* want bees?

If you have a colony of ground-nesting bees in an undesirable location, repeatedly dousing the area with water may be enough to persuade them to move.

- Place new nesting material (i.e., logs, dead wood, etc.) out before the ground has thawed, so the earliest emerging bees are aware of new potential nest sites.
- Give new nesting habitats at least a year after you've placed them for bees to find and colonize them.

BUILD A BEE BLOCK

- Use a piece of untreated lumber; it can be any size, but 5–9 inches long by 3–5 inches wide is ideal.
- Drill holes between 0.25 and 0.5 inch in diameter, and about 1.5–3 inches deep (respectively, narrower holes should not be as deep). Several can be put in one block of wood, placed about an inch apart.
- While bees will use any type of hole, you can increase nesting success by drilling holes that are smooth on the inside. This can be achieved by using a brad-point bit when drilling the holes. Another way to make the insides of the holes smooth is to insert rolled wax paper or printer paper into finished holes. You'll need to drill holes slightly larger for the latter alternative, so that with the inserts the hole is still the right diameter.
- Mount the block on a post, and place a "roof" over the top so rain can't fall directly into the bee nest holes. Keep your bee blocks off the ground just a little, to prevent extra moisture from seeping into the wood, and to minimize attack by ants and other nonflying enemies.
- Turn your nesting blocks so that the entrances face south or southeast. On cool days they will warm up earlier and your bees will be able to start gathering pollen sooner.
- Move your bee nests into a protected place for the winter months. You can move them to a shed or unheated garage, or just cover them with a tarp. It is important to note that if you choose to move them, wait until early November. Many wood-nesting bees emerge in the spring; consequently they overwinter as adults. These bees pupate and are then in their adult form by late fall. In this state they are much more stable and less inclined to suffer from the jostling that would come from moving their nesting blocks earlier in the fall. Don't forget to move the blocks back in the early spring.
- In areas of the southwest, adobe blocks can similarly be used. Drill holes in the adobe brick just as you would a piece of wood.

Making a bee block is one of the simplest ways to provide nesting habitat for many bees. Simply drill holes in a piece of wood and place it in a sunny location. Many kinds of cavity-nesting bees, like this leafcutter bee (*Megachile*) will take up residence.

Holes with a diameter of ⅜ inch or less drilled into a block of wood are perfect homes for many wood-nesting bees. Consider putting an eave over the top to prevent rainwater from seeping into the open cavities. Also, keep the block off the ground.

BUILD A BEE NEST BUNDLE

- Instead of drilling holes into a piece of wood, you can provide premade tunnels in twigs for bees.
- Cardboard straws, either prepackaged in a ready-made container and available from an insect supply store, or singly, can be purchased.
- Alternatively you can tie together bundles of hollow elderberry stems, pieces of bamboo, teasel, reeds, yucca stems, or other hollow stemmed plants. Roughly one to two dozen stems can be bundled together.
- The straws should be roughly 6–9 inches in length, and it is important that one end of the straw is closed. Ideally, cut your natural stems just below a node, so that one end is naturally sealed. Otherwise, a small dollop of caulk will work.
- Straws can be stuffed inside a piece of 2- or 3-inch PVC pipe with one end capped. Don't be surprised if some bees nest in the spaces between your straws too!

ABOVE: Making a bee nest bundle is another easy way to attract bees to your yard. This bundle was made from elderberry branches zip-tied together, then hung in a tree with a small cord. Entrances do not all need to face the same direction.

BUILD A BUMBLE BEE NEST

- Use wood that has not been pretreated, but do not use plywood. In total, the nest should be about 9 inches deep, 6 inches wide, and 6 inches tall. Within the nest, you will need to build two connected chambers: a vestibule, about 3 inches deep, and a nesting compartment, about 6 inches deep. When bumble bees enter their nest, they defecate in the vestibule before entering the nesting compartment.
- Drill two to six ventilation holes in the top of one of the side walls, and cover them with fine mesh or netting so that ants cannot get in. It is best to glue the netting in place, as ants will find their way around staples.
- Drill a large-diameter hole (0.5–0.75 inch) near the base of the vestibule as an entryway, and also one near the base of the inner wall, so the bees can walk from the vestibule into their nesting compartment.
- Once your nest is built, lay corrugated cardboard on the bottom of the nesting compartment to absorb extra moisture. Put wool, cotton, dried grass clippings, or moss in the nesting compartment for the bees to use.
- Put a short piece of PVC pipe into the opening between the vestibule and the outside, in order to minimize water getting into the vestibule.
- Place the nest on the ground near a tree, a fence, or a row of shrubs. Pick a location that is warm (preferably south-facing) and dry, but shaded from direct sunlight.
- Put a "roof" over the top of your nest box so it is sheltered from rain, strong wind, and snow.
- Place a log or large rock on top of the nest box so that the roof can't be carried off or blown away.

A bumble bee house sits next to a garden post. It may take bumble bees a year or more to find the house provided for them, but once they discover it, they can be excellent supplemental pollinators of garden plants.

Inside a bumble bee house there is a vestibule (on the left), where bees defecate before entering their living quarters. The main chamber has soft wool material in which to nest. The entrance is at ground level, through the black tube; a second entrance to the main nesting chamber is also at ground level but not visible.

- It may take a year for the nest to be "seasoned," discovered, and deemed acceptable by spring foundresses.
- Another way to build a bumble bee nest is to get a medium-sized flower pot (either clay or plastic is fine) and bury it upside down in the ground. The drain hole, now on the top, serves as an entrance. Place a thick layer of gravel on the bottom for drainage, put a roof over the top so rain doesn't enter the drain hole (a board, resting on a few rocks to elevate it just an inch off the top of the flower pot will suffice), and add cotton, wool, or grass clippings for nesting material. Bumble bees will use the covered outside area as the vestibule, before entering the nesting chamber inside the pot itself.

A FEW ADDITIONAL POINTERS

- In order to minimize parasites, change the nesting blocks, straws, or reeds every few years. Bee nests that are repeatedly reused tend to become popular with the many enemies of bees, diminishing their success.
- If possible, provide an area of packed dirt or mud nearby for the bees to use in modifying their nests.
- Broad-leaved plants can provide important nesting materials for leaf-cutting bees.
- Rock walls are another attractive habitat for nesting bees, and many will nest in the nooks and crannies between rocks.

Rock walls are excellent bee habitat as they provide cracks and crevices in which many bees can build their nests. Note also the unpaved area. Compacted earth is a favorite among many species of ground-nesting bees.

2.3 PROVIDING FOOD

In addition to bee nesting habitat, the other necessary resource for bees in your yard is flowers, for both pollen and nectar.

We have provided lists of flowering plants that can be planted in your yard that are known to be attractive to both specialist and generalist bees. Our lists are divided by region, with regions referring *not* to USDA hardiness zones, but to natural delineations in bee species ranges and seasonality across the year. These regions are fairly coarse; for example, several bees are found only in Florida, which could justifiably be placed in a region by itself. However, from the point of view of growing a flower garden for bees, dividing the regions this finely would serve no purpose. These regions are the same as the ones we use to illustrate the active period for each bee genus throughout this book. An additional note: in this book we use the phrase North America to refer to the United States and Canada.

Tit for tat

It is clear that bees are good for our gardens, but recent studies have shown that our gardens are also good for bees. Neighborhoods with more backyard gardens had higher bee diversity in adjacent natural areas than did neighborhoods with fewer gardens. Gardens can provide reliable pollen and nectar sources for bees living nearby.

ABOVE: Growing a field of mint is not difficult. Experienced gardeners will tell you to plant it in a pot because it can so quickly overwhelm a garden. Not that the bees will mind—mint flowers are a favorite with many kinds, especially Anthidiini.

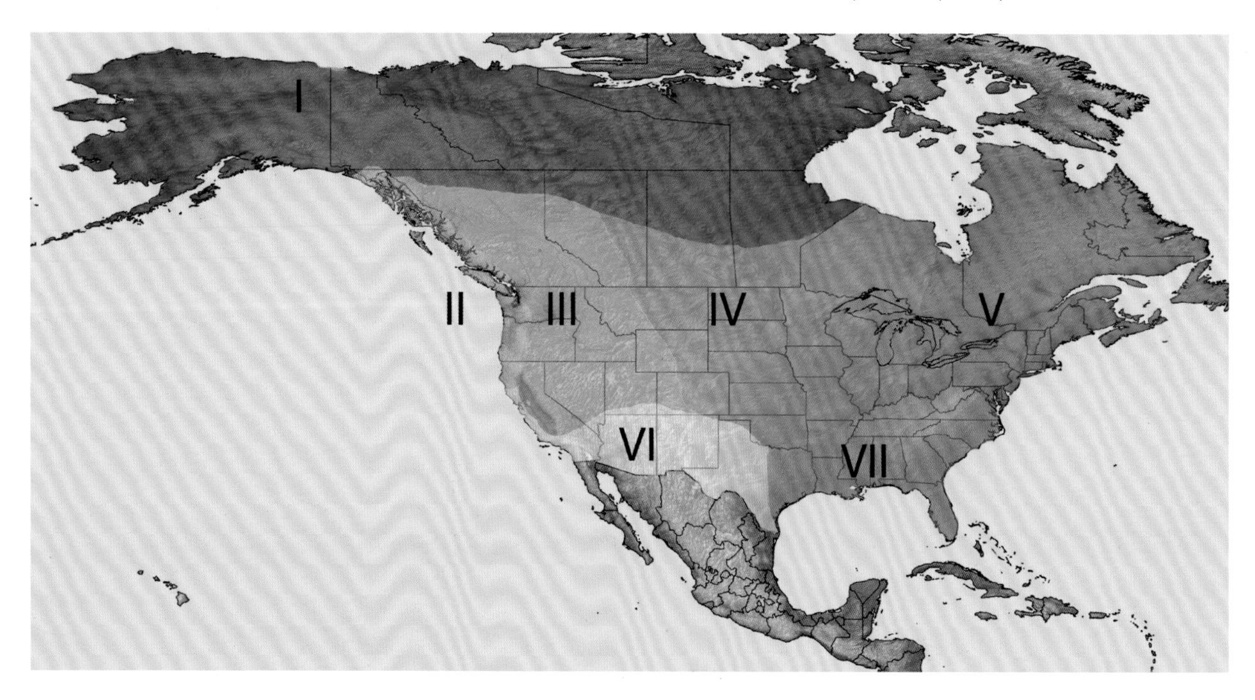

In order to provide useful information about what plants you can use to attract bees to your yard, we have divided the United States and Canada into seven different regions. These regions correspond roughly to natural delineations in the ranges of many bee species. In this book we use the phrase "North America" to refer to the United States and Canada.

For the beginner, easy to grow plants that attract bees (all regions)

Flower color	Scientific name	Common name	Family	Habit[1]
MAY				
	Trifolium	Clover	Fabaceae *(Peas)*	H
	Monarda	Bee balm	Lamiaceae *(Mints)*	H
	Stachys	Lambs ear	Lamiaceae *(Mints)*	H
JUNE				
	Perovskia	Russian sage	Lamiaceae *(Mints)*	H
	Rudbeckia	Black-eyed Susan	Asteraceae *(Sunflowers)*	H
	Echinacea	Coneflower	Asteraceae *(Sunflowers)*	H
	Helianthus	Sunflowers	Asteraceae *(Sunflowers)*	H
JULY				
	Agastache	Hyssop	Lamiaceae *(Mints)*	H/S
	Salvia	Sage	Lamiaceae *(Mints)*	H
	Hibiscus	Hibiscus	Malvaceae *(Mallows)*	H
	Cosmos	Cosmos	Asteraceae *(Sunflowers)*	H
	Nepeta	Catnip	Lamiaceae *(Mints)*	H/S

[1]H = herbaceous; S = shrub; T = tree

Region 1 (Alaska, Northwest Territories, and Nunavut)

Flower color	Scientific name	Common name	Family	Habit
MAY				
	Mertensia	Bluebells	Boraginaceae *(Borages)*	H
	Cornus	Dogwood	Cornaceae *(Dogwoods)*	S
	Fragaria	Strawberry	Rosaceae *(Roses)*	H
	Salix	Willow	Salicaceae *(Willows)*	S/T
	Rubus	Raspberry, Blackberry	Rosaceae *(Roses)*	S
	Symphoricarpus	Snowberry	Caprifoliaceae *(Honeysuckles)*	S
	Polemonium	Jacob's ladder	Polemoniaceae *(Jacob's ladders)*	H
JUNE				
	Amelanchier	Serviceberry	Rosaceae *(Roses)*	S
	Vaccinium	Blueberry, Cranberry	Ericaceae *(Heathers)*	S
	Hedysarum	Sweet pea	Fabaceae *(Peas)*	H
	Lupinus	Lupine	Fabaceae *(Peas)*	H/S
	Rosa	Rose	Rosaceae *(Roses)*	S
	Arnica	Arnica	Asteraceae *(Sunflowers)*	H
	Potentilla	Cinquefoil	Rosaceae *(Roses)*	H/S
	Senecio	Ragwort, Groundsel	Asteraceae *(Sunflowers)*	H
	Delphinium	Larkspur	Ranunculaceae *(Buttercups)*	H
JULY				
	Epilobium	Fireweed	Onagraceae *(Evening primroses)*	H
	Solidago	Goldenrod	Asteraceae *(Sunflowers)*	H
	Erigeron	Fleabane	Asteraceae *(Sunflowers)*	H
	Symphyotrichum and *Aster*	Aster	Asteraceae *(Sunflowers)*	H

Region 2 (British Columbia and the Pacific Northwest)

Flower color	Scientific name	Common name	Family	Habit
APRIL				
	Ribes	Currant	Grossulariaceae *(Currants)*	S
	Amelanchier	Serviceberry	Rosaceae *(Roses)*	S
	Mahonia	Oregon grape	Berberidaceae *(Barberries)*	S
	Balsamorhiza	Balsamroot	Asteraceae *(Sunflowers)*	H
MAY				
	Eschscholzia californica	California poppy	Papaveraceae *(Poppies)*	H
	Helianthus	Sunflower	Asteraceae *(Sunflowers)*	H
	Sidalcea	Checkermallow	Malvaceae *(Mallows)*	H
	Clarkia	Farewell-to-spring	Onagraceae *(Evening Primroses)*	H
	Gaillardia aristata	Blanket flower	Asteraceae *(Sunflowers)*	H
	Collinsia	Blue-eyed Mary	Plantaginaceae *(Plantains)*	H
	Eriogonum	Buckwheat	Polygonaceae *(Buckwheats)*	S
	Penstemon	Beardtongue	Plantaginaceae *(Plantains)*	H
JUNE				
	Lupinus	Lupine	Fabaceae *(Peas)*	S
	Veronica	Speedwell	Plantaginaceae *(Plantains)*	H
	Monardella	Mountain balm	Lamiaceae *(Mints)*	H
	Rosa	Rose	Rosaceae *(Roses)*	S
	Holodiscus discolor	Oceanspray	Rosaceae *(Roses)*	S
JULY				
	Solidago	Goldenrod	Asteraceae *(Sunflowers)*	S
	Asclepias speciosa	Showy milkweed	Apocynaceae *(Dogbanes)*	H
	Erigeron	Fleabane	Asteraceae *(Sunflowers)*	H
	Symphyotrichum and *Aster*	Aster	Asteraceae *(Sunflowers)*	H

Region 3 (Columbia Plateau, Great Basin, Snake River Basin, and Canadian Great Plains)

Flower color	Scientific name	Common name	Family	Habit
APRIL				
	Salix	Willow	Salicaceae *(Willows)*	S/T
	Arctostaphylos	Manzanita	Ericaceae *(Heathers)*	S
	Astragalus	Milkvetch	Fabaceae *(Peas)*	H
	Mahonia	Oregon grape	Berberidaceae *(Barberries)*	S
MAY				
	Oenothera and *Camissonia*	Evening primrose	Onagraceae *(Evening primroses)*	H
	Penstemon	Beardtongue	Plantaginaceae *(Plantains)*	H
	Sphaeralcea	Globe mallow	Malvaceae *(Mallows)*	H/S
	Allium	Onions	Amaryllidaceae *(Amaryllises)*	H/S
	Phacelia	Scorpionweed	Boraginaceae *(Borages)*	H

Flower color	Scientific name	Common name	Family	Habit
MAY *CONTINUED*				
	Opuntia	Prickly pear	Cactaceae *(Cactuses)*	S
	Lupinus	Lupine	Fabaceae *(Peas)*	H
JUNE				
	Cleome	Bee plant	Cleomaceae *(Bee plants)*	H
	Asclepias	Milkweed	Apocynaceae *(Dogbanes)*	H
	Dalea	Prairie clover	Fabaceae *(Peas)*	H
	Mentzelia	Blazing star	Loasaceae *(Blazing stars)*	H
	Helianthus	Sunflower	Asteraceae *(Sunflowers)*	H
JULY				
	Solidago	Goldenrod	Asteraceae *(Sunflowers)*	S
	Symphyotrichum and *Aster*	Aster	Asteraceae *(Sunflowers)*	H
	Erigeron	Fleabane	Asteraceae *(Sunflowers)*	H
	Grindelia	Gumweed	Asteraceae *(Sunflowers)*	H
AUGUST				
	Chrysothamnus	Rabbitbrush	Asteraceae *(Sunflowers)*	S

Region 4 (Great Plains)

Flower color	Scientific name	Common name	Family	Habit
APRIL				
	Salix	Willow	Salicaceae *(Willows)*	S/T
	Amelanchier	Serviceberry	Rosaceae *(Roses)*	S
MAY				
	Coreopsis	Tickseed	Asteraceae *(Sunflowers)*	H
	Mirabilis	Four o'clocks	Onagraceae *(Evening primroses)*	H
	Monarda	Bee balm, Firecrackers	Lamiaceae *(Mints)*	H
	Lupinus	Lupine	Fabaceae *(Peas)*	S
	Opuntia	Prickly pear	Cactaceae *(Cactuses)*	S
	Oenothera and *Camissonia*	Evening primrose	Onagraceae *(Evening primroses)*	H
	Penstemon	Beardtongue	Plantaginaceae *(Plantains)*	H
JUNE				
	Dalea	Prairie clover	Fabaceae *(Peas)*	H
	Hydrophyllum	Waterleaf	Boraginaceae *(Forget-me-nots)*	H
JULY				
	Liatris	Blazing star, Gayfeather	Asteraceae *(Sunflowers)*	H
	Solidago	Goldenrod	Asteraceae *(Sunflowers)*	S
	Symphyotrichum and *Aster*	Aster	Asteraceae *(Sunflowers)*	H
	Erigeron	Fleabane	Asteraceae *(Sunflowers)*	H
	Asclepias	Milkweed	Apocynaceae *(Dogbanes)*	H
AUGUST				
	Helianthus	Sunflower	Asteraceae *(Sunflowers)*	H

Region 5 (Northeast)

Flower color	Scientific name	Common name	Family	Habit
APRIL				
○	*Salix*	Willow	Salicaceae (*Willows*)	S/T
○	*Amelanchier*	Serviceberry	Rosaceae (*Roses*)	S
MAY				
●	*Hydrophyllum*	Waterleaf	Boraginaceae (*Forget-me-nots*)	H
○ ●	*Vaccinium*	Blueberry, Cranberry	Ericaceae (*Heathers*)	S
JUNE				
○ ● ●	*Monarda*	Bee balm, Firecrackers	Lamiaceae (*Mints*)	H
○ ●	*Lupinus*	Lupine	Fabaceae (*Peas*)	S
○	*Eupatorium*	Boneset	Asteraceae (*Sunflowers*)	H
○ ● ●	*Asclepias*	Milkweed	Apocynaceae (*Dogbanes*)	H
○ ● ●	*Penstemon*	Beardtongue	Plantaginaceae (*Plantains*)	H
JULY				
●	*Liatris*	Blazing star, Gayfeather	Asteraceae (*Sunflowers*)	H
●	*Solidago*	Goldenrod	Asteraceae (*Sunflowers*)	S
AUGUST				
○ ●	*Erigeron*	Fleabane	Asteraceae (*Sunflowers*)	H
●	*Helianthus*	Sunflower	Asteraceae (*Sunflowers*)	H
○ ●	*Symphyotrichum* and *Aster*	Aster	Asteraceae (*Sunflowers*)	H

Echinacea (purple coneflower) is an easy-to-grow perennial composite (Asteraceae) that thrives in a number of midwestern and eastern habitats, as well as wetter areas in the west. It is very popular with insects, including bees, and the seed heads are enjoyed by birds in the fall and winter.

Region 6 (Southwest Deserts)

Flower color	Scientific name	Common name	Family	Habit
FEBRUARY				
	Salix	Willow	Salicaceae (*Willows*)	S/T
	Layia	Tidytips	Asteraceae (*Sunflowers*)	H
MARCH				
	Eschscholzia californica	California poppy	Papaveraceae (*Poppies*)	H
	Nemophila	Ninespot, Baby blue eyes	Boraginaceae (*Forget-me-nots*)	H
	Amsinckia	Fiddleneck	Boraginaceae (*Forget-me-nots*)	H
APRIL				
	Collinsia heterophylla	Chinese houses	Plantaginaceae (*Plantains*)	H
	Robinia neomexicana	Locust	Fabaceae (*Peas*)	S/T
	Ferocactus	Barrel cactus	Cactaceae (*Cactuses*)	S
	Sphaeralcea	Globe mallow	Malvaceae (*Mallows*)	H/S
	Lupinus	Lupine	Fabaceae (*Peas*)	S
	Penstemon	Beardtongue	Plantaginaceae (*Plantains*)	H
MAY				
	Malacothamnus	Bush mallow	Malvaceae (*Mallows*)	S
	Eriogonum	Buckwheat	Polygonaceae (*Buckwheats*)	S
	Opuntia	Prickly pear	Cactaceae (*Cactuses*)	S
JUNE				
	Helianthus	Sunflower	Asteraceae (*Sunflowers*)	H
	Cleome	Bee plant	Cleomaceae (*Bee plants*)	H
	Asclepias	Milkweed	Apocynaceae (*Dogbanes*)	H
	Gaillardia	Blanket flower	Asteraceae (*Sunflowers*)	H
JULY				
	Erigeron	Fleabane	Asteraceae (*Sunflowers*)	H
	Solidago	Goldenrod	Asteraceae (*Sunflowers*)	S
	Hemizonia	Tarweed	Asteraceae (*Sunflowers*)	H
	Symphyotrichum and *Aster*	Aster	Asteraceae (*Sunflowers*)	H
AUGUST				
	Chrysothamnus	Rabbitbrush	Asteraceae (*Sunflowers*)	S

A variety of flowering plants create a beautiful garden and also attract equally beautiful bees, like this *Agapostemon* on blanket flower (*Gaillardia*).

Region 7 (Southeast)

Flower color	Scientific name	Common name	Family	Habit
MARCH				
	Vaccinium	Farkleberry, Blueberry	Ericaceae *(Heathers)*	S
APRIL				
	Mertensia	Bluebells	Boraginaceae (*Forget-me-nots)*	H
	Cercis canadensis	Redbud	Fabaceae *(Peas)*	T
MAY				
	Monarda	Bee balm	Lamiaceae (*Mints*)	H
	Dalea	Prairie clover	Fabaceae *(Peas)*	H
	Rosa	Rose	Rosaceae (*Roses*)	S
	Penstemon	Beardtongue	Plantaginaceae (*Figworts*)	H
JUNE				
	Gaillardia	Blanket flower	Asteraceae *(Sunflowers)*	H
	Rudbeckia	Black-eyed Susan	Asteraceae *(Sunflowers)*	H
	Asclepias	Milkweed	Apocynaceae *(Dogbanes)*	H
	Erigeron	Fleabane	Asteraceae *(Sunflowers)*	H
JULY				
	Helenium	Sneezeweed	Asteraceae *(Sunflowers)*	S
	Solidago	Goldenrod	Asteraceae *(Sunflowers)*	S
	Pycnantheum	Horsemint	Lamiaceae (*Mints*)	H
AUGUST				
	Eupatorium	Pieweed	Asteraceae *(Sunflowers)*	S
	Liatris	Blazing star, Gayfeather	Asteraceae *(Sunflowers)*	S

LEFT: Small asters, tansyweeds, and fleabanes (*Aster*, *Erigeron*, *Machaeranthera*, *Chaetopappa*, and others) can provide ground cover that is unobtrusive and easy to maintain, especially in arid areas of the west, and bees will be common visitors.

RIGHT: It is not just the flowers that are important for bees. *Megachile* use leaves to line their nest, chewing off rounded pieces and leaving behind plants that look as though they've been hit with a hole-punch. Planting leafy herbs and shrubs, especially roses, can also encourage bees.

Garden plants that attract bees

Scientific name	Common name	Family
HERB PLANTS		
Ocimum	Basil	Lamiaceae *(Mints)*
Origanum	Oregano	Lamiaceae *(Mints)*
Thymus	Thyme	Lamiaceae *(Mints)*
Rosemarinus	Rosemary	Lamiaceae *(Mint)*
Mentha	Mint	Lamiaceae *(Mint)*
Allium	Chives	Amaryllidaceae *(Amaryllises)*
Anethum	Dill	Apiaceae *(Carrots)*
BUSHES AND SHRUBS		
Vaccinium	Blueberry	Ericaceae *(Heathers)*
Rubus	Blackberry, Brambleberry, Raspberry	Rosaceae *(Roses)*
Sambucus	Elderberry	Adoxaceae *(Adoxa)*
TREES		
Malus	Apple	Rosaceae *(Roses)*
Pyrus	Pear	Rosaceae *(Roses)*
Prunus	Almond, Cherry, Peach, Apricot, Plum	Rosaceae *(Roses)*
Diospyros	Persimmon	Ebenaceae *(Ebonies)*
VINES AND HERBACEOUS PLANTS		
Cucurbita	Zucchini, Squash, Pumpkin, Watermelon	Cucurbitaceae *(Squashes and Melons)*
Cucumis	Cucumber, Cantaloupe, Honeydew	Cucurbitaceae *(Squashes and Melons)*
Citrullus	Watermelon	Cucurbitaceae *(Squashes and Melons)*
Solanum lycopersicum	Tomato	Solanaceae *(Nightshades)*
Capsicum	Pepper	Solanaceae *(Nightshades)*
Allium	Onion	Amaryllidaceae *(Onions)*
Fragaria	Strawberries	Rosaceae *(Roses)*
Phaseolus vulgaris	String bean	Fabaceae *(Peas)*
Lathyrus odoratus	Sweet pea	Fabaceae *(Peas)*

Osmia are important visitors to many orchards, increasing fruit set in industrial-sized orchards and backyards alike.

ADDITIONAL POINTERS FOR DEVELOPING A BEE GARDEN

- Don't be overwhelmed. Beginning a bee garden can seem intimidating, so start small. Just plant something. Anything that flowers. One fast-growing plant, adored by bees, is better than nothing.
- Over time, add to your garden. While one year you may plant a small patch of cone-flower, the next year you can add a mint or some catnip, and a sunflower the following year. The number of bees will increase exponentially.

LEFT: A bee-friendly landscape doesn't have to be separate from your living space. You can seamlessly blend bee habitat with your usable space by lining decks, driveways, or sidewalks with a variety of plants. Don't worry, if you don't bother the bees, they won't bother you. PHOTO BY LINDSEY E. WILSON.

BELOW: A front yard in a western climate that receives moderate rain and snowfall. It is landscaped with many native plants. The central area is planted with two kinds of thyme (Thymus) that do not require much water, bloom at different times, and are well liked by bees. Plants of various heights line the edges and include evening primrose (*Oenothera*), milkweed (*Asclepias*), willow (*Salix*), prickly poppy (*Argemone*), beardtongues (*Penstemon*), yarrows (*Achillea*), globe mallows (*Sphaeralcea*), and mints (*Mentha*).

- To attract the greatest diversity of bees, plant a variety of flowering plants; mix up flower families, flower colors, and flower shapes (e.g., tubular, star-shaped, sunflowers, bowl-shaped)
- Bees prefer large clusters of flowers. The relationship is not linear; whereas planting 1 flower may draw in (hypothetically) 3 bee species, planting 2 flower species will draw in 8, and 4 will draw in 20.
- Plant flowers that bloom at different times of year, so that something is always in bloom.
- Avoid using pesticides in your garden as much as possible. If they must be used, use the minimum amount necessary. Keep the spray contained, hitting only the target areas. Try to spray in the evenings, when humidity is low so the pesticide doesn't linger in the moisture, and when most bees are asleep or in their nests.
- Choose plants that haven't been treated with pesticides. You might need to ask at the garden shop before you buy. Pesticides will often be sprayed on flowers while they sit at nurseries and garden shops to keep destructive insects at bay. These pesticides can be absorbed into the nectar (and perhaps even the pollen) of these flowers. Even though you may choose not to use pesticides at your house, you can still end up with pesticides in the nests of your bees.
- Select native plants if you can; otherwise, cultivars of native plants are best. Your last choice should be ornamentals that have had their flower shape modified. Doubled petals, though beautiful, are especially spurned by bees because the extra petals come at the expense of nectar.
- Plant in the sun.
- Intersperse decorative flowers with vegetables in your garden to increase yield.
- If planting herbs, let a few bolt or go to flower.

Squash bees (*Xenoglossa*—pictured here—and *Peponapi*) are specialists on squash flowers, which cannot set fruit without them. You may not see these bees unless you're looking early in the morning, when they are busy foraging. PHOTO BY JILLIAN H. COWLES.

You can attract a variety of bees to your yard by cultivating plants of different colors, shapes, and blooming periods. Also, leaving bare dirt patches or rock walls can provide nesting habitat for bees. PHOTO BY LINDSEY E. WILSON.

3

ANDRENIDAE

Bees in the family Andrenidae come in a wide range of sizes, shapes, and colors. Here, a collection of bees belonging to this family shows some of the diversity. These bees are presented roughly to scale, from the large *Protoxaea* to the tiny *Perdita*.

Andrenidae is the largest of all the bee families, with in excess of 4500 species categorized into more than 40 genera. These bees can be found on all continents except Australia. They are numerous in the Western Hemisphere, particularly in drier areas and deserts. North of Mexico, there are over 1200 species in 13 genera. *Andrena* and *Perdita* are especially large groups; together these two genera comprise 80% of all Andrenidae species.

From nearly an inch long to smaller than a mosquito, the Andrenidae display an incredible diversity of sizes and colors. Some of the smallest bees in the world are Andrenidae, and these are found in the United States. *Perdita minima* (section 3.3) is gnat-sized and so petite that an entomological pin won't fit through its thorax (back). On the other end of the size spectrum, the Oxaeinae (section 3.5) subfamily includes brawny beasts that can be heard coming long before they're seen. Andrenidae are also impressively garbed, with hues from every arc of the rainbow. Rust-reds, albino-blonds, gunmetal blues, and vivid oranges can all be found in the glut of species in the Andrenidae.

The smallest bees in North America belong to the genus *Perdita*. While the bee pictured here is not the smallest of all the *Perdita* species, you can see how one could easily overlook these tiny bees as they hover around flowers.

Smaller than a mosquito

The smallest bees in North America are in the Andrenidae bee family. Specifically, bees in the genus *Perdita* can be smaller than 0.1 inch.

Specialists are well represented in the Andrenidae family, in large part because of the genus *Perdita*. Many Perdita have strict dietary preferences, collecting pollen from only certain flowers. Their bodies have evolved to support their

A *Calliopsis* takes a break from foraging on gumweed (*Grindelia*) to bask in the late afternoon sun.

ABOVE: A small male *Perdita* sits on a flower petal waiting for a potential mate.
RIGHT: An *Andrena* rests on a flower after a late summer rain.

lifestyle, and many have unique pollen-collecting hairs sized to meet the specs of the pollen grains associated with their preferred floral host. Several fly at dawn or dusk, when their favorite floral hosts are in bloom.

Sometimes called mining bees, all Andrenidae nest in the earth. Some nest shallowly—just an inch or two below the surface—but a few nest a foot or more below ground level.

All Andrenidae line their nests with a waterproof substance secreted by the female to protect her progeny from soil moisture and soil bacteria. For one rugged species of Andrenidae, this waterproof substance is so effective that it allows the still-developing bees to survive under water in seasonal lake beds.

Andrenidae are divided into four subfamilies. Three of the four are found in the United States and Canada but one (Alocandreninae) is found only in the Andes of Peru, and there is only one known species in this rare subfamily.

Journey to the center of the Earth

Mining bees dig their nest tunnels straight down. Nests commonly measure between 6 inches and a few feet deep, but at least one bee species can be an overachiever. Scientists researching *Andrena haynesi* in the deserts of Utah dug up a nest that ended 9 feet under the ground! This is the equivalent of a 6-foot-tall human using his hands to dig the depth of four football fields into the earth. Why the bee chose to dig this deeply is a mystery.

The three subfamilies in the United States and Canada are Andreninae, Panurginae, and Oxaeinae. Here, we cover each subfamily in one section, with the exception of the more diverse Panurginae. For that subfamily, we devote sections to the tribes: the Protandrenini and Panurgini (together), the Calliopsini, and the Perditini.

A *Protoxaea* forages on a legume flower.

IDENTIFICATION TIPS

Classifying a bee as belonging to Andrenidae while it is pilfering pollen from a flower in front of you is difficult. Without a still specimen, it is perhaps easiest to identify the genus and thus arrive at the conclusion that the bee is in the Andrenidae bee family. For example, *Perdita* are distinct because of their minute size and yellow bodies. By definition, all *Perdita* are Andrenidae.

IDENTIFYING FEATURES OF THE ANDRENIDAE BEE FAMILY

- Andrenidae can be identified by the two grooves below each antennal socket (subantennal sutures). If the face has a lot of punctation (small pock marks), this can be difficult to see, and there are a few species with only one.
- All Andrenidae have short tongues (see section 1.8).

Examples of diagnostic features for different genera can be seen below.

ANDRENINAE (SECTION 3.1)

Megandrena

Ancylandrena

Andrena

Though three genera within the Andreninae subfamily are found in the United States and Canada, *Andrena* are by far the most common. *Megandrena* and *Ancylandrena* may occasionally be encountered in the southwest. Because several genera in other bee families can look similar to *Andrena*, identifying them on the wing can be difficult. See below for identifying features, best seen under a microscope.

Identifying features of the Andreninae

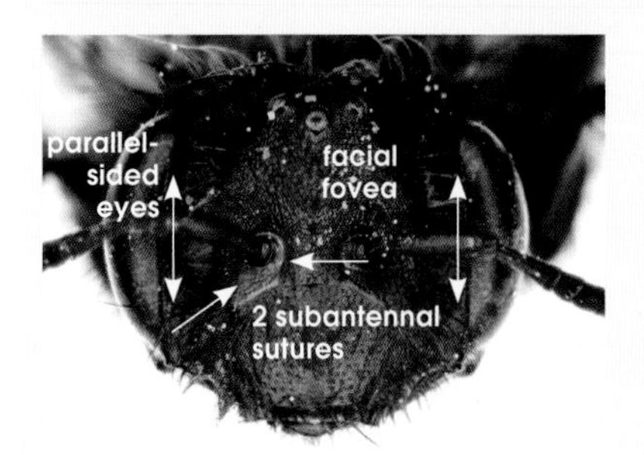

Looking head-on at an *Andrena* (or *Ancylandrena* or *Megandrena*) face, you can see a number of important features. With a microscope the two subantennal sutures are evident. The eyes are mostly straight up and down, so that the inner margins are close to parallel, and the face is usually wider than it is long; nearly oval-shaped. Beside each eye are "facial fovea," thick dense hairs inside the indentations that run next to each inner eye margin. They can often be seen without a microscope; turn the bee in the light or look down at the face from above.

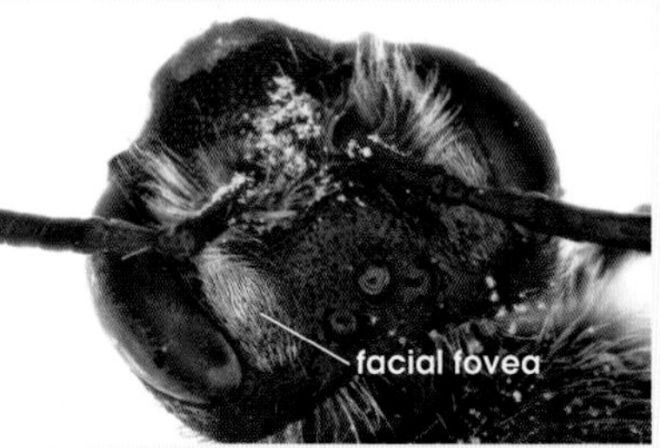

Seen by looking down on the head of an *Andrena*, the facial fovea (think of them as up-and-down eyebrows) are fairly obvious; thick patches of hairs inside slight depressions next to the eyes.

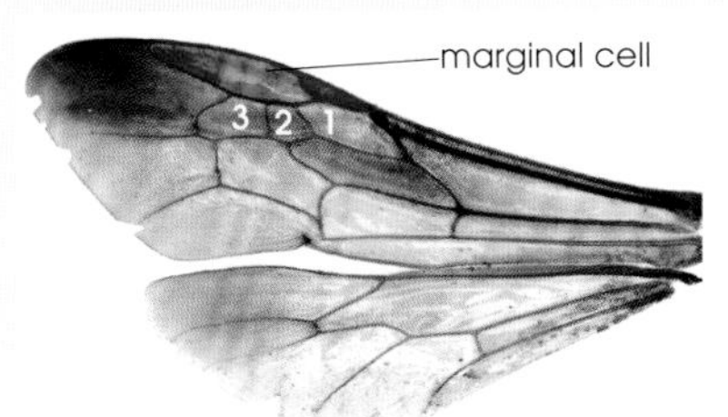

An *Andrena* wing. Note the three submarginal cells typical of almost all *Andrena*, as well as *Ancylandrena* and *Megandrena* (a few subgenera of *Andrena* have two). The marginal cell (in red) is relatively long and has a rounded tip. Look also at the basal vein and second recurrent vein; neither is strongly curved. Finally, the second submarginal cell is much shorter than the other two.

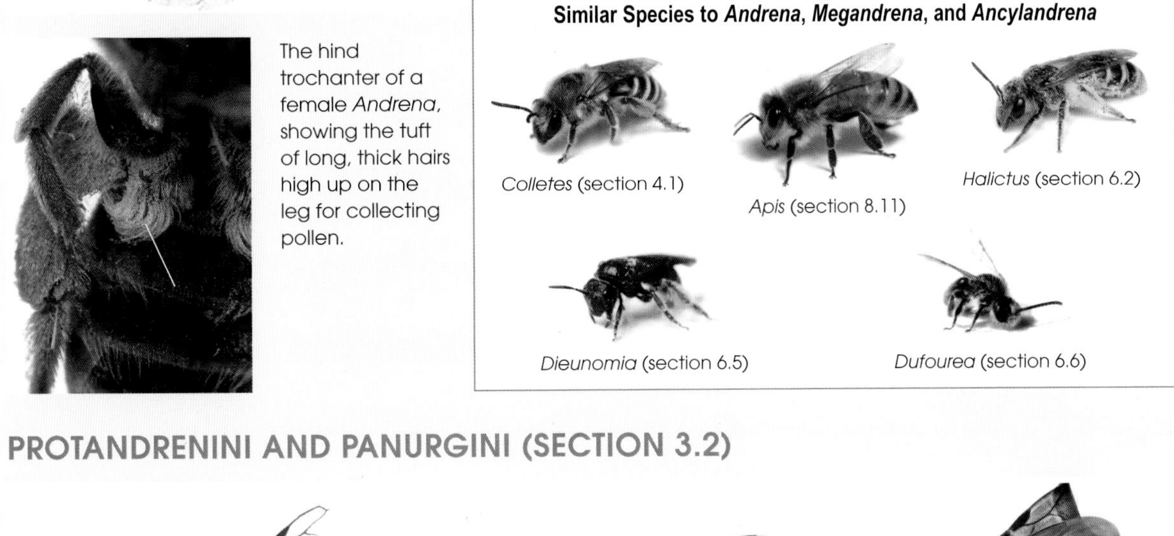

The hind trochanter of a female *Andrena*, showing the tuft of long, thick hairs high up on the leg for collecting pollen.

Similar Species to *Andrena*, *Megandrena*, and *Ancylandrena*

Colletes (section 4.1)

Apis (section 8.11)

Halictus (section 6.2)

Dieunomia (section 6.5)

Dufourea (section 6.6)

PROTANDRENINI AND PANURGINI (SECTION 3.2)

Panurginus *Protandrena* *Pseudopanurgus*

Three of the four North American bees in the Protandrenini and Panurgini bee tribes: *Panurginus, Protandrena,* and *Pseudopanurgus* (*Anthemurgus* is not pictured). All four of these bee genera have yellow markings on their black bodies, generally more on the males than females. They are slender and may be short or long. All have relatively round faces, which can be seen while the bees visit flowers. *Pseudopanurgus* represent the largest of these bees and have less yellow on their legs—some are completely black, though the face and areas of the thorax retain spots of yellow.

Identifying features of the Protandrenini and Panurgini

The face of a *Protandrena* (left) and a *Panurginus* (right). Most species have two subantennal sutures, though a few have lost the outer one. When there are two, the inner one is relatively long—longer than the diameter of the antennal socket (compare with *Calliopsis*). Note that the yellow markings are prominent but more or less limited to the lower half of the face. Both groups have tiny grooves (technically facial fovea), but it is never as dense or evident as in *Andrena*.

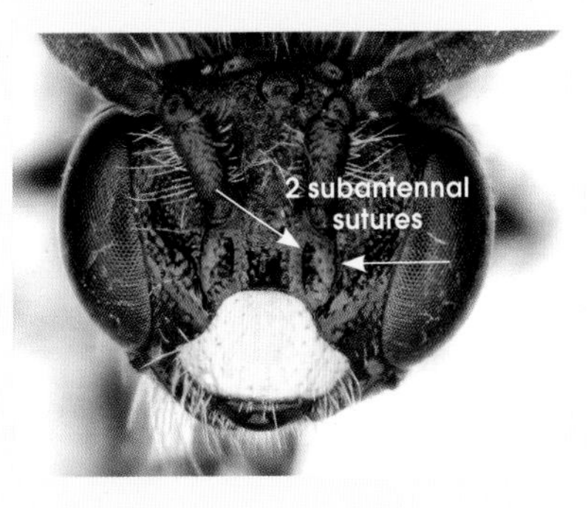

continued overleaf

Identifying features of the Panurgini and Protandrenini *continued*

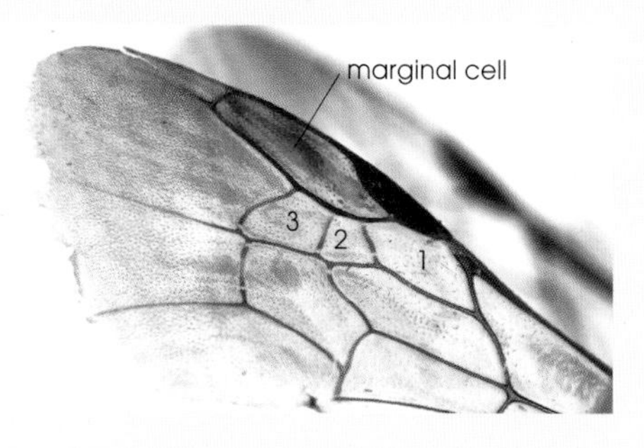

The wing of a *Protandrena* (left) and a *Pseudopanurgus* (below left). As with all bees in this section, the tip of the marginal cell looks snipped off. Some of the bees in these genera have two submarginal cells, while others have three (they are numbered here).

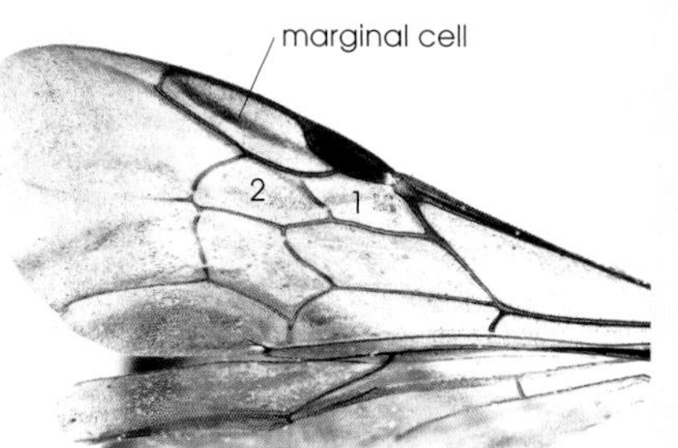

Similar Species to the Protandrenini and Panurgini

Hylaeus (section 4.2)

Ceratina (section 8.2)

Calliopsis (section 3.4)

PERDITINI (SECTION 3.3)

Perdita

Perdita

Macrotera

Perdita and *Macrotera* are very tiny bees—when males fly in swarms near a flowering plant, they look remarkably like gnats. Even the relatively larger Perditini have a slight build. Most are easy to recognize because of their yellow markings and miniscule size. Though some *Perdita* and *Macrotera* have a black abdomen, the majority have some yellow.

Identifying Features of Perditini

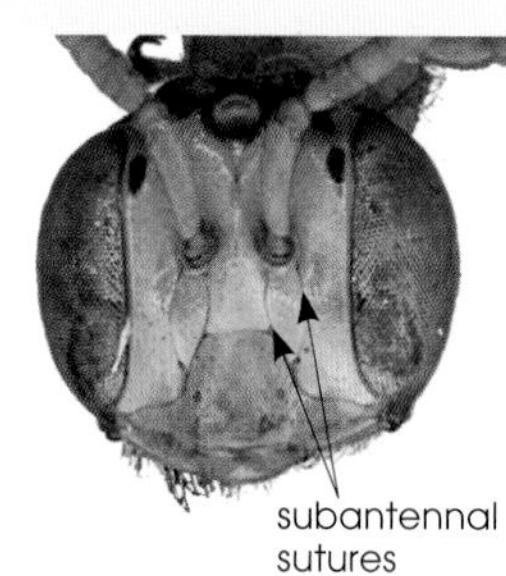

The face of a *Perdita*. *Perdita* faces tend to be very round, about as broad as they are long. Note the two subantennal sutures, typical of Andrenidae. Photo courtesy of USDA-ARS Bee Biology and Systematics Laboratory.

Similar Species to *Perdita* and *Macrotera*

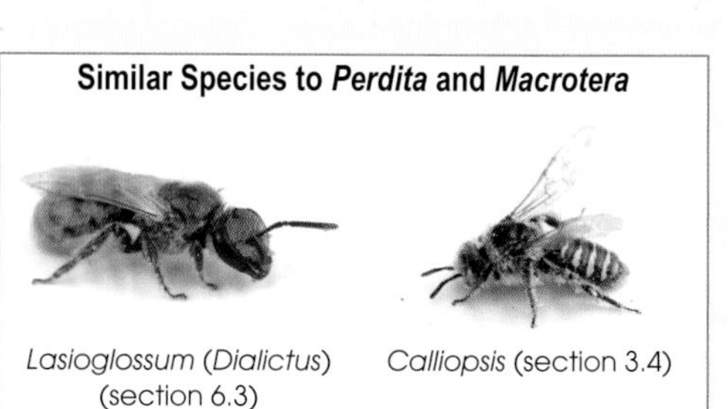

Lasioglossum (*Dialictus*) (section 6.3)

Calliopsis (section 3.4)

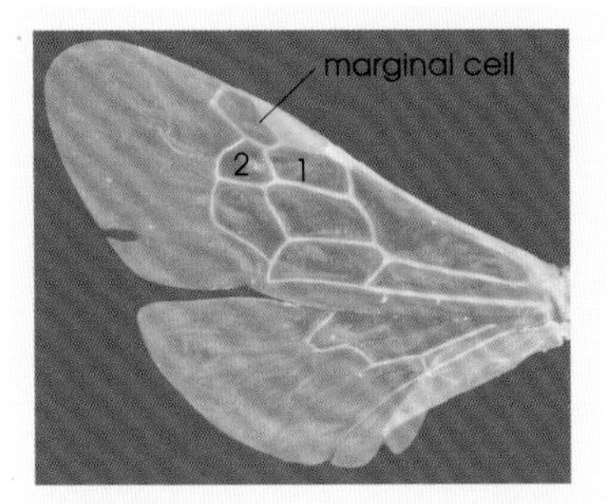

A *Perdita* wing, showing the existence of only two submarginal cells. The second submarginal cell is more or less triangular in shape. This image also shows the shape of the marginal cell (in pink). Rather than tapering to a point, it looks as though it were snipped off abruptly just past the stigma. This makes the upper margin of it (along the wing's edge) about the same length as the stigma's length, or even shorter. Some species have milky white veins, a feature not found in other bees.

CALLIOPSIS (SECTION 3.4)

Calliopsis are striking bees, and many have yellow or ivory markings on the face and legs. They tend to look flat, compared with many bees, as if the air had been sucked from their abdomen. Generally the face tends to be oval-shaped and the thorax shiny. Several species have a light steel-blue tint to their bodies, and even their eyes.

Identifying Features of *Calliopsis*

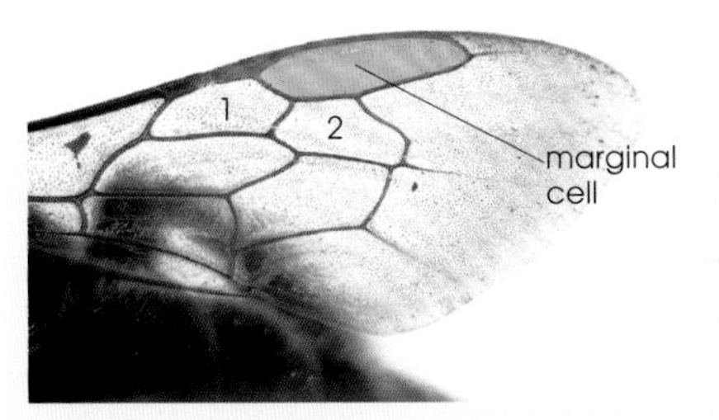

The wing of a *Calliopsis*, always with two submarginal cells. The marginal cell bends gently away from the wing margin, ending in a blunt or softly rounded tip.

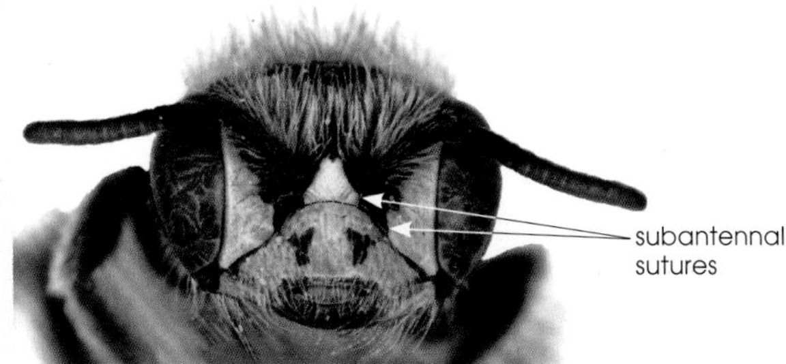

Like most bees in the Andrenidae bee family, *Calliopsis* have two subantennal sutures. Though you will need a microscope to see it, the inner of the two subantennal sutures is quite short, only as long as the diameter of the antennal socket, or shorter.

Similar Species to *Calliopsis*

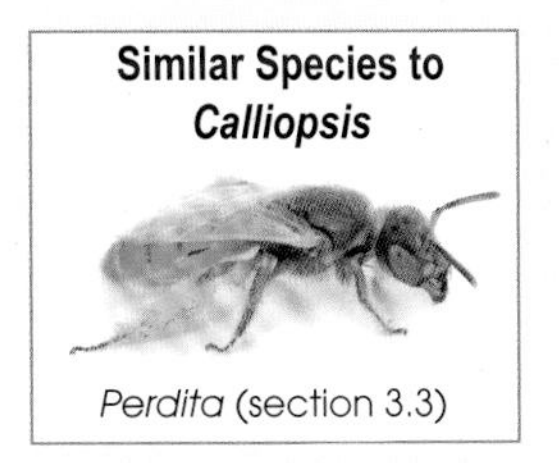

Perdita (section 3.3)

continued overleaf

OXAEINAE (SECTION 3.5)

As beautiful as these burly beasts are, they are seldom seen by most bee enthusiasts because they live only along the Mexican border.

Mesoxaea

Protoxaea

Mesoxaea (left) and *Protoxaea* (right) are the two genera in the Oxaeinae subfamily found in North America. Note that for both genera, the scopa is dense and runs from the hind coxa all the way to the basitarsus. *Mesoxaea* are extremely rare and seldom seen in the United States. Large, with a fuzzy thorax, they have a dark, relatively hairless abdomen. Though both are rare, *Protoxaea gloriosa* extends further north than any of the *Mesoxaea*, and it is the more likely of the two to be seen.

Identifying Features of Oxaeinae

LEFT: Both *Mesoxaea* and *Protoxaea* have two subantennal sutures. In profile, the clypeus can be seen to protrude from the face. For males, the eyes converge near the top, sometimes almost touching, which pushes the simple eyes (ocelli) low on the front of the face.

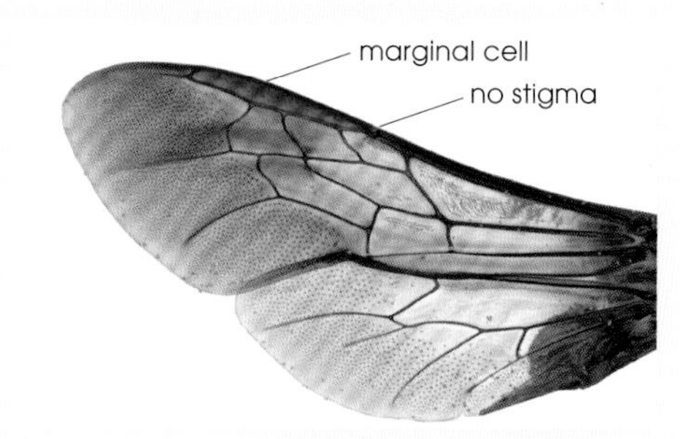

On the wing of any Oxaeinae, there is no stigma, and the marginal cell is long and very skinny. Notice also that the vein marking the edge between the second and third submarginal cells meets with the first recurrent vein, a good character for distinguishing it from *Caupolicana* (section 4.3).

Similar Species to Oxaeinae

Caupolicana and *Ptiloglossa* (section 4.3)

Centris (section 8.9)

Svastra (section 8.7)

3.1 ANDRENINAE

ANDRENA, *ANCYLANDRENA*, AND *MEGANDRENA*

This section includes three bee genera. All are close relatives of each other, belonging to the subfamily Andreninae, but *Andrena* is significantly more common than the other two genera.

Andrena

The name Andrena is Greek for "buzzing insect" (sharing the same root with drone).

DESCRIPTION

Andrena are small to medium-sized, lightly fuzzed bees ranging in color from gray and brown to bright red.

DISTRIBUTION

Andrena is one of the largest bee genera in the world, with nearly 1400 species. They are especially well represented in Europe and northern Asia, where there are an estimated 700 species. Considering their diversity, it is surprising that none occur in South America or Australia. *Andrena* are apparently indifferent to the cold, being common at high elevations, and occurring even in Alaska and northern Canada, where other bees are rare. These abundant flower lovers represent an extraordinarily diverse group of bees in North America, and they are ubiquitous across the United States and Canada. Roughly 550 species of *Andrena* occur in the United States, and about 100 of those species also occur in Canada.

A few species of mining bee are found naturally in both Europe and North America, including *Andrena clarkella* and *A. barbilabris. Andrena wilkella* occurs in the eastern United States and in Europe but was probably introduced into the United States.

DIET AND POLLINATION SERVICES

The floral preferences of *Andrena* species span the range of bee diets; some are broad generalists, and a number are strict specialists, while some fall between. Specialists visit an incredible diversity of flowers. For example, in the

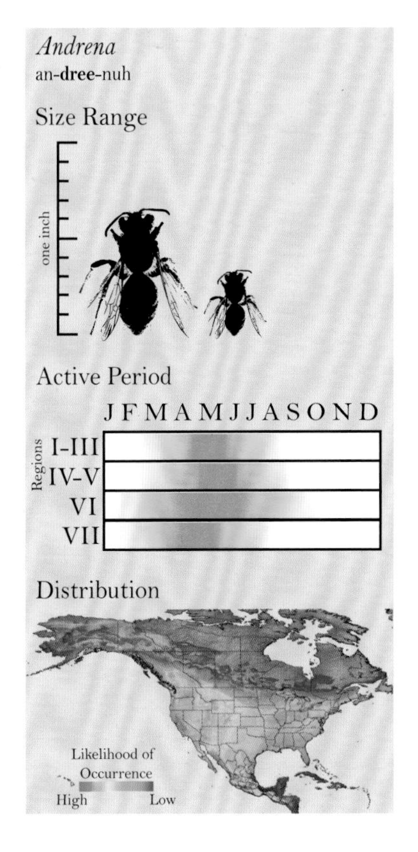

ABOVE: An *Andrena* gathering pollen from scorpionweed (*Phacelia*).

LEFT: Some *Andrena* will visit only specific flowers, and others, like this *Andrena prunorum*, can be found on a wide range of plants.

What's in a name?

The poorly named *Andrena astragali* specializes not on *Astragalus* (milkvetch) as its name suggests, but on death camas (*Toxicoscordion*), a plant that is poisonous to humans, cattle, horses, dogs, cats, and other creatures, including most bees. All parts of the plant, nectar included, contain neurotoxic alkaloids, and at least one species of death camas has been shown to be toxic to honey bees. It appears that *Andrena astragali* is immune to these toxins—possibly the only bee that is!

deserts of the southwest, there are several specialists that visit only sun cups (*Camissonia*). These flowers bloom very early in the day, perhaps in order to avoid losing precious nectar to evaporation. The *Andrena* that specialize on these flowers also fly very early, and they have harvested pollen and retired to their nests by 10 o'clock in the morning. The closely related evening primrose (*Oenothera*) also hosts specialist bees. It blooms at dusk, and you can find its specialists harvesting pollen late into the evening.

Other *Andrena* specialize on springtime tree species, especially maples (*Acer*) and willows (*Salix*). The sunflower family (Asteraceae) is another popular host plant for *Andrena*, including all 79 species in the subgenus *Callandrena*. The degree of specialization differs significantly between species, however. Some sunflower specialists visit only one genus of sunflowers—for example, blanket flowers (*Gaillardia*), coneflowers (*Echinacea*), black-eyed Susans (*Ratibida* and *Rudbeckia*), or dwarf dandelions (*Krigia*)—while others will visit any kind of sunflower.

Andrena are important for the success of some commercial crops, including blueberries, cranberries, and apples. One species, *Andrena prunorum,* is a frequent pollinator of wild onions *(Allium)*, and has been found to be better at pollinating cultivated onions than the honey bee.

Fooling around

In various areas of the world, complicated relationships exist between some species of bees (including *Andrena*) and orchids. The orchid petals resemble the "back" of a female bee, complete with little wings. They even emanate a floral scent like that of a female bee. Males are tricked into visiting the flower. They pounce on the backs of the female "bees," thinking only of mating, but leave frustrated and covered with pollen from the orchid. Not being fast learners, these hedonists often make the same mistake over and over with other orchid flowers, thereby unwittingly effecting pollination.

A male *Andrena* rests in a globe mallow flower (*Sphaeralcea*).

Other Biological Notes

Andrena are among the first bees to fly in the spring and can be seen in some areas before snow has melted and flowers have started to bloom. Other *Andrena* species emerge only in the fall, and a handful of species produce two generations in a year (see voltinism, p.85).

Bees that emerge during the brisk temperatures of early February are atypical. How they are able to withstand the chill is not yet clear to scientists. While bees can survive during remarkably cool temperatures, in order to fly they need to attain a minimum body temperature well above freezing. Newly emerging *Andrena* adults crawl from their underground lairs toward the surface at temperatures between 40 and 50 °F, but they can't fly until their body temperatures reach 50–60 °F. They rely on the warmth of the sun to get them up to speed, and it is not uncommon to see them warming up on leaves or rocks. This cold-hardiness makes them excellent pollinators of crops that bloom early in the season—especially before honey bees are active. Scientists have found that during the winter months, when the ground may be moist and

Many *Andrena* are active early in the year and can be found sunning themselves on leaves before foraging or looking for a mate. Photo by Susanna M. Messinger.

An *Andrena* on a mesquite flower (*Prosopis*). Notice the "facial fovea," the two vertical lines of short hairs just inside both eyes, a key characteristic shared by all *Andrena*. PHOTO BY JILLIAN H. COWLES.

soft because of moisture percolating from above, some *Andrena* species leave their nest cells and crawl through the soil toward the surface. They wait just below ground for the first stretch of warm days in the early spring to crawl the last distance and emerge. This behavior may help to explain how they can be among the earliest bees to appear in the spring.

Sometimes early spring *Andrena* emerge before any of the flowers in their habitat have started blooming. When this happens, they rely on their fat reserves to sustain them until blooming begins. During this waiting period, pairs will mate and the females will build nests, staying busy for up to a week while pollen sources become available.

All *Andrena* nest in the ground but the specifics vary between species; some prefer sand while others prefer

ABOVE LEFT: While many *Andrena* are various shades of gray or brown, they can be a wide variety of colors, including this black-and-red *Andrena prima*. PHOTO BY JILLIAN H. COWLES.
ABOVE RIGHT: A male *Andrena* drinks nectar from a manzanita (*Arctostaphylos*) flower. PHOTO BY SUSANNA M. MESSINGER.

Andrena are often called mining bees because they nest in the ground. Often many individuals will nest in the same area, preferring bare patches, like trails.

Andrena sphaeralceae, a specialist on globe mallow (*Sphaeralcea*).

Male *Andrena*, like this one, visit a wide variety of flowers for nectar. You can tell this is a male by the lack of pollen-carrying hairs (scopa) on the hind legs.

clay. Some seem not to care and can be found nesting in urban neighborhoods and lawns. Commonly, nests are at the base of rocks or under layers of fallen leaves, making nest entrances difficult to find. In contrast, nests on bare earth are often easy to see because they have piles of dirt around them (called tumuli, singular tumulus), resembling miniature gopher mounds.

Nest aggregations are not uncommon, with as few as 5 to upward of 2000 individuals nesting close together. Nest entrances in these cases are just inches from each other, and observers have reported seeing several hundred entrances per square yard. These aggregations might persist for decades, or they may be present for only one or two years before the bees disband and establish in new areas.

When nests are in large aggregations, every nest looks the same. To avoid mistaking another nest for her own, each time a bee emerges from her nest, she will spend several minutes flying in figure eights around the nest entrance from greater and greater distances in order to get a good sense of where her "door" is in the context of its surroundings. Because dozens of bees may be simultaneously orienting, the ground near these aggregations can appear to be moving.

Other species of *Andrena* nest communally. Between 2 and 50 females will share the same nest entrance but have individual tunnels inside the main entrance (see section 1.4). In turn, nests may have more than one entrance, with the entrances connected by underground tunnels.

ABOVE: A female *Andrena* takes a break from foraging to clean pollen off its antennae. PHOTO BY HARTMUT WISCH.

The metallic green-blue hue of this *Andrena cerasifolii* is not uncommon among *Andrena*, and gives it a slight resemblance to other metallic bees such as *Osmia*. PHOTO BY ALICE ABELA.

Ancylandrena

Ancylandrena means "curved Andrena," referring to the shape of the first tergite (section) of the abdomen.

Ancylandrena are striking bees, often with a light reddish hue to their abdomens, that look similar to *Andrena*. Only five species of *Ancylandrena* are known, all found in the southwestern United States. Most species are very rare, and all of them are active only in the early spring. *Ancylandrena* are specialists on desert shrubs like indigo bush (*Psorothamnus*) or creosote bush (*Larrea tridentata*). Little is known about the nesting habits of *Ancylandrena,* but it appears they have preferences for certain soil textures. For example, the females of some species in this group sort the grains of soil inside their nest according to coarseness. They use the finest particles to create very smooth walls around the cell containing an egg. In an antechamber leading to the nest cell, the bigger pebbles are stored; these are used to block the nest entrance at night.

Ancylandrena are rare and found only in the southwestern parts of the United States. PHOTO BY MARGARETHE BRUMMERMANN.

Megandrena

Megandrena means "big Andrena" in Greek.

Megandrena are larger than the average *Andrena*, and slightly hairier. Only two species of *Megandrena* are known, and one of them is very rare. *Megandrena mentzeliae* has a bright red abdomen with broad white stripes. It has been collected only in southern Nevada on blazing star flowers (*Mentzelia).* The other species, *Megandrena enceliae* has white stripes on its dark abdomen. It is relatively common in southern California, Nevada, and Arizona, where it can be found flying in the early spring. Like *M. mentzeliae, M. enceliae* is a specialist, and visits only creosote bush (*Larrea tridentata*) for pollen, although it visits other flowers for nectar. While it is known that *Megandrena* nest in the ground, the particulars of their nesting habits have not been studied.

A male *Megandrena* finding shelter inside a desert lily (*Hesperocallis*). PHOTO BY MICHAEL ORR.

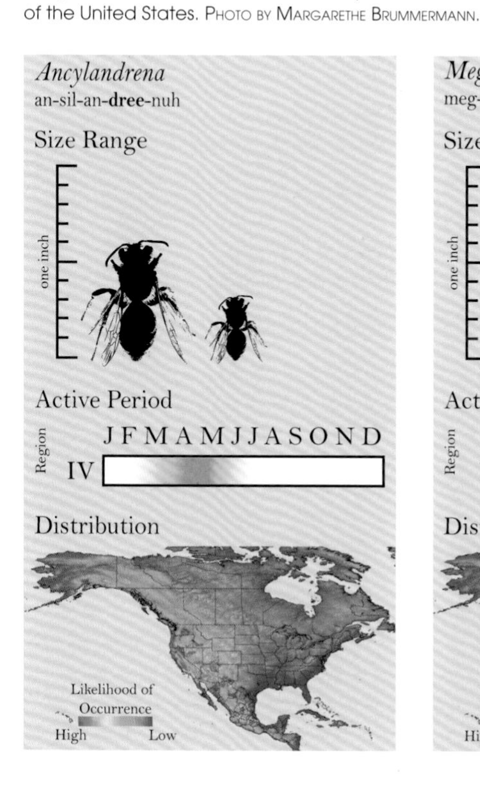

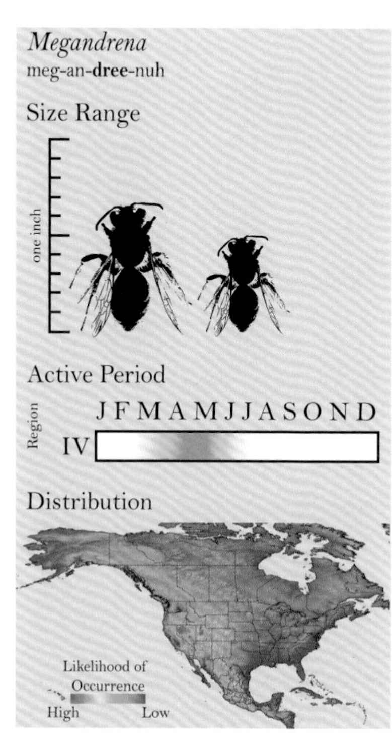

Look closely inside these cactus flowers; male *Megandrena* can be seen sleeping inside. PHOTOS BY MICHAEL ORR.

3.2 PROTANDRENINI AND PANURGINI

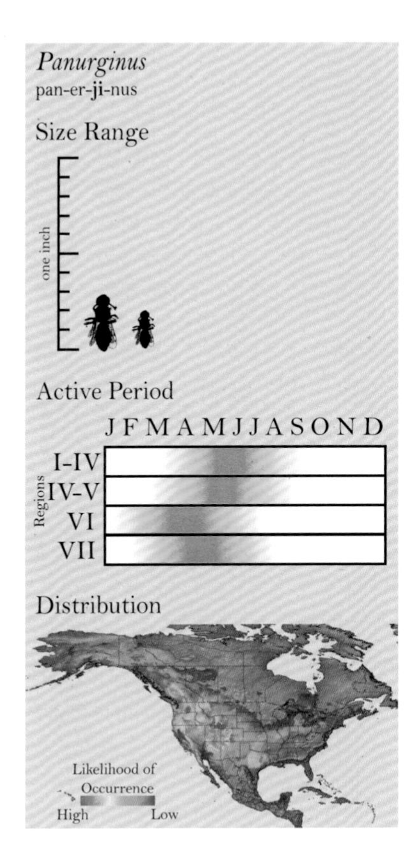

PROTANDRENA, *PSEUDOPANURGUS*, *ANTHEMURGUS*, AND *PANURGINUS*

These four groups of small to medium-sized bees are all in the subfamily Panurginae, with *Panurginus* in the tribe Panurgini, and the other three genera in the Protandrenini tribe. The Panurginae subfamily includes other tribes as well, but they are more common, and each has its own section following this one. Relative to other bees, very little is known about the life history of the Panurgini and Protandrenini, with the exception of *Anthemurgus*. There is only one species of *Anthemurgus*, and it visits only yellow passionflowers (*Passiflora lutea*) for pollen.

The Panurgini
Panurginus

Panurginus is Latin for "pertaining to Panurgus," a reference to their similarity to the European bees in the related genus Panurgus.

Lost in place

Bees fly long distances in search of flowers and other resources. A female's nest may be one small hole in the ground, perhaps surrounded by countless other bee nest holes. Or it may be a small tunnel in one of many nondescript pieces of wood in a large forest. Scientists are still learning how a female finds her way home again, and how she distinguishes her nest from all others. Females use the simple eyes on the tops of their heads to navigate, keeping track of the angle of the sun to locate home (i.e., "When I left, the sun was on my right side. To get home, it should be on my left side.") They also use landmarks, especially near the nest, to find their particular entrance. It is not uncommon to see a female leave her nest and fly in ever-wider figure eights in front of the entrance, presumably memorizing how it looks compared with all the features around it. Finally, she uses scent cues. A female marks her nest with a mix of chemical compounds that are unique to her particular body chemistry. This can help her hone in on her nest from a distance and also distinguish it from any of the others in close proximity.

Description

Only one genus from the tribe Panurgini is found in the Western Hemisphere: *Panurginus*. It is a small dark bee, commonly with yellow markings and very little hair.

Distribution

Panurginus are widely distributed in Europe and Asia. In North America they are found from Alaska south to northern Mexico, east to the Atlantic coast, and south as far as Georgia. There are 18 species in the United States, only 3 of which are found east of the Mississippi River. Three species are found in Canada.

Diet and Pollination Services

Some species of *Panurginus* are floral generalists, while others are specialists, visiting, for example, only mustards (Brassicaceae), or only roses (Rosaceae), or only buttercups (*Ranunculus*) for pollen.

Other Biological Notes

The nesting specifics of most *Panurginus* are not known. We can guess that their habits are probably similar to each other, although there are likely subtle differences as well. Like all bees in the family Andrenidae, *Panurginus* nest in the ground. The few nests that have been studied have been very shallow (less than a foot deep). They are typically quarter-inch holes without the characteristic mounds found around the nests of *Andrena.* Similar to her cousins, when a female *Panurginus* is in the nest, the entrance is plugged. When she is out foraging, she leaves the door open.

Panurginus found in the United States nest both individually and communally. At least one species, *Panurginus polytrichus*, is highly gregarious, sometimes with more than 100 bees nesting in one square yard! The same nesting site may be reused year after year; one site in Texas has been used for over 20 years. Male *Panurginus* hover hopefully above the emergence sites of females, aggressively attempting to copulate with anything that moves—even other males. Often three or more males will all attack a female at once, creating a massive ball of bees; only one male actually mates with the female, however. He pins her wings to her body until copulation is complete, then she kicks him loose and flies away. Once a female has mated, other males leave her alone.

The Protandrenini
Protandrena

The name Protandrena means "basic Andrena," referring to the similar familial characteristics of Andrena and Protandrena.

Description

Protandrena are slightly larger than *Panurginus*, but similarly hairless and with yellow markings on the face, legs, and thorax.

Distribution

These bees occur only in the Western Hemisphere: North, Central, and South America. They are not common in the tropics, but numerous species can be found in Mexico. In the United States and Canada, 80 species have been identified. They can be found from coast to coast but are more common in the west.

Diet and Pollination Services

Protandrena includes both specialists and generalists. *Protandrena abdominalis,* for example, will only visit bee balm (*Monarda*). Other species can be found foraging on many kinds of plants.

Other Biological Notes

All members of this genus are solitary, with each female maintaining her own nest, though some species may nest in aggregations. Nests are often located in areas with sparse vegetation and abundant sunshine, where females place their nest entrances next to, or just under, a small stone or clump of vegetation. It is possible that such an overhang conceals the entrance from potential intruders. Another possibility is that the marker, once memorized, aids the bee in finding the nest quickly when returning from visits to flowers. Bees that enter their nests speedily are less likely to be nabbed by predators, and they also lose parasites that have been trailing them to see where their nests are.

Inside the nest, each cell is lined with a waterproof substance secreted by the bees and painted onto the walls. Scientists have tried melting this "wallpaper" and dissolving it in various chemicals with no success. This material creates a strong, impenetrable barrier between the growing bee and soil bacteria and fungi.

When the egg hatches, most *Protandrena* appear to develop all the way through the various larval stages, and then overwinter as mature larvae. They do not spin a cocoon as other bees do, but develop hard waterproof skins instead.

A female *Protandrena* foraging on milkweed (*Asclepias*).

A male *Protandrena* foraging on milkweed (*Asclepias*).

Pseudopanurgus

The name Pseudopanurgus refers to their rough resemblance to the closely related Panurgus, which are found in Europe.

Pseudopanurgus often have less yellow on their bodies, especially their legs. Despite a broad range from the Great Plains of the Midwest across to California, and south to Costa Rica, these bees are relatively uncommon. Only 12 different kinds are known in the United States and 6 in Canada. A number of *Pseudopanurgus* are specialists on flowers in the sunflower family (Asteraceae). All species appear to be solitary. Similar to *Protandrena*, these bees nest in open ground, often in fairly dense aggregations, line their few nest cells with a waterproof substance, and don't spin cocoons.

Pseudopanurgus
soo-doh-pan-**er**-jus

Size Range

one inch

Active Period

Regions	J F M A M J J A S O N D
III-IV	
IV-V	
VI	
VII	

Distribution

Likelihood of Occurrence
High Low

A mating pair of *Pseudopanurgus*. PHOTO BY JILLIAN H. COWLES.

Pseudopanurgus sipping nectar from a mustard flower (Brassicaceae).

Pseudopanurgus with a large pollen load. PHOTO BY JILLIAN H. COWLES.

A male *Pseudopanurgus* resting on a composite (Asteraceae).

A female *Pseudopanurgus* forages on a composite (Asteraceae). PHOTO BY JILLIAN H. COWLES.

Anthemurgus passiflorae

Anthemurgus means "flower worker" in Greek.

Anthemurgus are relatively rarely seen. Mostly black with yellow on the lower halves of their faces, they are restricted to yellow passionflowers (*Passiflora lutea*). This is the only species of *Anthemurgus* anywhere in the world. It occurs from Texas, east to North Carolina, and north to Illinois. As its name implies, *Anthemurgus passiflorae* is extremely specialized and will collect pollen only from the yellow passionflower; however, *Anthemurgus* isn't a particularly good pollinator of this plant. Unlike other bees, the *Anthemurgus* female harvests pollen with her mandibles, not her legs. She very thoroughly collects all pollen from passionflower anthers, then coats it with a little regurgitated nectar before passing it to her legs for transport. The nectar coating prevents pollen transfer to receptive passionflower stigmas. *Anthemurgus passiflorae* flies for about three weeks in the late spring and early summer and is bivoltine, meaning that it produces two generations every year (see voltinism, below). Like other Panurgini bees, *Anthemurgus* dig their nests in bare ground, often in trails or other areas with little vegetation. *Anthemurgus* nest singly and are solitary bees. Several females may nest in the same area, but they are only loosely aggregated.

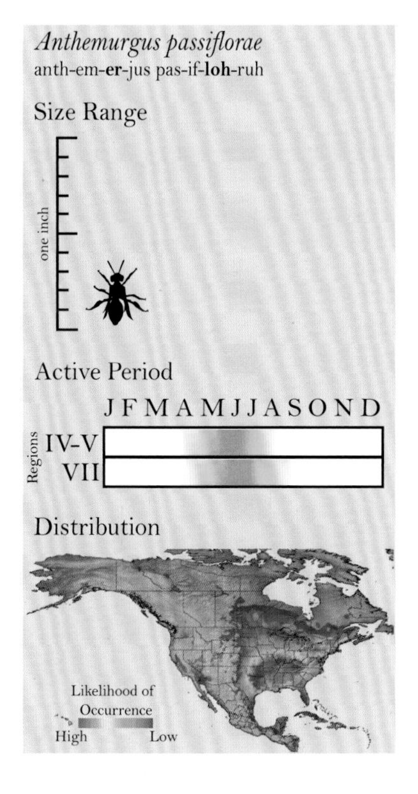

Voltinism: A second go

Volitinism refers to the number of generations produced by an organism in one year. Bees can be "univoltine," meaning that only one generation is produced each year; for the majority of the year univoltine bees are in diapause (i.e., dormant), waiting for the height of flowering season before beginning the process of producing offspring that will emerge the following year. Other bees are "bivoltine," meaning that two generations are produced in one year, usually one spring and one fall generation. Finally, some bees are "multivoltine," meaning they produce numerous generations per year. Multivoltine bees are generally social species, like honey bees, with a queen that continuously lays eggs. What determines voltinism is not entirely clear, but temperature plays a large role. Cold temperatures appear to trigger diapause, so bees at northern latitudes and higher elevations are often univoltine. However, there are bees in the deserts, where temperatures are warm for a long period of time, that are also univoltine. Photoperiod, or the length of the day, also seems to play a role, and this may be what is important to bees in desert regions.

3.3 PERDITINI

Home bodies

With their little wings and diminutive bodies, small Perditini don't fly very far. It is thought that most species never venture further than 200 feet from their nest entrances when looking for pollen. The smallest obstruction can thus change the landscape dramatically for these bees. Roads have been found to completely isolate populations of *Perdita* that at one time may have interbred, and to separate these bees from plants that rely on them for pollination.

PERDITA AND *MACROTERA*

The Perditini bee tribe (*Perdita* and *Macrotera*) includes the smallest bees in the United States and Canada. When flying, they are often hard to see and look more like gnats than bees. The Perditini are in the Panurginae subfamily.

Perdita

Perdita was named in 1853 by Frederick Smith, an entomologist who cataloged all the insects in the British National Collection. His thought process while naming this bee isn't known, but perdita is Latin for "lost"; it may be that the diminutive size of this petite bee made it easy to lose in those extensive collections.

DESCRIPTION

Though minute, *Perdita* are strikingly colored, often pure yellow, burnt orange, or even blue-tinted chrome.

DISTRIBUTION

More than 650 species of *Perdita* have been named, and there are probably several hundred more that have not been discovered, or have yet to be officially named. *Perdita* are found only in North America, ranging from Canada (12 species) south through Costa Rica. They are especially common in the desert areas of the southwestern United States and northern Mexico. In the east, they are fairly rare, with only three species in the New England area, and fewer than 30 species east of the Mississippi River.

Perdita
per-did-uh

Size Range

one inch

Active Period

J F M A M J J A S O N D

Regions: II-III, IV-V, VI, VII

Distribution

Likelihood of Occurrence
High Low

Not for the faint of heart

At just under 0.1 in long, *Perdita minima* is the smallest bee in North America, where it flies in July and August in the Sonoran Desert. *Perdita minima* works tirelessly in these sweltering conditions to prepare nests for her offspring (with temperatures often higher than 100 °F). The photographer who took this amazing image, Jillian Cowles, spent several hours on her stomach, at the level of the bee, in order to photograph it in its natural habitat. This species rarely lands for very long, so getting a good image was difficult. By the time Jillian had a picture she was happy with, she was so dehydrated from lying in the hot sun that she couldn't stand and had to crawl to her backpack to get some water.

PHOTO BY JILLIAN H. COWLES.

Diet and Pollination Services

Nearly all *Perdita* are specialists, collecting pollen from specific blooming plants and ignoring all the rest. These specializations have led to some unique traits. For example, some species of *Perdita* in the subgenus *Xerophasma* only fly very early in the morning and at dusk or after dark. They have enormous simple eyes on the top of their heads (called ocelli), and collect pollen from evening primroses (*Oenothera*), which bloom at dusk. Other species specialize on (for example) prickly pear cactus (*Opuntia*), globe mallow (*Sphaeralcea*), spurges and sandmats (*Euphorbia* and *Chamaescyce*), gilias (*Gilia*), borages (*Coldenia* and *Heliotropium*), trefoils (*Lotus*), or plants in the sunflower family (Asteraceae). Some species visit just one genus in the sunflower family and others will visit several different genera.

Other Biological Notes

Perdita can be seen during the warm months of the year, from early spring in the deserts through late fall. Because so many are floral specialists, their emergence is often timed to coincide with the bloom of the flowers on which they rely.

A small *Perdita* collects green pollen (visible on her legs) from the anthers of a Rocky Mountain bee plant (*Cleome*). Photo by B. Seth Topham.

Perdita forage well on shallow flowers like this desert marigold (*Baileya*), because their short tongues can reach the nectaries at the base of the flowers.

A mating pair of *Perdita* on a spring mustard (Brassicaceae).

A *Perdita* exits a flower with a full load of pollen on her back legs.

A male *Perdita* prepares to fly from a desert marigold (*Baileya*).

A *Perdita* in a desert dandelion (*Malacothrix*).

Can you see the bee in this flower? This bee blends in well with the yellow reproductive parts of the globe mallow flower (*Sphaeralcea*). Species of *Perdita* that are completely pale yellow are often called "blond *Perdita*."

He's got a big head

The males of several species of Perditini are unique in that their head size can be highly variable; one brother may have a "normal-looking" head, and the other may have gigantic mandibles and a huge head—so big, in fact, that he is unable to fly. Scientists think the enlarged heads make it easier for these males to grasp females when mating. Those males that are so weighed down they don't fly spend their lives inside the communal nests in which they were born. In the dark tunnels they mate with females that are often their sisters. The advantage may be that these males are the first to mate with the females before they lay their eggs, ensuring that the offspring are theirs. However, this phenomenon, combined with the fact that such tiny bees can't fly very far, can lead to a lot of inbreeding in populations of *Perdita* and *Macrotera*.

While the daunting task of studying all species of *Perdita* has yet to be completed, it does appear that most nest in the ground. One, *Perdita opuntiae*, is known to nest in sandstone. Generally *Perdita* make one main tunnel with many branches, each branch of which ends in a nest cell for one tiny little egg. Nest entrances are frequently in bare ground where vegetation is sparse. The entrances are so very small that it is difficult to distinguish the nest opening from other irregularities in the ground's surface. Some *Perdita* nests have miniscule mounds of dirt ringing them. This can make them easier to spot, though these mounds erode away after a few days. Several species of *Perdita* nest gregariously, meaning that multiple females will nest in close proximity to one another; there may be more than 10 nests in a square yard. As one would imagine with a bee so small, the nests are not very deep, often just an inch or two below the surface (though some may be a foot deep).

Perdita may be very specific about the soil they choose when digging a nest. Some in Texas will nest only in limestone-derived soil. A few species nest only in soft sand dunes, and others nest inside the cracks that form as mud dries in intermittent washes. At least one species nests at the bottom of seasonal lake beds. The lake bed is dry in the summer when the bee is active, but during the winter and spring months, the nests will be under water, sometimes for five or six months!

The majority of *Perdita* nest alone; however, a few species nest communally, with up to 30 females using the same tunnel entrance. Inside the nest, each female digs her own offshoot from the main tunnel, and also provides for her own eggs. These nests are often used by successive generations, though each female digs new nest cells each year for her offspring.

Perdita appear to be very sensitive to the humidity in their nests, emerging only when it reaches certain levels indicative of a significant rain event above ground. This may

Perdita come in a variety of colors. Most are yellow or have a yellow abdomen, but some, like this species, are black and red.

A *Perdita* cleaning its abdomen with its back legs as it rests on an indigo bush flower (*Psorthamnus*).

A *Perdita* foraging on a mesquite flower (*Prosopis*). Photo by Jillian H. Cowles.

be because rainfall events in the deserts are correlated with abundant floral blooms.

It takes between 4 and 10 trips to get enough pollen to provide for one small *Perdita* larva. The mother first piles all the pollen in a heap in the cell where she will lay her egg. Once she has enough, she'll work slowly with her front and midlegs to pack the pollen into a perfect sphere. While most bees line their nests with some sort of water-repellent substance, *Perdita* do not. Instead, they cover the pollen ball itself with a waterproof coating. A *Perdita* can usually prepare the nest, and gather enough pollen, for one egg per day.

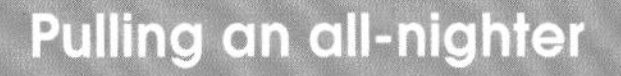

Pulling an all-nighter

Most species of Perditini have not been studied in detail, and how they construct their nests is not well understood. One species of *Macrotera* has been studied at length, however: *Macrotera portalis*. The scientist who studied this bee dug into the ground beside the nests, creating an observation chamber 6 feet deep, where he could stand and watch the bees as they worked underground (like a walk-in ant farm). Interestingly, he found that these little bees do much of their nest construction at night, often working until 1:00 in the morning!

Macrotera

Macrotera was named by the same entomologist who named Perdita. The name means "large monster," which is ironic for such a tiny bee. It is likely a reference to the larger size of these bees compared to Perdita, with which they were once grouped.

Around 30 species of *Macrotera* are known. They are found only in the southwest and western United States and northern Mexico, ranging as far east as Oklahoma and as far north as North Dakota. The life habits are extremely similar to *Perdita*, and many species were once classified as the latter, and may be again someday; it could be that *Macrotera* would best be considered a subgenus of *Perdita,* rather than its own genus. Most *Macrotera* are specialists, and all nest in the ground, either communally or solitarily. In fact, the one notable difference between *Perdita* and *Macrotera* is that, in contrast with *Perdita, Macrotera* line their nests, rather than the pollen loaf, with a waterproof substance. For other details on the *Macrotera* lifestyle, read the section on *Perdita* above.

A pair of *Macrotera* mating inside a cactus flower. Notice the extra large head of the male. PHOTO BY JILLIAN H. COWLES.

A female *Macrotera* collecting pollen from a globe mallow flower (*Sphaeralcea*).

A male *Macrotera* on globe mallow (Sphaeralcea).

Many *Perdita* prefer to forage in high temperatures and are active in the hottest parts of the summer. This *Perdita* was photographed when the air temperature was more than 100 °F.

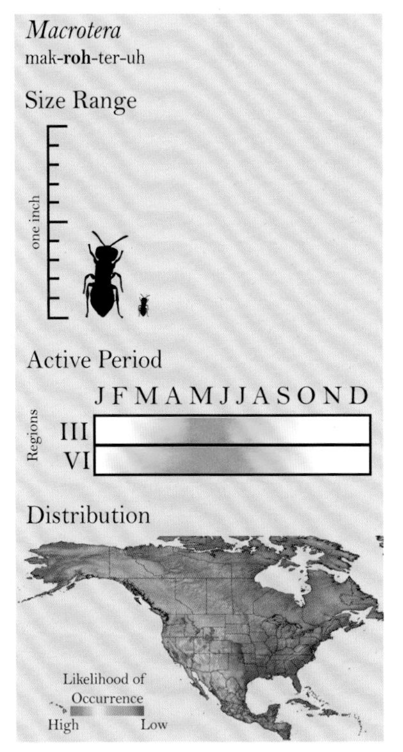

3.4 *CALLIOPSIS*

Calliopsis is the sole representative of its tribe (Calliopsini) in the United States and Canada. The Calliopsini are in the Panurginae subfamily

In Latin, Calliopsis is derived from the word calliope, which means "beautiful," as these bees surely are.

Description

Calliopsis are small to medium-sized bees that often have eye-catching yellow bands on the abdomen, yellow markings on the face, and a shiny black thorax under a light fuzz of hair.

Distribution

Calliopsis are found only in the Western Hemisphere (North and South America) with nearly 90 species known; roughly 70 species occur north of Mexico, with less than 6 in Canada. They are common in dry areas of the southwestern United States and some prairies.

Diet and Pollination Services

Almost all *Calliopsis* are specialists, though there are also "picky" generalists in this group. Seven subgenera are found in the United States and Canada, and several of the subgenera comprise species that all specialize on the same plant group. Hosts for different subgenera include spurges (Euphorbiaceae), legumes (Fabaceae), globe mallow (*Sphaeralcea*), various flowers in the potato family (Solanaceae), and the sunflower family (Asteraceae). Bees in the subgenus *Verbenapis* visit only *Verbena*. The *Verbena* specialists even have special adaptations to maximize their efficiency in the form of hooked hairs on their front legs. They reach into the flower's narrow corolla tube and "hook" the pollen to pull it back out. *Nomadopsis* and *Micronomadopsis* are the only subgenera in which all species do not specialize on the same type of plant.

Calliopsis are not used to pollinate any important crops, but they are very likely important pollinators of the plants on which they specialize. As an example, one researcher found that *Calliopsis* are essential to the pollination of at least one species of *Nama* (fiddleleafs).

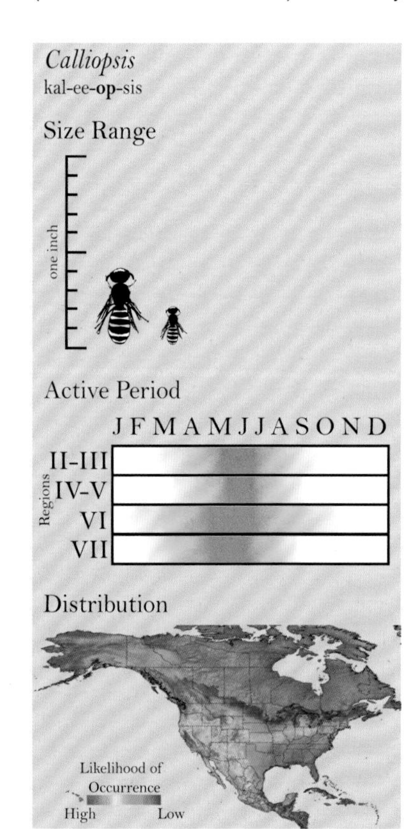

A male *Calliopsis* waits in a globe mallow (*Sphaeralcea*) flower for a receptive female. Note the lack of yellow markings on the abdomen—not all *Calliopsis* have yellow bands.

A *Calliopsis* foraging on a desert dandelion (*Malacothrix*) in April in the Sonoran Desert.

Calliopsis foraging on a desert dandelion (*Malacothrix*) in March in the Mojave Desert.

A *Calliopsis* foraging on a globe mallow flower (*Sphaeralcea*). Photo by Hartmut Wisch.

Attached at the hip

Calliopsis have a fairly unique mating ritual. Once a male mounts a female, he will stay attached as she flies from flower to flower until she returns to the nest. If you see two bees "attached" to one another in this way, consider that they might be *Calliopsis*. PHOTO BY ALICE ABELA.

OTHER BIOLOGICAL NOTES

Calliopsis are most often seen in the summer months and early fall. In desert areas where most flowers are blooming in the spring, spring-flying species also occur. For those *Calliopsis* that live in habitats subject to drought, if the flowers don't bloom, the bees may not emerge until a more promising year (see bet-hedging, p.17).

All *Calliopsis* (like all other bees in the Andrenidae bee family) nest in the ground, and their tiny nests are often found in hard-packed trails or where solid layers of clay or alkaline soil occur. If they are specialists, they nest in close proximity to their preferred pollen source.

Some *Calliopsis* nest in very dense groups, with in excess of 1500 nests per square yard; however, all of them are solitary, and each female builds and provisions her own nest.

With so many bees nesting so close together, competition by males for newly emerged females is intense. Males have been observed fighting for airspace around the nesting site, squaring off in the air and flying in an upward spiral until they are as much as 6 feet above ground before attacking and tumbling to the ground, biting at each other's legs and bodies as they fall.

The nests are very short tunnels (less than 6 inches deep). Inside are side passages ending in one to many nest cells. Depending on the species, one female bee may lay just a few eggs in a nest, or nearly two dozen. The walls of the nest cells are lined with a waterproof substance painted on by the female. They also paint the pollen and nectar loaves they leave for their offspring with the same waterproof lining. Interestingly, *Calliopsis pugionis* nest on floodplains, and the nests may be underwater for more than three months. The waterproof lining keeps the growing bee inside the nest dry and safe until the soil dries out and the water above ground evaporates during the dry season.

A *Calliopsis* resting on a gum weed flower (*Grindelia*).

Buzzing in

Unlike other bees that close their nest entrances when they're home but leave them open when they're gone, many *Calliopsis* hide their nests underneath a loose pile of dirt through which they must dig both coming and going. In the most extreme case, *Calliopsis larreae* nests in areas where sandy soil covers a hard layer of clay. The bee vibrates her way through the sand until she reaches the hard clay, where she digs her nest. Somehow she is able to relocate her nest every time she returns, though the nest entrance may be covered by 3 inches of sand! Even when the female is carrying large loads of pollen on her legs, she manages to keep the loads as she burrows through the sand to her nest in the clay underneath. Oddly, *Calliopsis* males also burrow and will often dig shallow nests to sleep in at night.

3.5 OXAEINAE

MESOXAEA AND *PROTOXAEA*

These are the black sheep of the Andrenidae bee family; *Mesoxaea* and *Protoxaea* look nothing like other Andrenidae in the United States and Canada. They are, in fact, so different that they were once placed in their own bee family. Oxaeinae are found only in the Western Hemisphere, and 19 species are distributed among 4 genera. They are most abundant in the tropics, though a few species range north, to the southern border of the United States.

Protoxaea gloriosa

Oxaea means "pointy" in Latin and likely refers to the pointy glossa (tongue) of these bees, while proto- means "first" or "basic," probably a reference to the fact that this bee shares similarities with bees in the genus Oxaea.

DESCRIPTION

Protoxaea are fast-flying, very hairy, stocky bees.

DISTRIBUTION

Three species of *Protoxaea* are known in the world, but only one occurs in the United States. *Protoxaea gloriosa* can be found in southern Arizona, southern New Mexico, and southwestern Texas; it ranges just a few hundred miles north in each state and also south through central Mexico.

DIET AND POLLINATION SERVICES

Protoxaea are generalists, though with slight preferences when it comes to gathering pollen. While they visit a wide variety of blooming plants, they appear to favor flowers in the pea family (Fabaceae), the potato family (Solanaceae), and flowers of creosote bush (*Larrea tridentata*). Some scientists believe this bee has a preference for any plant that can be buzz pollinated (p.21).

These bees must often contend with other large, bullying bees as they forage for pollen and nectar. *Caupolicana* and *Ptiloglossa* (section 4.3) will both knock *Protoxaea* off flowers when they land to forage. To avoid being displaced, *Protoxaea* time their foraging trips for later in the morning, after the other early rising bee ruffians have mostly finished.

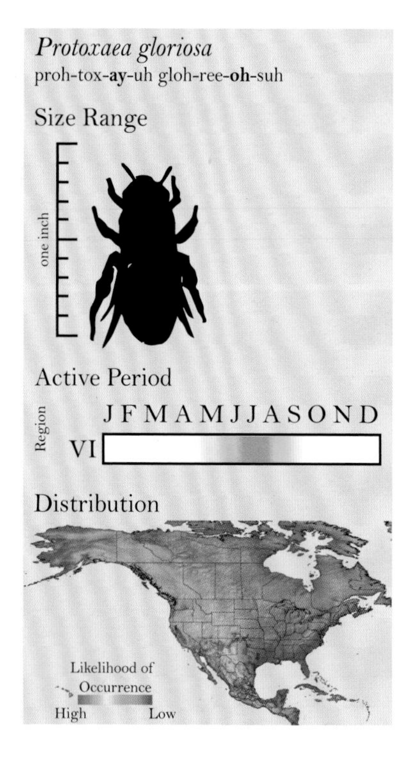

A male *Protoxaea* drinking nectar from a flower in southern Arizona. PHOTO BY LON BREHMER AND ENRIQUETA FLORES-GUEVARA.

Frenemies

Though they battle during the day, at night the gloves are put away. Male *Protoxaea* huddle together to sleep in large aggregations of up to 300 individuals! There does not appear to be a favorite plant for aggregation; they congregate on bushes or shrubs like jimson weed (*Datura*) or Mormon tea (*Ephedra*), or even deserted bird nests. It is thought that huddling together provides the bees some warmth during the cold desert nights. Photos by Jillian H. Cowles.

Protoxaea fly during the summer and early fall months, following the monsoon rains that scour the southwest during these seasons.

Other Biological Notes

The males of these bees are what one is most likely to see, as they hover around flowers waiting for nubile females. They are territorial and will engage other *Protoxaea* males as well as males of other bee species. Some have been observed dislodging another male *Protoxaea* while he is in the act of mating.

Nests are most often built in flat, hard-packed ground with sparse vegetation. There is typically a mound of freshly excavated dirt around the rather large nest entrance. While the nesting habits for all of *Protoxaea* aren't yet known, for the ones that have been studied, the nests range from a foot to more than 2 feet deep. There is usually one long vertical tunnel, with horizontal tunnels radiating from the main shaft. Each horizontal shaft ends in a single cell, again dug vertically into the soil. It appears that some species have soil preferences when building their nests.

A female *Protoxaea*.

A female *Protoxaea* preparing to take flight after foraging. Note the distinctive tuft of hair near the tip of the abdomen.

A male *Protoxaea* on a mesquite flower (*Prosopis*). Photo by Jillian H. Cowles.

A loving death grip

To avoid being knocked off a female, a mating male hangs on tight, jamming his head between the thorax and abdomen of his partner and clamping down with his mandibles. Once a female has been mated, other males seem to recognize that she is no longer fecund and will avoid her.

These bees often nest in large aggregations of several thousand individuals. Each female builds and maintains her own nest, often in close proximity to other females. In Texas, in 1932, the Reverend Birkmann observed an incredibly large aggregation and wrote an account for the local newspaper, the *Gidding News*. The article read, in part, as follows:

> My son, with whom I am now staying, tells me that he has seen a lot of bees or wasps in the air and they are making a great ado and hubbub flying in all directions, and not only over a small area, but for quite a distance the noise is heard and the ground is full of newly dug holes. Next morning when the day grows warm we go out and I found all this to be true, and not the half told. I could not see the insects on account of my poor eyesight, but my son tells me all I want to know. He tells me the bees are flying, some of them only a half foot from the ground, others fly higher, so as to get over the weeds, still others are pretty high up, all of them flying fast and like so many little furies, making all the noise possible to them. You could hear them at some distance and the sound was similar to that heard in the telegraph wires at times, but much louder, and to listen to them for awhile would make me a little nervous.

Mesoxaea

Meso- is Latin for "middle," likely a reference to the distributional center of this group in Central America; oxaea means "pointy," a reference most likely to the tongue.

Though darker in coloration, *Protoxaea* and *Mesoxaea* look remarkably similar, with eyes that converge near the top of the head in males, and fuzzy fast-moving bodies. These bees are extremely rare, but seven species of *Mesoxaea* can be occasionally seen in the United States, along its southern border. *Mesoxaea* was formerly considered a subgenus of *Protoxaea,* and their biologies are remarkably similar. North American species of both genera hover around the same types of flowers, and they nest in a similar manner. For details on the nesting and foraging biology of these species, read the *Protoxaea gloriosa* section above.

A *Protoxaea* foraging on a plant in the pea family (Fabaceae).

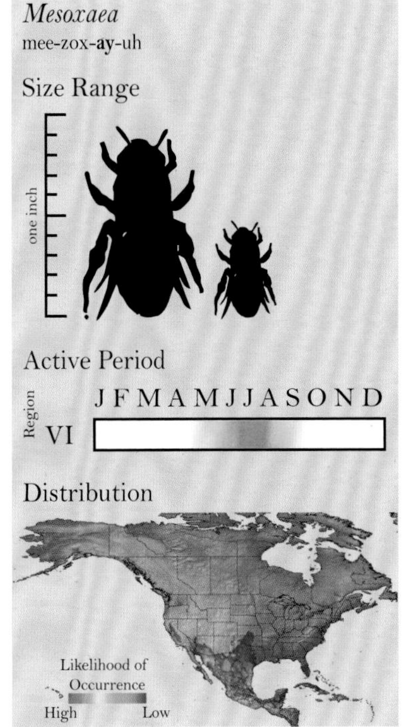

4

COLLETIDAE

Bees in the family Colletidae range from scrawny to burly, as illustrated here. The bees presented above are shown roughly to scale, from the large and brightly colored *Caupolicana* to the matte-black and wasplike *Hylaeus*.

The Colletidae are a diverse family with many striking and easily recognized bees. The group includes agile fliers, nocturnal bees, extreme specialists, and bees that carry pollen internally, instead of on their legs or abdomen. One entire genus is nearly hairless, resembling wasps more than bees. At the other extreme, several genera are robust, loud buzzers with enough hair to make a bear envious.

North of Mexico only 5 genera are known, but worldwide there are 56, and they are best represented in South America and Australia. Of the North American species, two genera (*Colletes* and *Hylaeus*) are widespread, occurring in every U.S. state and nearly every Canadian province. One of these, *Hylaeus*, is even relatively well represented on Pacific islands where native bees are generally rare. It is the only native bee genus living in Hawaii. The other three Colletidae genera found in the United States and Canada are most common in Central and South America, with some species extending just over the border into the southwestern United States.

The bee family Colletidae includes generalists and specialists, and they are likely important pollinators of many wildflowers. None have been developed as commercial pollinators in North America, but their abundance in Australia makes them potential candidates there.

Ready for a close-up

Some species of *Colletes* found in South America are especially colorful, with a deep sea-blue abdomen or bright-red hair on the thorax.

All Colletidae in the United States and Canada are solitary nesters, and there is no evidence of sociality, meaning these bees do not cooperate in making a nest and caring for their young. However, some species nest in large groups, called aggregations, and they may even share a nest entrance. The range of nesting habits among Colletidae is impressive. A large suite are ground nesters, digging tunnels of varying depths in the soil and building

Many *Colletes* are gray or brownish, with contrasting stripes on their abdomen. Several species prefer flowers in the pea family (Fabaceae).

Caupolicana are large beautiful bees with a thick coat of orange hair and black-and-white stripes on their abdomen.

lateral offshoots for their nest cells. *Hylaeus*, in contrast, are all twig nesters and burrow into the hollow stems of perennial bushes and other stout plants. Interestingly, among the ground nesters, many secrete a remarkable waterproof substance that looks very much like cellophane wrap. They "paint" this substance onto the walls of nest cells using their unique tongues, thus protecting their young from moisture as well as from oft-associated soil fungi.

Five subfamilies have been identified worldwide, including three in the United States and Canada: Colletinae, Hylaeinae, and Diphaglossinae. We have devoted a section to each subfamily.

Hylaeus are very unique-looking bees. They are nearly hairless and thus easily confused with wasps. Unlike other bees, they carry pollen in their crop (like a stomach) instead of externally, and regurgitate it when they get back to their nest.

Many bees in the family Colletidae secrete a special cellophane-like substance they use to line their nest cells. Here, a female *Colletes* sits next to the clear nest cell she is filling with a mixture of nectar and pollen. Photo by Dana Atkinson.

IDENTIFICATION TIPS

With bees on the wing, identification is difficult at best. With Colletidae, it is often easier to deduce the genus, then work out the family. *Hylaeus* bees, for example, are among the most easily identifiable genera in North America. Rather than working out the family as Colletidae, it is more straightforward to identify them as *Hylaeus* and know, therefore, that they are Colletidae. If the genus isn't obvious, collect a specimen (or take photographs), and use the characters given below as a guide. A microscope, hand lens, or several macrophotographs might be necessary.

IDENTIFYING FEATURES OF THE COLLETIDAE BEE FAMILY

- Colletidae have few unifying features to aid in their identification. The most obvious shared characteristic is also the hardest to see: All bees in the Colletidae bee family are short-tongued, and the glossa especially is extremely short. It is frequently broader than it is long, and the end is split.
- There is only one subantennal suture beneath each antennal socket.

Diagnostic features for different genera can be seen below.

If you can see it, the tongue of *Colletes* is unique: short, fat, and split at the tip like a snake's, though the ends are rounded or squared off, instead of pointed. Also note the labial palps—all four segments are the same length, indicating a short-tongued bee. Photo courtesy of USGS Bee Inventory and Monitoring Lab.

COLLETINAE (SECTION 4.1)

Eulonchopria *Colletes*

The Colletinae are united in some of the features of the tongue, but the two genera found in the United States and Canada look very little alike. *Colletes* are by far the more common. They are gray, hairy, medium-sized bees. Unfortunately, many other genera also include gray, hairy, medium-sized bees. Look for the heart-shaped face and slightly flattened body. *Eulonchopria* are only occasionally seen along the Mexican border. They are medium-sized black bees with extremely coarse punctation, little hair, and broad integumental yellow bands at the end of each tergal segment.

Identifying Features of *Colletes*

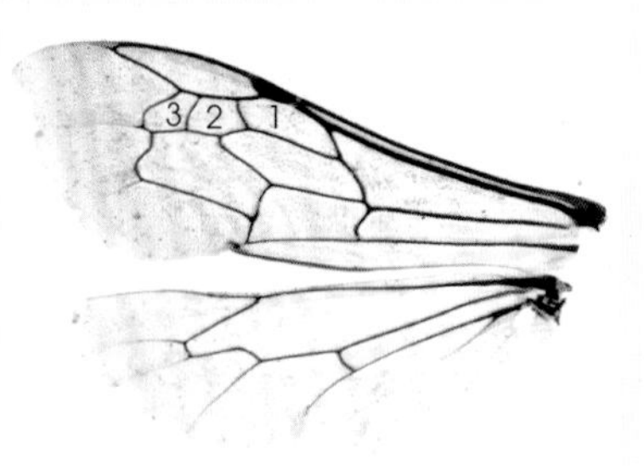

Right: A *Colletes* wing, showing the three submarginal cells and the curved, almost S-shaped, second recurrent vein (in red). The basal vein is straight, not curved. The second and third submarginal cells are about the same width (compare to *Andrena* wing, section 3.1).

Identifying Features of *Colletes* continued

RIGHT: The face of a *Colletes*. Note that the eyes are angled (rather than being parallel), making the face slightly heart-shaped. Under each antennal socket there is just one subantennal suture (hard to see here), and there is no hair near the compound eyes (facial fovea) as in *Andrena*.
FAR RIGHT: *Colletes* lack a pygidial plate (p.34) and the associated fimbriae (tufts of hair).

Similar Genera to *Colletes* and *Eulonchopria*

Andrena (section 3.1)

Halictus (section 6.2)

Anthophora (section 8.8)

Melitta and *Hesperapis* (chapter 5)

HYLAEUS (SECTION 4.2)

Hylaeus are conspicuous. The specific pattern of yellow markings on the face that extend up beside the eyes are seen in no other bee. There is very little hair, no metallic sheen, and there are no pollen-collecting hairs. They in fact look more like a wasp than a bee. From above they appear narrow, almost as though pinched.

Identifying Features of *Hylaeus*

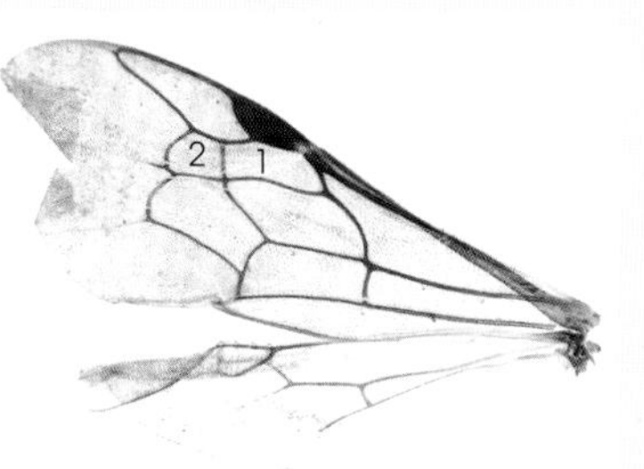

The wing of a *Hylaeus*, showing the two submarginal cells, and no especially curved wing veins.

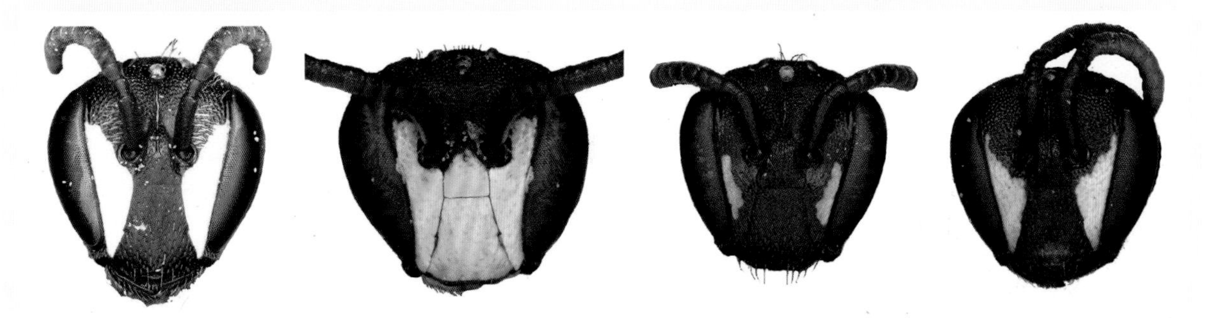

Faces of four species of *Hylaeus* showing a range of facial patterns. *Hylaeus* faces tend to be longer than wide, or round, and they have only one subantennal suture. The patterns seen here can even differ between individuals of the same species. PHOTO COURTESY OF USDA-ARS BEE BIOLOGY AND SYSTEMATICS LABORATORY.

continued overleaf

HYLAEUS (SECTION 4.2) *continued*

Similar Genera to *Hylaeus*

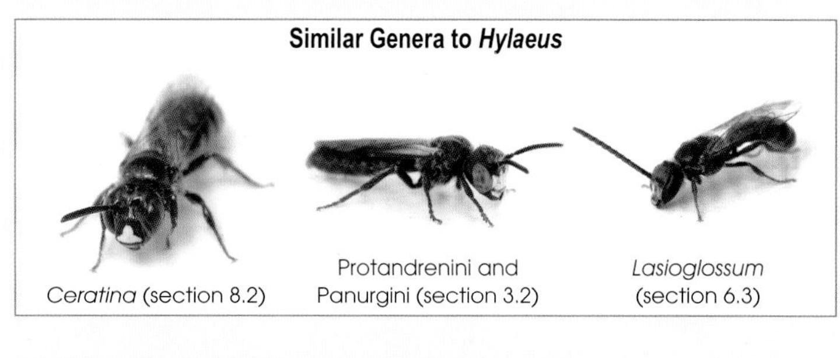

Ceratina (section 8.2) Protandrenini and Panurgini (section 3.2) *Lasioglossum* (section 6.3)

DIPHAGLOSSINAE (SECTION 4.3)

Ptiloglossa *Caupolicana*

LEFT: The large eyes, large bodies, and fast flight can provide gestalt identification clues for these two Diphaglossinae. At first, they may seem similar to many of the large hairy bees in the family Apidae, but there are notable differences. In addition to the characters shown below, these bees are often among the earliest fliers each day, up and busy a half-hour to an hour before the sun. On the left is *Ptiloglossa*. This genus has a slight metallic sheen and no white hair bands. This is the rarer of the two. On the right is *Caupolicana*. The striking white bands on the thorax are distinctive. *Caupolicana* are matte black (underneath an auburn coat). This genus is fairly widespread in the United States.

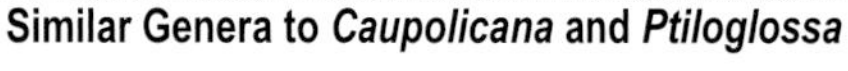

Similar Genera to *Caupolicana* and *Ptiloglossa*

Protoxaea (section 3.5) *Centris* (section 8.9)

Identifying Features of the Diphaglossinae

LEFT: The Diphaglossinae have three submarginal cells (pictured here is *Ptiloglossa*), and the middle one is usually very small. Note that the beginning of the submarginal cell (nearest the stigma) has a "hook" to it, extending like a finger pointing toward the stigma. Though this is not as pronounced in *Caupolicana*, it is still visible and can help in distinguishing these bees from large hairy bees in other families.

MIDDLE AND RIGHT: In the middle, the face of a *Caupolicana* head-on. On the right, a *Ptiloglossa* from above. Note the large ocelli on both faces, between equally large eyes. The ocelli face more forward than up as they do in most bees. In males they are even further down on the face.

4.1 COLLETINAE

COLLETES AND *EULONCHOPRIA*

In North America, the Colletinae subfamily is represented by two bee genera. *Colletes* can be found across the continent. Its closest cousin, *Eulonchopria,* occurs rarely along the border with Mexico.

Colletes

The genus name Colletes means "one who glues," referring to their habit of applying a glue- or cellophane-like lining to the walls of nest cells, using their specialized tongues. This lining gives rise to their nicknames: cellophane bees, polyester bees, and plasterer bees.

Description

Colletes are medium- to large-sized bees that are generally hairier than many bees of similar size. Though it is hard to see without a microscope, *Colletes* have a distinctive tongue that is short and forked.

The right tool for the job

Most bees that nest in the ground have a specialized structure on the tip of their abdomen (a pygidial plate) with a brush of hairs around it. This structure is thought to be useful for sculpting dirt when making ground nests. Surprisingly, *Colletes* have neither a pygidial plate nor the surrounding hairs, even though they nest in the ground. One reason may be that the first *Colletes* nested in woody stems (as some species still do in South America, the ancestral place of origin for *Colletes*). Researchers propose that over time they lost the pygidial plate because it wasn't needed—there is no soil to tamp down inside a woody stem. Later, the species reverted to ground nesting, but they never re-evolved this structure, and all *Colletes* species today simply do without.

Saved by the bug

Colletes was named by Pierre Andre Latrielle in 1802, after he was released from prison. As a member of the clergy, he was imprisoned during the French Revolution. His skill in identifying a beetle in his dungeon cell caught the attention of a visiting physician, who used his connections to save this impressive scientist from the guillotine. He went on to name thousands of insect species.

Many *Colletes* are specialists and will collect pollen only from specific plants. This *Colletes*, for example, collects pollen only from prairie clovers (*Dalea*).

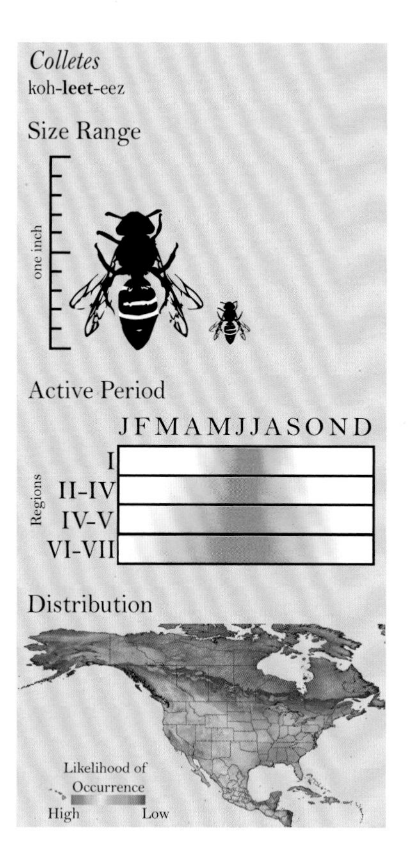

A *Colletes* collecting the bright orange pollen from an indigo bush flower (*Psorthamnus*).

Distribution

Colletes are prevalent in North and South America, as well as in Europe and Africa. Interestingly, though the bee family Colletidae is profuse in Australia, the genus *Colletes* is entirely absent there.

In North America, *Colletes* are found throughout Mexico and as far north as Alaska. East to west, *Colletes* are found in every state except Hawaii; 99 species are found in the United States with 25 of those also in Canada. *Colletes* occur in all environments, from grasslands to forests, from swamps to tundra, and even in backyards in Manhattan and San Francisco.

Home is where the habitat is

Though the genus thrives in many environments, several *Colletes* species have specific habitat preferences. For example, certain Florida scrub habitats are home to *Colletes francesae,* a rare bee found nowhere else in the United States. This bee collects pollen exclusively from the bully tree (*Sideroxylon*). *Colletes validus* is a species that is active in the spring and can be found in the wet woods and swampy areas of New England and the east coast, but only as far west as Michigan. *Colletes impunctatus* is found only north of 42°N latitude (Maine, New Hampshire, Michigan, Wisconsin, Minnesota, Montana, and north to Alaska). Another species, *Colletes stepheni,* only lives on sand dunes in the Mojave Desert and collects pollen primarily from creosote bush flowers (*Larrea tridentata*). In order to avoid the high heat in these desert habitats, this bee is most active at sunrise.

Diet and Pollination Services

A number of *Colletes* are specialists, foraging for pollen on only one group of plants. Some specialize on all members of a plant family (often Fabaceae or Asteraceae), while others will visit only one genus. Oddly, *Colletes linsleyi* appears to prefer collecting pollen from salt cedar (*Tamarix*). While the bee is native to the United States, salt cedar is an invasive species imported from the Mediterranean less than 200 years ago.

Not all *Colletes* are floral specialists, however, and a few species will visit a taxonomic smorgasbord of blooming flowers. Some species, even among the generalists, will collect pollen from only one type of plant for each egg, so that each egg develops on a different type of pollen. Because the ins and outs of manipulating flowers can be difficult to retain for an insect with a brain smaller than a popcorn kernel, it is possible that *Colletes* stick to one flower type at a time because it is all they can remember. Alternatively, it may be more efficient (see constancy, p.203).

Mi casa, su casa

In Europe, one species (*Colletes daviesanus*) nests in sandstone embankments, with thousands of nests in one area. The villages nearby are frequently "modified" by bees that branch out from natural embankments to nest in the walls of brick and mortar buildings.

Other Biological Notes

In warmer areas, *Colletes* can be found from March through September, but in most regions of the United States and Canada, there are more species active in fall than in spring. Although much is not known about this genus, there is evidence that some species of *Colletes* bee may be bivoltine, meaning they produce two broods in one year (see voltinism, p.85).

Colletes usually nest in the ground in areas with little vegetation. Some species are very particular about the soils in which they nest, for example preferring sand dunes over silty soils, or loamy turf over sandy embankments. Many of the nesting sites are used for successive years, and over time giant nesting aggregations may form, sheltering up to several thousand bees.

Colletes gather pollen, water, and nectar (which is stored for transport in their crop) to provide food for eggs. Rather than the harder ball of pollen that most bees leave for their offspring, *Colletes* leave a soupy mass for them to devour. Perhaps because the provision is so soggy, *Colletes* bees

Some *Colletes* species are commonly found on plants that are not native to North America, like this *Colletes* on sweet clover (*Melilotus*), a flower native to Europe and Asia.

A mating pair of *Colletes*; the male is on the right. Photo by Alice Abela.

A *Colletes* foraging on a fleabane (*Erigeron*).

stick their eggs to a side wall of the cell chamber rather than leaving them on top of the pollen and nectar mixture.

Colletes line their nests with a distinctive cellophane-like substance made from saliva and secretions from the Dufour's gland (a small gland on the abdomen). Using flat, short, forked tongues, they paint the walls with saliva, then, with secretions from the Dufour's gland, they add a coat of varnish. This creates a clear covering that is strong, durable, and resistant to mold and water. Because it is so strong and does not degrade easily, scientists have recently begun analyzing this natural polymer (silk) with the hope that it may someday be synthetically manufactured for use in a variety of applications.

Colletes are often called cellophane bees or polyester bees because they line their nest cells with a clear, cellophane-like substance. They then fill these nest cells with a liquid mixture of nectar and pollen. Here, a female *Colletes* deposits some pollen into her transparent nest cell. PHOTO BY DANA ATKINSON.

A *Colletes* resting on a globe mallow flower (*Sphaeralcea*).

Eulonchopria punctatissima

Eulonchopria means "well-serrated spear," most likely a reference to the inner hind tibial spur of females, which is long and has a row of very fine teeth.

Eulonchopria is a small to medium-sized bee. Just one species is found in the United States, along the border with Mexico; the other half-dozen species are found in South America. Interestingly, they are not found in wet rainforests, but occur on either side of it. This distribution pattern is different from the majority of Earth's organisms, which have continuous geographical ranges (see amphitropical distributions, p.212). The species found in the United States appears to be a specialist on *Acacia*. Its specific nesting habits are not known, but observations suggest that this species follows the pattern of other bees in the Colletidae family and nests in deep burrows with several branching tunnels for nest cells.

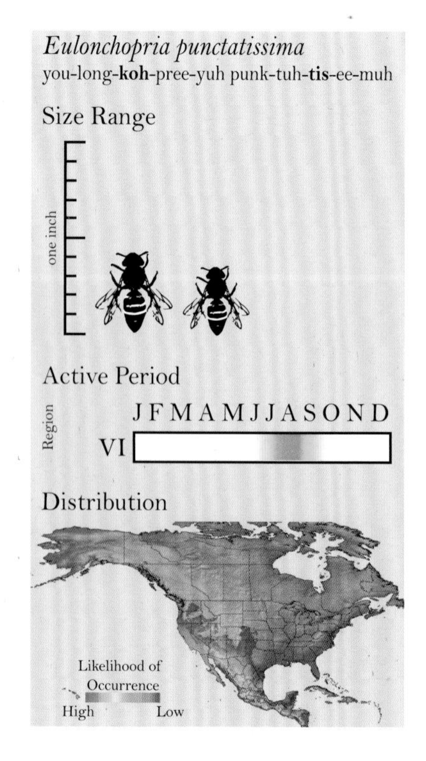

4.2 *HYLAEUS*

Hylaeus means "of the woods" and refers to the observed preference of these bees for nesting in woody materials; also, Hylaeus are relatively abundant in forests, a habitat where bees are generally not common.

Hylaeus are the only member of the Hylaeinae subfamily that resides in the United States and Canada, though 13 other genera have been identified around the world.

Description

Though small and rather slender, *Hylaeus* are one of the most distinctive bee groups in North America. They are dark, mostly black, but with white or yellow markings on their faces and legs. This, and the fact that they have very little hair, makes them look more like wasps than bees.

Distribution

Hylaeus are found around the world, and (notably) on many islands; native species have even been found in Hawaii. Worldwide, scientists have identified a total of 700 species of *Hylaeus*, and they are particularly diverse in Australia. On mainland North America, approximately 50 *Hylaeus* species occur, ranging in latitude from the Arctic Circle in Alaska south to Florida and in longitude from coast to coast.

Many species of *Hylaeus* are found in the mountains, where woody materials abound, though a few species are restricted to deserts. Because their nesting preferences are flexible (see below for nesting details), *Hylaeus* are also quite common in urban/suburban areas.

Hawaiian native

Interestingly, *Hylaeus* are the only bees native to Hawaii, historically home to 60 different *Hylaeus* species (more than are found in mainland United States and Canada combined). The Hawaiian *Hylaeus* are found nowhere else in the world, and many species are restricted to a few habitat fragments on a single island of the archipelago. Researchers believe these 60 species all evolved from a single ancestor that may have come to Hawaii from Japan or mainland Asia. Recent surveys of Hawaiian bees have failed to find several of these *Hylaeus* species, and scientists hypothesize that many of them might have gone extinct. The limited distribution of the Hawaiian species, along with the rapid rate of habitat destruction, has led to a few rare species becoming candidates for federal protection.

In Canada and mainland United States, two species of *Hylaeus* are thought to be relatively recent introductions. Both are native to Europe. *Hylaeus punctatus* was first seen in the United States in the early 1980s and is rare. The other, *Hylaeus leptocephalus,* is well established and common across the United States.

The fact that *Hylaeus* nest in wood (instead of in the ground) might explain their abundance on islands around the world (which are notoriously lacking in bees).

A male *Hylaeus*.

A female *Hylaeus* resting on a flower.

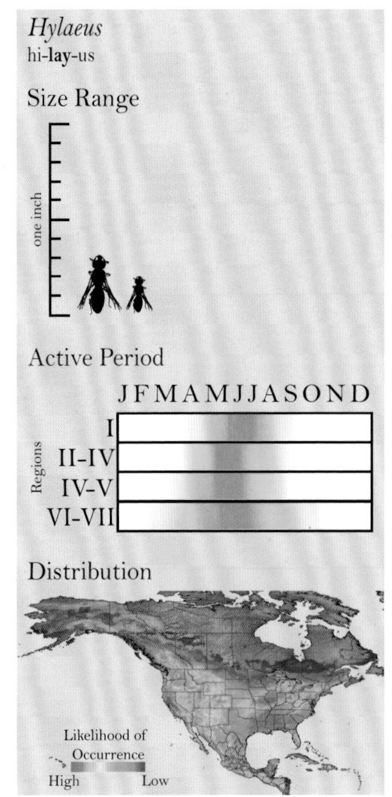

Hylaeus nesting in pieces of driftwood may emerge to find themselves on new shorelines. This may also explain the non-native species found in the United States. It is likely that a piece of wood containing a European *Hylaeus* nest was shipped across the sea. When the bees emerged, they easily established in the New World.

Diet and Pollination Services

Most nonparasitic female bees have pollen-collecting hairs (called scopa) somewhere on their bodies for carrying pollen back to their nests. *Hylaeus* are unique among females in that they have no scopa. Instead, *Hylaeus* eat the pollen and nectar they find in flowers and regurgitate it when they return the nest. It is thought that all *Hylaeus* species in North America are generalists, collecting floral resources from any number of blooming plants; however, because they carry their pollen internally, it is hard to determine the flowers from which they have collected. One scientist has found three *Hylaeus* species that appear to visit flowers only in the rose family (Rosaceae). It is possible, therefore, that some specialist *Hylaeus* occur in North America. The fact that several Australian species are specialists supports these findings.

While the pollen preferences of most *Hylaeus* are unknown because they carry their pollen internally, researchers suspect that some species prefer to forage on plants in the rose family (Rosaceae).

Other Biological Notes

Hylaeus fly mainly in the spring, but some species emerge at the end of summer or early fall. Several species are bivoltine, with a generation emerging in the spring and another in the fall (see voltinism, p.85).

Hylaeus are often confused with wasps because of the hairless, rough integument (skin), and lack of pollen-collecting hairs (scopa) on the legs of females. Photo by Alice Abela.

A *Hylaeus* resting on a flower bud. PHOTO BY JILLIAN H. COWLES.

BELOW: A foraging *Hylaeus*. Notice the white markings on the face and legs. This is a key characteristic of *Hylaeus*. PHOTO BY JILLIAN H. COWLES.

Hylaeus nest in premade holes; they do not excavate their own nests. Suitable holes include the hollow stems of dead plants, beetle burrows, earthworm holes, holes in wasp nests, nail holes, natural holes in rocks, and even the holes made by other bees of about the same size. They may use preexisting holes because their dainty mandibles (mouthparts) are not strong enough to dig nests in hard materials like wood and hard-packed earth.

Like their cousins *Colletes* (section 4.1), *Hylaeus* line their nests with a transparent cellophane- or silk-like substance that is waterproof and contaminant-proof, using their tongues to paint the substance on the cell walls. Also like *Colletes*, *Hylaeus* secrete chemicals from the base of their mandibles that help prevent the spread of fungi and bacteria in the nest. These chemicals are identical to those produced by many plants and smell like citrus.

Stronger than silk?

Curious scientists have tried any number of methods to destroy the silklike lining created by *Hylaeus* for their nest cells. They have found that it doesn't melt when heated, and it doesn't dissolve when infused with strong chemical solvents.

Many small bees, like *Hylaeus*, gather pollen from the anthers of a plant without contacting the plant's pollen receptor (stigma); in those instances they are poor pollinators. Here a small *Hylaeus* collects pollen from, but does not pollinate, a yellow spider flower (*Cleome*).

4.3 DIPHAGLOSSINAE

CAUPOLICANA AND *PTILOGLOSSA*

Caupolicana and *Ptiloglossa* are two of the rarer genera in the Colletidae bee family, and the only genera in the United States representing the Diphaglossinae bee subfamily.

Caupolicana

Caupolicana is named after a fierce warrior of the Mapuche people of Chile. He led an army against the Spanish conquistadors in the 1500s. The first specimens to be labeled with the name Caupolicana were collected in Chile.

DESCRIPTION

Caupolicana are burly, extremely fast-flying bees with large eyes. In the males these eyes curve up around to the tops of their heads.

DISTRIBUTION

Found throughout the Midwest, they range as far east as Florida, and west through Arizona and Utah. Only five species are known in the United States; in the east are *Caupolicana electa* and the very similar-looking *C. floridana*, and in the west are *C. yarrowi* and *C. ocellata*. The fifth, *C. elegans,* is in a subgenus known as *Zikanapis*, which some scientists believe should be classified as a separate genus.

DIET AND POLLINATION SERVICES

While specialization has not been verified, *Caupolicana* appear to have a strong preference for flowers in the pea family (Fabaceae) and the mint family (Lamiaceae), though they are also common on creosote bush (*Larrea tridentata*). One rare species that lives only in the east, *Caupolicana floridana,* has never been collected with pollen, but this species has been observed visiting mints, which they often nectar rob (p.54). This *Caupolicana* species lands on the back end of the mint flower and makes a quick slit with its sharp tongue. Prying open the slit, the bee then sips out the nectar without ever coming in contact with the pollen presented at the mouth of the flower. Interestingly, these bees often visit flower buds that are ready to open the following day. It seems the flowers begin producing nectar before they actually bloom, and the bee steals the reward from other nectar-sipping insects.

Some scientists consider this *Caupolicana* species (*Caupolicana elegans*) to be in a different genus (*Zikanapis*); regardless, they show the typical rusty red thorax and large eyes common among *Caupolicana*. PHOTO BY MARGARETHE BRUMMERMANN.

Caupolicana
koh-poh-lik-**ay**-nuh

Size Range

one inch

Active Period

J F M A M J J A S O N D

Regions

IV

VI

VII

Distribution

Likelihood of Occurrence

High Low

A count who counts

Caupolicana was named by Maximilian Spinola, a well-to-do count in Italy who lived in the 1800s. It was fashionable for gentlemen of the time to maintain a worldly insect collection. With his wealth he often bought showy beetles and other insects from collectors abroad. *Caupolicana* is common in South America, where Spinola owned property, and it was from here that he received many specimens, including bees of this genus.

Caupolicana can often be found on creosote bush (*Larrea*). Photo by Jillian H. Cowles.

A *Caupolicana* foraging on a mint flower (Lamiaceae).

Other Biological Notes

To see *Caupolicana,* plan to be out just after sundown or just before the sun comes up, as they prefer not to be out midday. In fact, you can often hear *Caupolicana* buzzing around flowers before light is sufficient.

Caupolicana nest in the ground, and many species prefer sandy soils. While only a few nests have been studied in detail, the general design of a *Caupolicana* nest is a relatively deep main tunnel, extending straight down as much as 3 feet, with upward-slanting side tunnels that at some point drop abruptly down again, forming a final vertical nest cell. In the manner of other bees in the family Colletidae, each *Caupolicana* nest cell is lined with a cellophane-like substance that is waterproof. The up-and-down positioning of the cell and the waterproof lining are both ideal for holding the almost-entirely liquid food resource that the mother leaves for her offspring. After excavation of the nest cell is complete, she makes a cocktail of mostly nectar, with a little pollen, then lays an egg on the top, where it floats until it hatches.

Ptiloglossa

Ptiloglossa means "feathered tongue" and refers to the brushy tongue common in many bees in the family Colletidae.

Like *Caupolicana*, *Ptiloglossa* are brawny, fast-flying, and furry. They are found only along the southern borders of Arizona, New Mexico, and Texas. Two species occur in the United States and Canada, *Ptiloglossa jonesi* and *Ptiloglossa arizonensis*. *Ptiloglossa* fly when there is very little light and are seldom seen when the sun is at full force. To aid in these predawn flights, the simple eyes on the top of their heads are abnormally large, which helps them orient when landmarks are almost completely invisible. Flower preferences are similar to those of *Caupolicana* (see above). These bees nest in the ground and, like their relatives, line the nest walls with a clear membrane. Inside, the bee leaves a liquid provision of a little pollen and a lot of nectar.

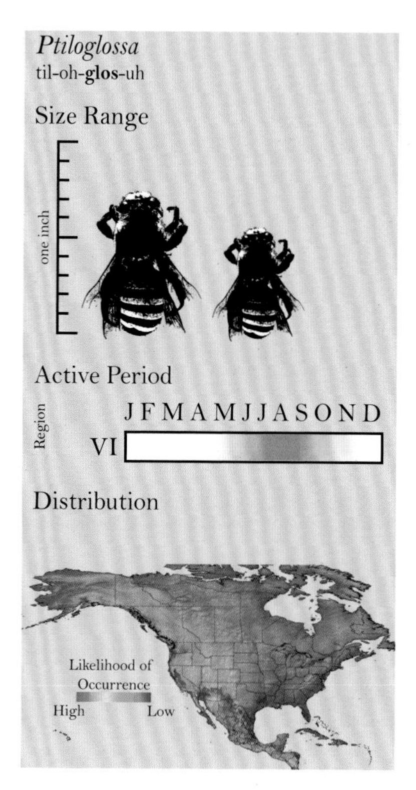

Bee beer!

Bees need protein to develop, and nectar is mostly sugars and water. The larvae of *Ptiloglossa* get their protein from the yeasts common in flower nectar, which ferment the almost-liquid provision left by the mother. That gives a new meaning to the phrase, "getting a buzz from your beer!"

5

MELITTIDAE

The family Melittidae is the least diverse of all the North American bee families. Most bees in this family are small and unassuming. The most common genus in the family, *Hesperapis*, is pictured here.

The bee family Melittidae is a small group of relatively uncommon bees. Though they are found across the Northern Hemisphere and in the southern part of Africa, they are completely absent from South America and Australia. Many bee genera in North America comprise more species than the entire Melittidae family, which contains only about 200 species worldwide.

In North America, only 33 Melittidae bee species occur, classified into 3 genera. Of these, 25 species are placed in the genus *Hesperapis*, 4 in the genus *Macropis*, and 4 in the genus *Melitta*. *Hesperapis* are relatively common in the deserts of the southwestern United States, though not as frequently seen as other desert-dwelling bees. *Melitta* and *Macropis* are rare across the United States.

All Melittidae are solitary bees that nest in the ground, and most species are pollen specialists, meaning they gather pollen from a select few plant species, even when the countryside is abounding with blooms. The hosts they prefer run the gamut, however, with closely related bee species often specializing on distantly related plant groups. In addition to their pollen specialization, *Macropis* species collect floral oils, which they use both to line their nests and also to mix with pollen to feed to their offspring.

Some researchers have suggested that these bees should actually be split into two distinct families, Melittidae and Dasypodaidae, but because this split has not been widely accepted, we consider them a single family here. In fact, though the three Melittidae genera are distinct from each other and have their own unique life histories, we have combined them into one section because they are relatively uncommon.

Ancient bees

Despite their relative rarity worldwide, Melittidae are ancient bees and were probably flying side by side with pterosaurs. The oldest known bee fossil is thought to be around 100 million years old, and it is a specimen that belongs to the family Melittidae. Recent genetic studies support the fossil record, suggesting that Melittidae may have appeared shortly after flowering plants evolved.

A *Hesperapis* drinking nectar from an indigo bush flower (*Psorthamnus*). Notice how she pushes the wing petals of this pea flower apart with her hind legs to access the pollen nestled inside the keel petal. The pollen will stick to her abdomen, but when she is finished drinking, she will pack the loose pollen grains onto her legs.

IDENTIFICATION TIPS

Hesperapis *Melitta* *Macropis*

Melittidae are probably the hardest of all the bees to identify; even determining that they are in the family Melittidae is difficult! The characteristics that separate them from other bees are found on the tongue and the reproductive parts—two areas of the bee that are difficult to see even with a microscope. In truth, the best way to identify Melittidae is probably to rule out other bee families: if it lacks the characteristics common to other genera (i.e., two subantennal sutures, arcuate basal vein, or some other telling feature), it is likely Melittidae. *Hesperapis* are common in western North America, whereas *Melitta* and *Macropis*, are more common in the east. Note that the abdomen of *Hesperapis* tends to look extremely flattened, especially on dead specimens. *Melitta* tend to be more solidly built than *Hesperapis*, but *Macropis* are chunkiest, with a notably rounded abdomen.

IDENTIFYING FEATURES OF MELITTIDAE

- All Melittidae have short tongues.
- The middle coxa (p.38) may be slightly longer compared with other bees, and thus more visible.

Diagnostic features for different genera can be seen below.

ABOVE: The tongue of all Melittidae is short and pointed, but not forked. PHOTO COURTESY OF USDA-ARS BEE BIOLOGY AND SYSTEMATICS LABORATORY.

HESPERAPIS

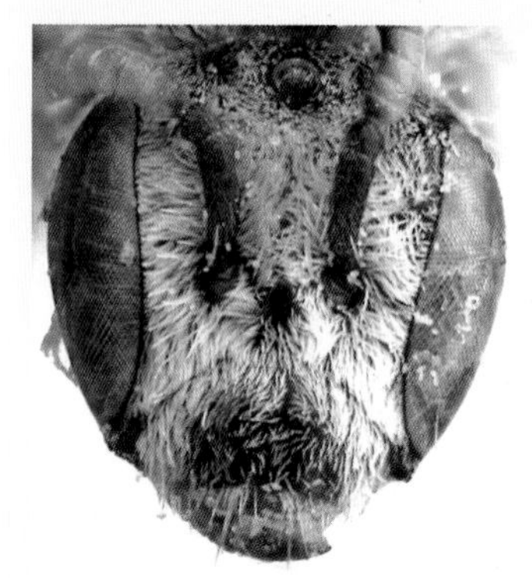

FAR LEFT: *Hesperapis* occur mostly in the western United States and up into western Canada. These petite bees have oval heads, slightly longer than wide, and covered with fine short hair. PHOTO COURTESY OF USDA-ARS BEE BIOLOGY AND SYSTEMATICS LABORATORY.

ABOVE LEFT: The basitarsus of *Hesperapis* is relatively long and slender, much longer than wide. Notice also that the wing has two submarginal cells—this is true of all *Hesperapis*.

MELITTA

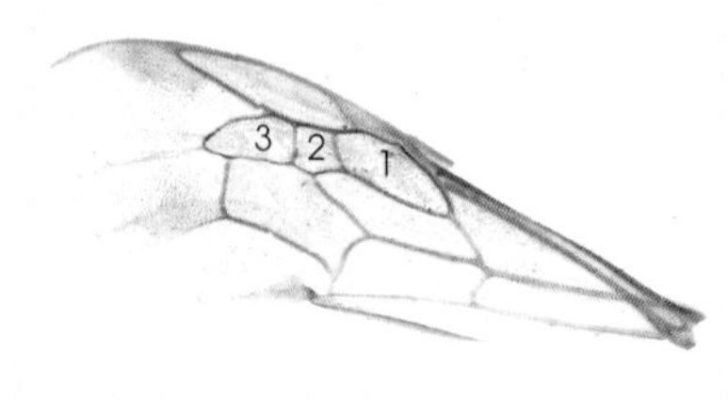

Melitta species have three submarginal cells on the wing, which separates them from *Hesperapis* and *Macropis*.

MACROPIS

LEFT: *Macropis* specialize in collecting oil from *Lysimachia* flowers and have highly modified back legs that aid in the task. As a result the basitarsus of *Macropis* is short and wide; this and the stout body separate it from *Hesperapis*.

RIGHT: *Macropis* is the other Melittidae with two submarginal cells (with *Hesperapis*). See the hind basitarsus character for help differentiating between the two.

Similar Species to the Melittidae

Andrena (section 3.1)

Halictus, or other large halictids (chapter 6)

Colletes (section 4.1)

Apidae (chapter 8)

HESPERAPIS

The name Hesperapis means "western bee." In Greek, hesperus means "evening star" and is a reference to the west, while apis is a Greek word for "bee."

DESCRIPTION

Hesperapis are petite bees, usually with gray or black hair on their abdomens.

DISTRIBUTION

Around the world, nearly 40 species of *Hesperapis* are known. This genus is found only in North America and, oddly, South Africa. In North America, *Hesperapis* are the most commonly found Melittidae. Twenty-five species occur in the United States, but none occur in Canada. Unlike the other Melittidae (*Macropis* and *Melitta*), which are largely restricted to the eastern United States, *Hesperapis*

A *Hesperapis* visits a pincushion flower (*Chaenactis*) in the Mojave Desert.

A *Hesperapis* resting on a desert dandelion (*Malacothrix*).

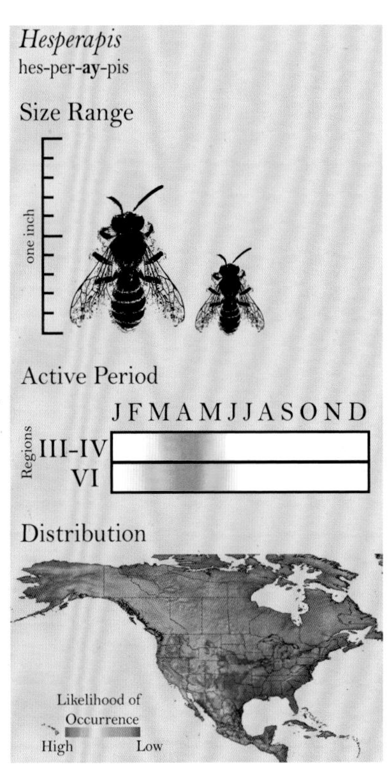

A *Hesperapis* eyes the photographer from inside a suncup (*Camissonia*). Photo by Alice Abela.

A *Hesperapis* foraging on a flower in the borage family (Boraginaceae). Photo by Hartmut Wisch.

are found mainly in the arid deserts and grasslands of the western United States; however, one species (*H. carinata*) can be found as far east as the Mississippi River, and another species (*H. oraria*) occurs only along the eastern Gulf Coast, including Florida.

Diet and Pollination Services

All *Hesperapis* species are pollen specialists. Different species, however, have evolved to specialize on entirely different kinds of plants, from weedy annuals to well-established perennials. For example, two *Hesperapis* species, *H. larreae* and *H. arida*, gather pollen primarily from the creosote bush (*Larrea tridentata*). In contrast, another species, *Hesperapis regularis,* collects pollen only from farewell-to-spring (*Clarkia*). Though the bees are close relatives, the plants they prefer are in separate taxonomic orders (for comparison, a bee and a grasshopper are also in separate taxonomic orders).

Other Biological Notes

Hesperapis may be seen from March through October. Many species in the deserts are bivoltine (having two generations per year), taking advantage of the blooms that follow spring runoff as well as those that follow summer monsoons (see voltinism, p.85).

Hesperapis are all ground nesters, and many prefer to dig their nests in loose sandy soil. In general the nests are burrows with several long side branches, each leading to a

A *Hesperapis rests* on an indigo bush flower (*Psorthamnus*).

A *Hesperapis* on a mariposa lily (*Calochortus*). Photo by Hartmut Wisch.

single nest cell. Unlike most ground-nesting bee species, which line their nest cells with a waterproof coating, *Hesperapis* species generally forgo the lining step. There is some evidence, however, that the *Hesperapis* larvae, instead of their parents, might apply a waterproof coating from the inside of the nest cell before they become pupae. A few species are known to line their nest cells with smooth, very fine soils mixed with saliva.

Nest entrances of *Hesperapis* are often surrounded by a small mound of soil (a tumulus), deposited during the excavation of the nest. Several species, particularly those that nest in sand, have specialized ridges and stiff hairs on their hind legs that are thought to help them dig in the loose sand. The nests of *Hesperapis* are relatively deep compared with those of other Melittidae, with many nests reaching three feet below the surface—impressive for a bee that is only half an inch long!

Weathering the storm

Several *Hesperapis* populations living in the coastal dunes of the northern Gulf of Mexico survived a Class 3 hurricane with sustained winds of 115 miles per hour, 5–10 inches of rain, and a 14-foot tidal surge that covered their dunes with seawater. We still don't know exactly how these bee populations survived a direct hit from a powerful hurricane, but a combination of factors might have played a role. Specifically, their deep nests (some *Hesperapis* dig as deep as 3 feet down) and waterproof nest cells may have helped them survive. Though this bee is able to endure hurricanes, it is still in danger of extinction because of urban development in its coastal habitats.

MELITTA

The name Melitta is a subtle reference to the fact that they are petite bees; meli is Greek for "honey," and itta means "diminutive" or "small."

Description

Light brownish-gray fur covers these slight and delicate-looking bees. When seen, they are often abundant, but seeing them is rare.

Distribution

Melitta species are found across Europe and Asia, in North America, and north and south of tropical (central) Africa. Worldwide, *Melitta* is the most diverse bee genus in the Melittidae bee family (with nearly 50 species), but only four species live in the United States and Canada. Three *Melitta* species are found east of the Mississippi River (*Melitta americana*, *M. eickworti*, and *M. melittoides*), where they range from Florida all the way to Canada. Only one species (*M. californica*) is found in the west, primarily in the Mojave and Sonoran deserts of California and Arizona.

Diet and Pollination Services

North American *Melitta* are all picky about the flowers from which they gather pollen, specializing on just one plant

Melitta are so rare that they are seldom photographed. Pictured here is *Melitta nigricans*, a European species. They look very similar to North American *Melitta* species. PHOTO BY DICK BELGERS.

family or genus. Like *Hesperapis*, closely related species of *Melitta* specialize on flowers that are not closely related and often look nothing alike. *Melitta americana*, for example, specializes on pollen from blueberry and cranberry flowers (*Vaccinium*). In contrast, *M. californica* visits flowers only in the mallow plant family (Malvaceae), primarily globe mallow (*Sphaeralcea*).

Studies have shown that *Melitta americana* is often the most frequent native pollinator of large cranberries, particularly in the northeast parts of North America. *Melitta* are especially effective pollinators of these and related berries because of their ability to buzz pollinate. Also, *M. americana* collects pollen almost exclusively from cranberries, whereas other bee visitors are generalists.

Eclectic tastes

The fact that several species in the Melittidae bee family, all closely related, specialize on such distantly related flowers is interesting to researchers. If we trace the family tree of Melittidae backward, all modern-day species share one common ancestor (the same way all your cousins share a common ancestor, your grandfather). That one bee species, presumably, was a specialist on just one plant group. Because today's species are found specializing on so many different plants, it seems likely that the descendants of that common ancestor switched from the plant on which it specialized to something new. Why? Scientists hypothesize that environmental changes may have changed available floral resources, forcing specialist bees to find new hosts. Another hypothesis is that the bee evolved over time so that it was better able to recognize other suitable hosts. These ancient changes in pollen preference likely led to the wide variety of plants on which we see *Hesperapis* and *Melitta* today.

By gathering what scientists call a "pure load," *M. americana* is more likely to transfer the appropriate pollen to receptive cranberry flowers; visiting only one plant species results in a better pollination rate and more fruit. Unfortunately, recent studies have shown that these important bees are disappearing as more potential nesting sites are converted to agriculture. When properly prepared, our yards and gardens can provide critical nesting habitat and food resources for native bees as their natural nesting sites vanish.

OTHER BIOLOGICAL NOTES

Melitta, like all North American Melittidae, are ground nesters. While little is known about the nesting habits of North American *Melitta*, we can infer some of their nesting behaviors based on studies of European species. These studies suggest that *Melitta* nest in flat or gently sloping ground and often build their nests in a location hidden by vegetation. *Melitta* give rise to a single generation per year and females probably dig a single nest in that time.

MACROPIS

It is unclear what the name Macropis means. Klug and Panzer named the genus in 1809, and their Latin description describes the bee as having a distinctly rounded abdomen. A likely translation of the name from Greek, then, is "large back end" (from macro- + opis).

DESCRIPTION

Macropis individuals are the stoutest of the Melittidae, with a broad thorax and beefy hind legs. Their floral-oil collecting habits restrict their distribution, and they are seen only where loosestrife, their preferred floral host, blooms.

Macropis europea, pictured here, is a European species, but it looks similar to North American *Macropis* species. PHOTO BY DICK BELGERS.

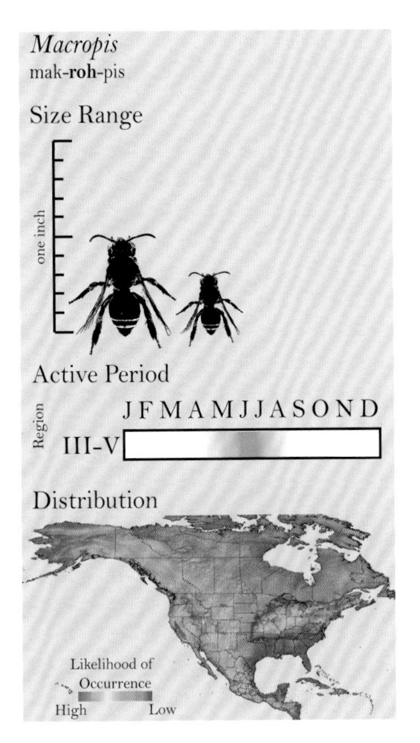

A female *Macropis* forages on a *Lysimachia* flower. Note the giant balls of moist pollen on her legs. The pollen is moist because it has been mixed with floral oils. Photo by Joel Gardner.

Distribution

Worldwide, 16 species of *Macropis* have been identified, all of which are restricted to temperate parts of the Northern Hemisphere. In the United States, there are four species, found mainly in eastern and northern states as well as in southern Canada. All are relatively rare and seldom seen. Recently, *Macropis* have been found as far west as Washington State, but they are generally absent from the desert southwest.

Diet and Pollination Services

Macropis in all parts of the world specialize on the flowers of loosestrife (*Lysimachia)*. Interestingly, *Macropis* is one of the few groups of bees that feed their offspring oils collected from these flowers, instead of nectar. In fact, *Lysimachia* flowers don't even produce nectar. Adults, which need nectar to fuel their activities, visit other flowers for their own nutrition, but females have specially modified hairs for transporting *Lysimachia* oil to their nests. The small tarsal (wrist) segments of the bee's legs have thick, almost spongelike hairs for sopping up oils in flowers. The hind legs have feathery hairs that make carrying the oil back to the nest easier. Most bees not only drink a plant's nectar, but also use it to moisten the pollen they are gathering for their young. *Macropis* bees mix the floral oils with the pollen instead. They also line their nest cells with it, giving each cell a thick, greenish, waxlike coating. The oil is water-resistant, thus protecting the growing larvae from getting too wet.

Biological Notes

Macropis nest in fine, well-drained soil, often on sloped banks with some plant cover to conceal nest entrances. The same nesting areas, and even the same nests, are used for consecutive years by successive generations. *Macropis* build relatively shallow, compact nests compared with those of most other ground-nesting bees. The deepest parts of the nest are generally only a few inches below ground. Despite the shallowness of the nests, these bees use the available nesting space quite efficiently, with one study finding 25 active nest cells in an area of 3 square inches.

A male *Macropis* rests on a leaf. Photo by Joel Gardner.

6

HALICTIDAE

Bees in the family Halictidae are some of the most common bees in people's backyards. Though there are few of them that are large bees, most are medium-sized to small beauties are variable in color and shape. The bees here are presented roughly to scale.

Halictidae can be found on every continent, and they are ubiquitous in habitats ranging from desert mountaintops to jungle basins. Not only are they widespread, they are often abundant, and it is not uncommon to see hundreds or even thousands of individuals in one area. More than 3500 species have been described in 76 genera in the Halictidae bee family. In the United States, 16 genera are known; 9 of these are also found in Canada.

The family Halictidae includes remarkable bees exhibiting widely differing social behaviors, including solitary, communal, semisocial, and primitively eusocial species. Like honey bees, bumble bees, and a few other bees not found in North America, many halictids have nests with queens and workers living together and dividing the labor involved in providing for bee larvae. Unlike any other bees, though, eusocial species of Halictidae are flexibly eusocial, and the degree of cooperation depends on the climate, elevation, seasonal variants, and other environmental factors. For example, a species may be entirely solitary at high elevations where climates are cooler and flowering

A *Halictus* foraging on a desert marigold (*Baileya*).

A male *Agapostemon* resting on a sunflower (*Helianthus*).

seasons are short. At low elevations, that same species may be primitively eusocial, with several overlapping generations and cooperative nesting behaviors. Within a communal nest, many halictids that take on the duties typical of a worker bee in a honey bee hive are still able to lay eggs. Should the opportunity arise (say, because the established queen dies), they can assume egg-laying duties. Because several varieties of "social lifestyle" occur in the Halictidae bee family, scientists eagerly study them, hoping to gain insight into the conditions under which eusociality may have evolved. The fact that these bees appear to retain some flexibility in their nesting preferences leads scientists to believe that they represent early stages in the evolution of eusociality. For Halictidae, the condition of eusociality appears to have evolved several times, in several different lineages, sometime around 22 million years ago, at roughly the boundary between the Miocene and Oligocene geologic time periods. This was a time of climate change, and some scientists think this may have been the catalyst for new social adaptations in Halictidae.

The bee family Halictidae includes both extreme specialists (almost the entire genus *Dufourea*), and broad generalists (some *Lasioglossum* collect pollen from over 50 different plant species). The specialist bees sometimes have unique adaptations to aid them in collecting pollen. For example, some *Halictidae* that specialize on pollen of evening primrose flowers (*Oenothera*) have widely spaced hairs in a single row on their back legs, making the collection of the threadlike pollen produced by this plant much more efficient.

Halictidae nest almost exclusively in the ground, though a few species nest in rotting wood instead. Their nests are often large complicated mazes of interconnecting tunnels, and they are sometimes used year after year.

The family Halictidae includes four branches in North America, the subfamilies Augochlorinae, Rophitinae, Nomiinae, and Halictinae. Halictinae is by far the largest, with more than 80% of the family's bee species. Among the Halictinae is one group that deserves special mention, the genus *Lasioglossum*. Including some 1800 species globally, *Lasioglossum* is the largest bee genus in the world. Individuals within this genus (at least north of Mexico) are minute to medium-sized, often lightly metallic colored, and inordinately abundant. The magnitude of the genus is due, in part, to historic difficulties encountered by scientists who looked to establish distinctions between genera. Finding physical features that unite one group, but are distinctly separate from another, has been difficult for *Lasioglossum*. Currently, in order to deal with the "exceptions" to every taxonomic rule, scientists have lumped many previously distinct genera into this one large and diverse genus.

A *Nomia* visiting on a flower in Arizona. PHOTO BY JILLIAN H. COWLES.

A female *Lasioglossum* collecting pollen and drinking nectar from a fleabane flower (*Erigeron*).

A *Dufourea* resting in a California poppy. PHOTO BY HARTMUT WISCH.

An augochlorine bee working on the turret surrounding its nest. Augochlorine bees, like all Halictidae, build nests in the ground. PHOTO BY DANA ATKINSON.

IDENTIFICATION TIPS

Though Halictidae include a diverse company of bees of various shapes and colors, they are distinctive, sharing a few unique characters that make identification relatively easy.

IDENTIFYING FEATURES OF THE HALICTIDAE BEE FAMILY

- All the Halictidae placed in the Halictinae subfamily can be distinguished from nearly all other bees in North America by the arcuate basal vein on the wing; this curve is seen in some other genera of the Halictinae, though it is not as strong (e.g. *Xeralictus*). Because the Halictinae are so common, this character can be extremely helpful.
- All Halictidae have short tongues.
- Halictidae have only one subantennal suture.

Diagnostic features for different genera can be seen below.

AGAPOSTEMON (SECTION 6.1)

Agapostemon, with their vibrant green thorax (and abdomen in many females), are among the bees easiest to identify in the United States and Canada. Males are especially noticeable because, in conjunction with the green thorax, the abdomen is strongly black-and-yellow striped—a pattern not seen in other North American bees. While most females have a green abdomen, this is not true of all species. Black abdomens and red-and-white striped abdomens also occur. When the abdomen is green, the colors usually vary across each segment, so that the end of each tergite is darker than the beginning. The eyes tend to be a light shade of grey, though this color fades after death. Females can be mistaken for the closely related Augochlorini bee tribe (section 6.4), but a few characteristics, outlined below, can be used to quickly separate the two.

Identifying Features of *Agapostemon*

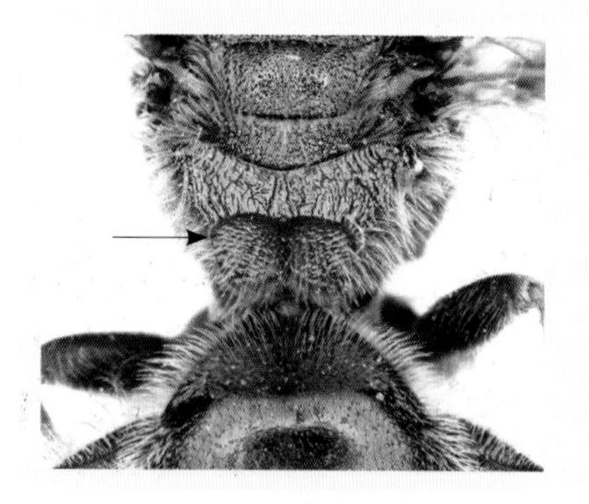

LEFT: Female *Agapostemon* can be identified by the raised ridge appearing all the way around the propodeum. The similar-looking Augochlorini tribe do not have this. (Remember that male *Agapostemon* can be identified by the unique combination of yellow-and-black abdomen and green thorax).

Similar Species to *Agapostemon*

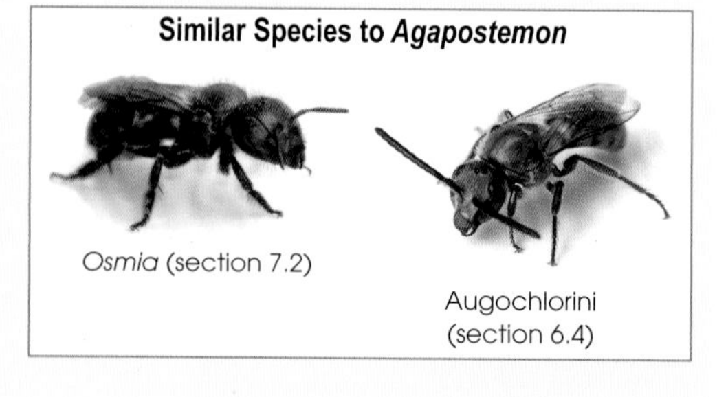

Osmia (section 7.2)

Augochlorini (section 6.4)

HALICTUS (SECTION 6.2)

For all their amazing social behaviors, *Halictus* are far from showy bees, with a brown to black body and white stripes across the abdomen. There is little that sets them apart, which, ironically, is what sets them apart. Without a microscope, look for the prominent bands of hair at the *end* of each tergal segment, and an often thick-looking head—there is much surface area behind the compound eyes of many species.

Identifying Features of *Halictus*

ABOVE LEFT: The wing of *Halictus* has several notable characters, some of which can be seen without a microscope. First, the marginal cell comes to a point (albeit not a sharp one) right along the margin of the wing. Second, the basal vein (labeled) has a distinct curve in it, like a sideways letter J. And finally the three veins that divide the submarginal cell are all dark and prominent; each side of each vein is a bold line (compare to some *Lasioglossum*).

ABOVE RIGHT: Members of a commonly seen group of *Halictus* species are relatively easy to identify, sometimes even when the bee is resting on a flower or near a nest (see images in the text). *Halictus ligatus/poeyi* has a very sharply defined "jaw"; that is, the gena has a strong hook to it, rather than being gently curved. This makes the head look especially big. *Halictus ligatus* and *Halictus poeyi* are nearly indistinguishable "sister species," best identified by DNA analysis; however, they appear to overlap very little in their geographic ranges. *Halictus poeyi* is found in Florida, along the Gulf coast, and north up the Atlantic coast. It seems to overlap only with *Halictus ligatus* up the Atlantic coast. Put another way, a bee with a strong genal angle in Florida is *Halictus poeyi*; in the north or west, it is *Halictus ligatus*, and along the Atlantic Coast, it could be either.

While many bees have stripes running across the abdomen, few have such distinct ones. *Halictus* hair bands are restricted to just the apical portion of each tergal segment, and even the very first segment (T1) has hair bands. The area between hair bands is largely devoid of hair.

Similar Species to *Halictus*

Lasioglossum (section 6.3)

Andrena (section 3.1)

continued overleaf

LASIOGLOSSUM (SECTION 6.3)

Lasioglossum (*Evylaeus*)

Lasioglossum (*Dialictus*), male

Lasioglossum (*Dialictus*), female

Lasioglossum can differ dramatically in appearance, depending on the subgenus. *Lasioglossum* (*Dialictus*) is the most common (and may be one of the most commonly seen bees, period!). *Lasioglossum* (*Dialictus*) have a subtle metallic sheen to their bodies—always the thorax and often the abdomen. A handful, like the female above, have a red abdomen. *Lasioglossum* (*Evylaeus*), which are slightly less common, are generally black and have more distinctive hair bands than do *Dialictus*. Finally, male *Lasioglossum* (*Dialictus*), look very different from their female counterparts, with a long abdomen and slender thorax.

One of these things is not like the others...

Identifying *Lasioglossum* to species can be very difficult. The subgenus *Dialictus* in particular has been referred to as "morphologically monotonous"—a scientist's way of saying "they all look the same." Quite often the differences between species are little more than the density of punctations on the thorax, and the surface texture between punctations.

Identifying Features of *Lasioglossum*

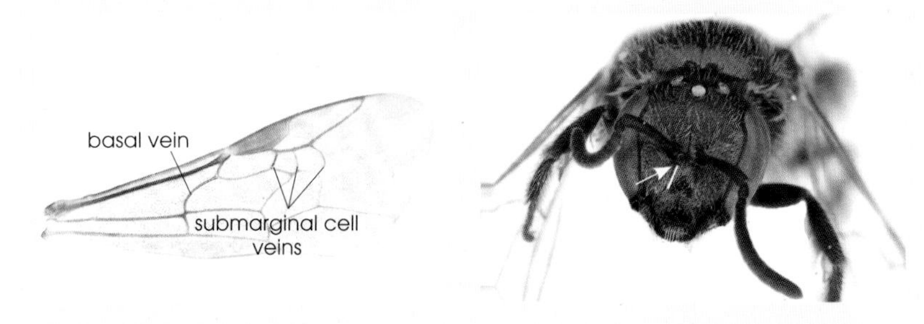

ABOVE LEFT: Like other bees in the family Halictidae, *Lasioglossum* have a distinctly inwardly curved basal vein (described as "arcuate"). There may be two or three submarginal cells (usually three), but the veins defining them are not generally as well defined as *Halictus*, with only the first appearing thick and dark.
ABOVE RIGHT: *Lasioglossum* antennal sockets are high on the face. If you measure the diameter of the antennal socket, you will find that the antennae are always more than one antennal socket-length above the top edge of the clypeus (usually two or three lengths), shown here with the white bar. This can separate them from *Dufourea* (section 6.5).

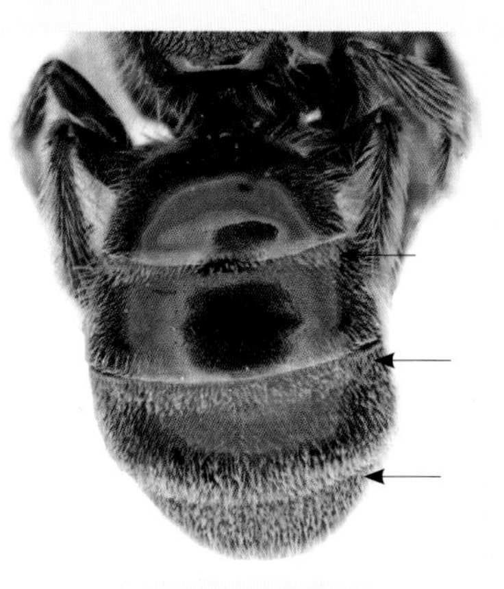

Lasioglossum have scant hair on their abdomens. Many are shiny and mostly bald, though a few subgenera have thick flat hairs along the top (the basal end) of each tergal segment. This is in contrast to *Halictus*, which have hair bands at the bottom (the apical end) of each tergite.

Similar Species to *Lasioglossum*

Halictus (section 6.2) *Perdita* (section 3.3) *Dufourea* (section 6.5)

AUGOCHLORINI (SECTION 6.4)

Augochlorella *Pseudaugochlora* *Augochlora* *Augochloropsis*

Vibrant and green, these bees are hard to miss as they forage on flowers. With a few key characteristics, they can be easily distinguished from other green bees (especially *Agapostemon*).

Identifying Features of Augochlorini

LEFT: In all Augochlorini, the inner margin of the compound eyes has an angle or a notch in it. In some genera it is quite strong; in others not as much, but it is always present.
MIDDLE: In all Augochlorini, the middle of the fifth tergal segment has a cleft or notch in it. Though this is a very reliable character, it can be hard to see because of the hair running across the end of this segment.
ABOVE RIGHT: The propodeum (the back face of the thorax) of Augochlorini is smooth, transitioning to the sides and the top of the thorax with no sharp ridges (compare to *Agapostemon*, section 6.1).

Similar Species to the Augochlorini

Osmia (section 7.2) *Agapostemon* (section 6.1)

NOMIINAE (SECTION 6.5)

Dieunomia and *Nomia* are distinctive, not only within the Halictidae bee family but also compared with most other bees in North America. Though they look little alike, both are larger bees with bold markings.

Dieunomia

Nomia

continued overleaf

NOMIINAE (SECTION 6.5) *continued*

Identifying Features of Nomiinae

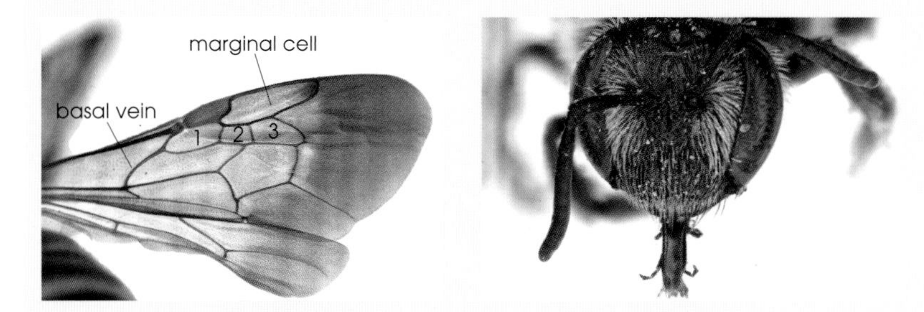

LEFT: For both *Nomia* and *Dieunomia* (pictured here), the marginal cell on the wing is gently rounded, not ending in a point on the edge of the wing. Also on the wing, the second submarginal cell tends to be much shorter than the first and third (both genera in the United States have three submarginal cells). The basal vein is not as strongly curved as it is in other species in the Halictidae bee family. Also, the wing of *Dieunomia* is often very dark, though some just have black tips.
MIDDLE AND ABOVE RIGHT: Both of the Nomiinae have very rounded faces, with the angle of the eyes melding into the lower margin of the clypeus so that the clypeus doesn't appear to extend abruptly below the face. Just visible in this picture of a *Dieunomia (on the right)*, the face also tends to be bulky from front to back, with a thickened gena (the area behind the compound eye), as well as extra surface area behind the simple eyes.

The hair band running across the last tergal segment (called the "prepygidial fimbria") does not appear to be divided. This separates the Nomiinae from many other Halictidae.

Because *Nomia* and *Dieunomia* look so different from each other, we also include some characteristics unique to each below.

Identifying Features of *Nomia*

The abdomens of all *Nomia* north of Mexico are marked with beautiful smooth bands of pearl-white or ivory, sometimes tinged with blue and green. These markings are so unique they can help distinguish *Nomia* from any other bee in the United States and Canada.

Identifying Features of *Dieunomia*

Unique to *Dieunomia*, the sides of the abdomen are lined with tufts of pollen-collecting hairs that often wrap up to the top face.

When viewed from the top, *Dieunomia* are distinct in that the first tergal segment of the abdomen (called T1) has a V shape because a long groove runs down its center. This can help distinguish it from *Andrena* species (section 3.1) that have a red abdomen.

NOMIINAE (SECTION 6.5) *continued*

Similar Species to *Nomia*

Halictus (section 6.2) *Calliopsis* (section 3.4)

Similar Species to *Dieunomia*

Andrena (section 3.1)

ROPHITINAE (SECTION 6.6)

Micralictoides *Dufourea* *Protodufourea* *Conanthalictus* *Xeralictus* *Sphecodosoma*

The Rophitini includes one very common genus (*Dufourea*) and several rather rare ones. While recognizing the Rophitini tribe is relatively easy, identifying the genera within can be tricky. The low antennae on the face are fairly unique among U.S. and Canadian bees. The body tends to be slender and long. *Dufourea* photo courtesy of USDA-ARS Bee Biology and Systematics Laboratory.

Identifying Features of Rophitinae

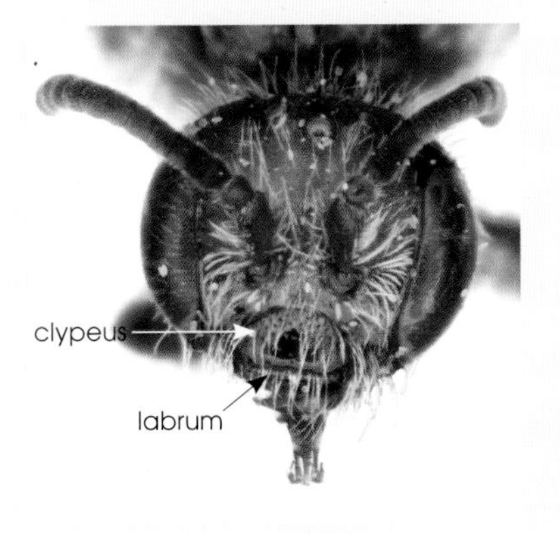

LEFT: Characteristic of all Rophitini, the antennae are set very low on the face (just one antennal socket length above the clypeus), and are joined to the clypeus by only one subantennal suture. The clypeus looks squashed compared with that of other Halictidae; it is much wider than it is tall. Also, the labrum is rounded at its apex (the outer margin), like a half-moon, and can be seen even when the mandibles are closed (it is hidden in other Halictidae as well as in many other bee families).

Rophitines differ from other tribes in the Halictidae because the females have scopa on the tibia, but not so much on the femur or the coxa, the two leg segments on either side of the tibia.

Similar Species to the Rophitinae

Male *Lasioglossum* (section 6.3)

Male *Andrena* (section 3.1)

Protandrenini and Panurgini (section 3.2)

6.1 *AGAPOSTEMON*

The name Agapostemon means "stamen loving" (stamens are the pollen-producing parts of plants).

Description

Agapostemon are beautiful bright metallic green or blue-green bees of medium size found throughout North America. Their brassy green coloration gives rise to the common name: metallic green bees.

Distribution

Agapostemon are found only in the Western Hemisphere (North and South America and associated islands). Approximately 45 species are found from northern Canada stretching south to Argentina, including many of the islands of the Caribbean. In the United States and Canada, 14 species are known.

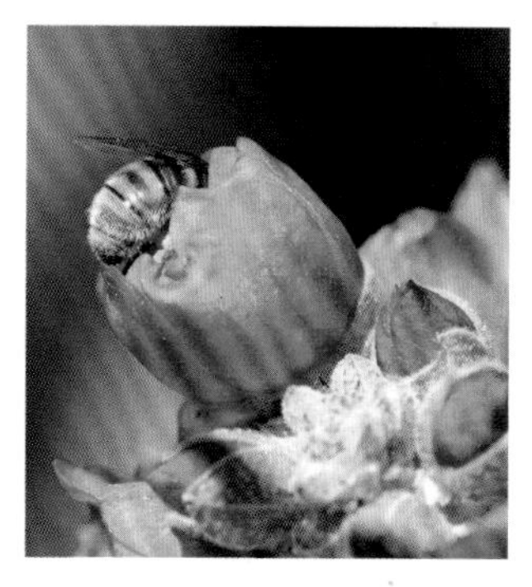

The abdomen of a foraging *Agapostemon* on a globe mallow flower (*Sphaeralcea*). Notice the slight change in green shading from the front to the back of each tergal segment.

Most *Agapostemon* species are either westerners or easterners. One species, though, *Agapostemon texanus*, can be found from the east coast to the west coast, and from southern Canada through Mexico. *Agapostemon virescens* is also widespread, but diminishes as one heads south. In the east (i.e., east of the Great Plains, but including its eastern edge), there are two additional species: *A. sericeus* and *A. splendens*. In the west, and especially the southwest, *Agapostemon* is particularly common, and about a dozen

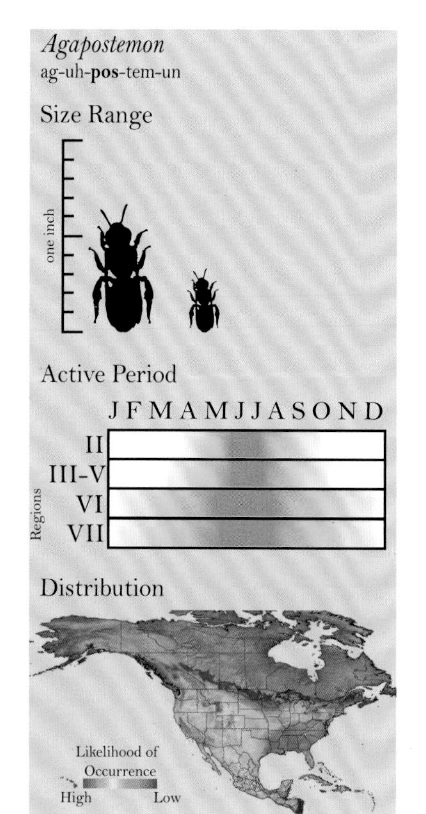

An *Agapostemon* visiting a composite flower (Asteraceae).

species can be found there. *Agapostemon* occur across a wide range of elevations from below sea level in Death Valley to well above timberline (over 10,000 feet).

Diet and Pollination Services

All *Agapostemon* are generalists with no particular floral preferences. They are frequently seen on plants in the sunflower family (Asteraceae), including garden varieties.

Other Biological Notes

All Agapostemon nest in the ground. Soil types span a broad range, from very fine sand to silt to sticky clay, and many species have even been found in lawns. *Agapostemon* do not often nest in large aggregations like other bees, so their widely scattered nests can be difficult to find.

Nests are built in vertical banks, on hillsides, or on flat level ground. Quite often the nests are surrounded by a tumulus, a mound of dirt similar to an anthill, with the nest entrance in the middle. Inside the nest, a tunnel extends downward (nests in vertical banks also have tunnels angling down), with branches off to the sides, each of which contains a single nest cell.

While the majority of *Agapostemon* do not nest in aggregations, there are species that nest communally, with as few as 2 or up to 20 females sharing the same nest entrance in the ground. Still, each female builds her own nest inside the burrow and provides for her own larvae (section 1.5). In North America, *Agapostemon nasutus, A. cockerelli, A. virescens, A. texanus,* and *A. angelicus* have all been observed to nest communally, though it appears to be an opportunistic decision—if other females aren't around, these bees will nest individually.

Multicultural neighborhoods

It is not uncommon for several different species of *Agapostemon* to nest in the same general area, or even for *Agapostemon* to nest in close proximity to bees of other genera. In fact, some may take over nests being built by other bees and remodel the cells for their own offspring. One scientist found an area with *Andrena, Halictus,* and *Agapostemon* all nesting in the same vicinity.

Like other bees in the family Halictidae, *Agapostemon* build nests in the ground. This female is starting a nest by chewing at the soil with her mandibles (jaw).

An *Agapostemon* nest in an open patch of dirt. Some *Agapostemon* nests are surrounded by a small mound of dirt (called a tumulus) resulting from the excavation of the nest. Incidentally, leaving bare patches of earth in your yard is an excellent way to encourage them to nest near you.

A female *Agapostemon melliventris,* showing a variation of the color patterns of the abdomen. Photo by Jillian H. Cowles.

A male *Agapostemon* patrolling near a sunflower (Asteraceae), looking for females. The yellow-and-black stripes on the abdomen are typical on males.

An *Agapostemon* collects pollen from a globe mallow flower (*Sphaeralcea*).

An *Agapostemon*, dusted with pollen, resting on a desert marigold (*Baileya*).

LEFT: Most female *Agapostemon* are completely green colored, but a few species, like this female *Agapostemon virescens*, have a green head and thorax with a black-and-white striped abdomen.

BELOW LEFT: *Agapostemon* are some of the most noticeable bees in your garden because their bright metallic green bodies often contrast strikingly with the flowers they are feeding on, like this *Agapostemon* on desert marigold (*Baileya*).

BELOW: A male *Agapostemon* drinking nectar from a mint flower (Lamiaceae). Note the yellow-and-black stripes on the abdomen and legs, indicative of a male. PHOTO BY SUSANNA M. MESSINGER.

6.2 *HALICTUS*

The name Halictus is a very old Greek word used as a general term for bees long before it was applied to this specific group. It is thought to have originated from the Greek term for "gathering" or "collecting," perhaps as a reference to females collecting pollen.

Description

Though they are fairly nondescript, *Halictus* have curious social behaviors for native North American bees. These medium-sized brown and black bees are ubiquitous and abundant and can be found year-round in some habitats. *Halictus* sometimes sip sweat from humans, and they, along with *Lasioglossum* (section 6.3), are commonly referred to as sweat bees.

Distribution

Halictus are found worldwide and on every continent except Australia. They are most common in the Northern Hemisphere. In North America they range from Alaska west across the continent to Nova Scotia and south through Central America. A few species have extensive ranges. *Halictus ligatus* ranges from the Arctic Circle all the way to Venezuela. Another species, *Halictus rubicundus,* is found in Europe, northern Asia, and across the United States and Canada.

Diet and Pollination Services

There are no known specialist species of *Halictus*; all are generalists. Considering that many are social and produce several generations per year, this is not surprising; most blooming plants are season-specific. A bee that requires pollen and nectar across multiple seasons would not thrive as a specialist.

Because of their relative abundance in any given area, they are often important pollinators of flowering plants. Researchers think *Halictus* may be key pollinators of some carrots (Apiaceae), onions (*Allium*), and sunflowers (Asteraceae).

Other Biological Notes

While considerable variation occurs between species, and even between populations of the same species, the general eusocial life cycle of a *Halictus* begins with a fertilized "foundress" that has spent the winter in hibernation. She emerges in the spring and digs a new nest. Sometimes she works side by side with another foundress. She, or they, lay a handful of eggs, all females, which develop into workers. In cases where two foundresses establish a nest together, one will eventually become subordinate to the other and assume the role of worker once the nest is developed. The workers, which are often slightly smaller than their foundress mother(s), gather pollen and nectar and otherwise care for the next round of bee larvae. The number of subsequent generations depends on the length of the flowering season, but colony sizes can reach up to

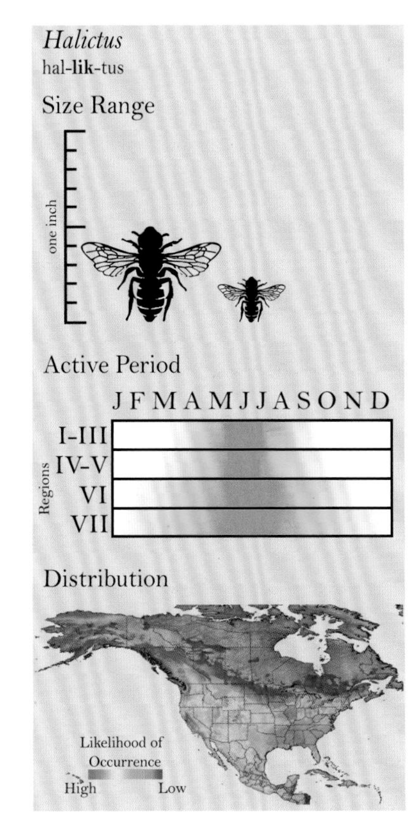

A *Halictus* on a yarrow flower (*Achillea*).

A *Halictus* on a small composite flower. Note the distinctive white bands on the abdomen.

A *Halictus* foraging on a sweet clover flower (*Melilotus*).

A *Halictus* with lots of pollen packed into her pollen-collecting hairs (scopa), foraging on a flower in the sunflower family (Asteraceae).

200 females. Regardless, for the last generation of the season, the brood comprises eggs that develop into males, and also the next generation of female foundresses.

Social *Halictus* species are remarkable for two reasons. First, all females retain the ability to lay eggs and can therefore act as queens. If a queen dies, any worker can take her place; further, a worker can leave the nest and establish her own separate colony. Workers in other social bee colonies (i.e., bumble bees and honey bees) are unable to lay eggs as their ovaries are never fully developed. In some species of *Halictus*, the queen may limit the egg-laying ability of her worker offspring by feeding them a smaller amount of pollen. Females that eat smaller amounts of pollen develop into smaller adults. Their subordinate role in the colony is maintained by their diminutive size, since queens, being larger, can bully them into obeisance. These smaller bees are also less likely to survive a winter and are therefore not well suited to establishing colonies in the following year.

Second, whether a female establishes a eusocial nest or a solitary nest appears to depend to some extent on the environment. Unpredictable weather leads to a greater occurrence of social behaviors. When floral resources are spotty because the weather is capricious, queens leave smaller pollen loaves for their offspring. The bees that develop from these smaller loaves are also small. It is easy for the queen to maintain her dominance over these petite workers, and sociality is promoted. In contrast, when the weather is cooperative and flowers abound, queens provision many nest cells with large pollen loaves. Larger workers are produced and, between the large brood size and the lesser comparative difference in body size, the queen loses control of her egg-laying monopoly. More of these workers lay eggs of their own, and sociality breaks down (remember that sociality in insects is defined, in part, by a clear division of labor).

The evolution of sociality

Because sociality is rare in nature, its evolution is of interest to scientists, and Halictidae are exceptionally interesting because they differ in many ways from other social organisms. Some wasps, termites, and ants are eusocial, and it seems they have been so for more than 65 million years. In contrast, sociality appears to have arisen within the Halictidae bee family only 20 to 22 million years ago. Among bees, eusocial behaviors are seen in a handful of groups: the Allodapines, stingless honey bees, honey bees, and bumble bees. In all of these cases, the condition of eusociality appears to have arisen just once, and it is obligatory. Within the Halictidae bee family, however, sociality has evolved several times. In other words, there is no one common social ancestor for all social Halictidae. Moreover, it is generally the case that once an organism has become social, there is no going back. It is not possible to revert (as it were) to a solitary life. *Halictus* appear to be the exception to this rule. Based on phylogenetic studies, scientists conclude that these bees have gone from solitary to primitively eusocial and back to solitary over evolutionary time.

A *Halictus* foraging on a snakeweed flower (*Gutierrezia*). Notice the sharp projection—almost a hook—on her "jawline" (the area known as the gena). In the west and north, this "tooth" is a defining characteristic of the female of the species *Halictus ligatus*.

Female (left) and male (right) *Halictus* often look remarkably different from each other. Males generally are more slender, with longer antennae, and more yellow markings on their legs and their face, especially their clypeus (just below the antenna).

Cool temperatures also tend to lead to more solitary behaviors, while warm temperatures lead to more social behaviors. Where temperatures are too cold for the flowering season to be very long, sociality is rare. This is because time is insufficient to produce more than one generation. Alternatively, larvae that develop in very warm climates go through each developmental stage faster and spend less time dining on pollen. As a result they tend to be smaller, which helps to reinforce sociality.

Regardless of social behavioral differences, nesting preferences among *Halictus* species are generally the same. All *Halictus* in North America nest in the ground, and they may nest in the same area for decades. Nest entrances can be identified by the little mound of excavated soil (called a tumulus) near each nest entrance. One creative scientist blew smoke into a *Halictus tripartitus* nest entrance and observed it coming out another hole nearby. This suggests that at least some of their nests have multiple entrances, so that what looks like many individual bee nests is in fact one large one. *Halictus* bees may also nest in aggregation, with separate nests congested in the same area. Underground, nests consist of one long relatively vertical tunnel. In solitary bee nests, side branches off this main chute contain individual nest cells that are sealed off once the egg has been laid. In social nests, these side branches have open brood chambers, with both nest cells and growing larvae. For eusocial bees, different generations are found in different brood chambers. The deeper the side branches, the younger the generations. Nests can be 8 inches to nearly 3 feet deep.

A male *Halictus* visiting rabbitbrush flowers (*Ericameria*) in California. Photo by Hartmut Wisch.

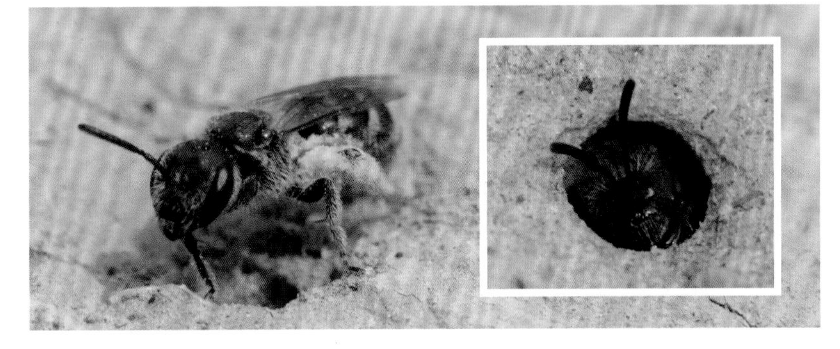

A female *Halictus*, returning home with a load of pollen, standing above her nest entrance in a bare patch of soil. As seen from above (inset), what is likely an elaborate underground maze of nest cells appears as a barely noticeable hole.

Other *Halictus* nests are surrounded by a small mound of dirt so they look like miniature anthills. Here, a female prepares to takes flight as she leaves her nest.

6.3 *LASIOGLOSSUM*

The genus *Lasioglossum* contains a grab bag of bees that have in the past been considered separate genera; thus a wide range of shapes and sizes is clustered under this one generic roof. However, they retain their old generic name as a subgenus name, which is put in parentheses between the genus and specific epithet (e.g., *Lasioglossum (Dialictus) tegulare)*. Like *Halictus*, they express a range of social behaviors that fascinate scientists.

The name Lasioglossum means "hairy tongue" in Greek.

Description

Lasioglossum are among the most abundant and commonly seen bees in all of North America, perhaps making up for their minute size with their incredible numbers. They range in color from solid black to shiny aquamarine. All are relatively petite, though they can be long. Like some of their relatives, *Lasioglossum* are known to lap sweat from human skin; they are therefore often called sweat bees.

Distribution

Lasioglossum species are found on every continent, and they constitute the largest bee genus, with close to 1800 species.

Five subgenera of *Lasioglossum* can be found in the United States and Canada. *Lasioglossum* (*Dialictus*) are the most likely to be seen; there are over 300 species of these tiny metallic bees north of Mexico (worldwide they are found throughout the Northern Hemisphere). *Lasioglossum* (*Evylaeus*) is just as ubiquitous, ranging from Alaska and Nova Scotia south through Florida and Baja Mexico. Roughly 70 species occur in the United States and Canada. *Lasioglossum* (*Lasioglossum*) is most common north of the equator. In the United States and Canada, it ranges from the Northwest Territories south to Panama, with about

Why are they called sweat bees?

Many of the bees in the Halictidae bee family appear to crave sweat; they will seek out sweltering humans and lick it off them. The unfortunate part is that these frequently tiny bees can get caught in clothing and skin folds and sting. Bees in the genus *Lasioglossum* and those in the Augochlorini tribe are particularly known for this behavior. It appears that when they find a sweaty human, they secrete a pheromone that attracts other bees, so 10 to 50 bees may eventually swarm to the source. The swarm can be annoying, but swatting one of the bees and killing it isn't the best deterrent—a dying sweat bee gives off one more burst of pheromone, drawing in even more of these little pests. Interestingly, a group of social bees in Thailand lap up not just the sweat of humans, but also their tears! They land gently on the lower eyelashes of their tearful host and suck up the tears, storing the liquid in their crop and taking it back to the nest, where they regurgitate it for growing larvae. Why do bees drink sweat and tears? Nobody knows for sure; sweat is high in salt, which the bees may need, and tears contain both salt and proteins (which sweat also contains in small doses). It may be these proteins that are important.

Lasioglossum are often called sweat bees because they have a habit of landing on people to lick sweat off their skin.

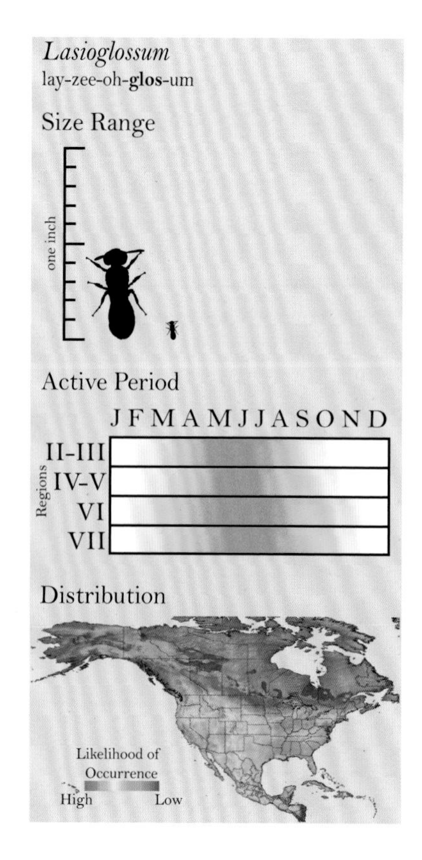

Even on flowers as strangely shaped as this one, a *Lasioglossum (Dialictus)* is able to find and collect pollen.

A *Lasioglossum* (*Dialictus*) foraging on a summer composite flower (Asteraceae).

A *Lasioglossum* (*Dialictus*) with a full load of pollen, foraging on a wild rose (*Rosa*).

30 species. Interestingly, *Lasioglossum (Lasioglossum) leucozonium* is one of the few bees that is more common in Canada than in the United States, dipping south only as far as Maine and New York.

Lasioglossum (*Sphecodogastra*) and *Lasioglossum* (*Hemihalictus*) are the two subgenera with the smallest ranges. *Lasioglossum* (*Sphecodogastra*) is only found in North America, where it stretches from southern Canada south all the way through central Mexico, and from coast to coast in the United States. There are eight species. There is only one species of the subgenus *Hemihalictus: Lasioglossum* (*Hemihalictus*) *lustrans.* It can be found in the eastern United States as far north as Michigan, south to Florida, and west as far as New Mexico.

A *Lasioglossum (Dialictus)* resting on a composite flower (Asteraceae). Many *Lasioglossum*, especially those in the subgenus *Dialictus*, are shiny gunmetal green.

Diet and Pollination Services

The majority of *Lasioglossum* are generalists, but there are a few exceptions. The subgenus *Sphecodogastra* comprises only specialists on evening primrose (*Oenothera*). Some *Sphecodogastra* have immense ocelli (simple eyes) on the top of their heads to aid them in flying at dawn and dusk when there isn't much light (but when evening primrose flowers are open). On full moon nights these bees will even fly all night long. The pollen of evening primrose flowers is held together in clumps by long sticky threads. *Sphecodogastra* are so specialized on these flowers that they have adapted a single row of stout bristles as their scopa, in place of the more typical bunch of long stiff whiskers. These specialized hairs are modified just for dealing with *Oenothera*'s strange pollen.

The bee *Lasioglossum* (*Hemihalictus*) *lustrans* visits mostly chicories and false dandelions (*Pyrrhopappus*). Other species of *Lasioglossum* are not specialists, likely because of their long flying season. Successive generations in the same year aren't likely to be able to forage on the same flowering plants, which don't bloom for as long.

Because *Lasioglossum* are so abundant throughout the flowering season, they are often important pollinators.

A *Lasioglossum (Sphecodogastra)*. Though hard to see from this angle, the ocelli (simple eyes on top of the head) are notably large, allowing these rarely seen bees to forage at dawn and dusk when there is very little light.

Ignoring efficiency, their sheer numbers are enough to achieve excellent pollination of many wildflowers, especially of plants in the sunflower family (Asteraceae), which have shallow flower tubes that are easily accessed by these minute bees.

Some of the subgenera of *Lasioglossum* have light and dark contrasting hair bands on their abdomens. Up close it would be evident that these hair bands sprout from the beginning of each tergal segment, not from the end, as they do in *Halictus*.

A *Lasioglossum (Dialictus)* resting after foraging on a desert marigold (*Baileya*).

Other Biological Notes

Lasioglossum can be found from the earliest of spring days through the fall. A few of the subgenera whose members specialize have narrower windows of adult activity. For example, *Lasioglossum* (*Sphecodogastra*) species, mentioned above, are seen only in the spring. *Lasioglossum* (*Hemihalictus*) *lustrans* is seen mostly in the summer when composites are in full flower.

Lasioglossum includes solitary, communal, semisocial, primitively eusocial, and even parasitic species. The diversity of social behaviors in this large bee genus makes it impractical to discuss each unique life cycle. We instead highlight a few exemplars that illustrate the amazing behavioral variety in this group.

Almost all *Lasioglossum* in the United States and Canada nest in the ground; however, a few species nest in rotting wood. *Lasioglossum* (*Dialictus*) *coeruleum* uses the old burrows of beetles or other insects in decaying stumps or logs as their nests. The preexisting beetle burrow is expanded and modified so that side chambers can hold nest cells. These nests are built in the spring by fertilized females (called foundresses) that spent the winter in hibernation. Each foundress lays a batch of eggs (a brood) that develops into males and females. The females of that second generation become either new foundresses, "understudy" queens in their mother's nest, or workers, helping their mother to raise successive broods. The nest grows with each additional generation of bees, and tunnels eventually wind around and connect together, creating an elaborate maze of intersecting hallways off which are many nest cells.

A small *Lasioglossum* (*Dialictus*) feeding on a mustard flower. Many *Lasioglossum (Dialictus)* are completely gunmetal green in color, but some species have a red or rust-colored abdomen, like this one.

Another species, *Lasioglossum* (*Dialictus*) *laevissimum* is a more typical ground nester. A few fertilized females

What makes a queen a queen?

For *Lasioglossum*, it appears that any bee can become a queen, and that behavior in the nest is the only thing that separates a queen from a worker. A future queen will nudge the other bees in a nest more often, somehow communicating her role as "queen bee" through these nudges. If she dies, another bee that was once a worker will begin nudging her coworkers until her role as queen has been established.

Lasioglossum (*Dialictus*) are among the smallest bees in the United States and Canada. Here, a male feeds on a miniscule mustard flower. Male *Lasioglossum* are more slender and have much longer antennae than the females.

A secret scent

In another primitively eusocial bee, *Lasioglossum (Dialictus) zephyrum*, one female actually guards the nest entrance, letting in only those bees that are relatives. Amazingly, it appears that these bees distinguish between relatives and strangers by scent.

Lasioglossum (*Dialictus*) are some of the most abundant bees in many backyard gardens in North America. These small bees are often overlooked, but they can be important pollinators of both native and cultivated plants.

overwinter in seclusion—separate from nesting sites where they may later raise their broods. In the spring, these fertilized females swarm in the area where they intend to nest, blurring the ground beneath them. Sometimes several thousand individual females will nest in the same area, so dense that 100 nests may be found in a square yard. These first females excavate one main tunnel about

RIGHT: A *Lasioglossum* (*Dialictus*) visiting a desert poppy flower (*Arcetmecon*).

BELOW LEFT: A small *Lasioglossum* feeding on a manzanita flower (*Arctostaphylos*); only tiny bees can fit into the narrow floral tubes of such a flower. Note the small holes at the base of some of these flowers where larger bees have chewed through the flower base to steal nectar (p.54).

BELOW RIGHT: A *Lasioglossum* (*Lasioglossum*) on a desert marigold (*Baileya*). Notice how the hair bands on the abdomen are at the start of each segment of the abdomen (see identification section for more details on the differences in hair bands between *Lasioglossum* and their relatives).

A male *Lasioglossum* (*Dialictus*) on a dandelion (*Taraxacum*). Males have much longer antennae than females.

A *Lasioglossum* (*Lasioglossum*) resting on a desert marigold (*Baileya*).

A *Lasioglossum (Evylaeus)* foraging on a composite flower. PHOTO BY MARGARETHE BRUMMERMANN.

5–7 inches deep. Each lays a brood in her nest, in cells built in a concentric ring off of the main tunnel, and then seals herself and her eggs inside. She waits with them for four to six weeks until they mature. Her first brood is all females. When they emerge, they begin gathering pollen and nectar for the second brood, also laid by the mother. This brood also make the nest deeper, digging down an extra 5 inches and building a second set of concentric rings for their younger siblings. While this second generation is developing, the nest is reclosed. The second brood consists of both males and females (mostly males). They mate, and at the end of the season these fertilized females dig a deep burrow from whence they emerge the following spring.

There are also solitary species of *Lasioglossum*. *Lasioglossum* (*Hemihalictus*) and *Lasioglossum* (*Sphecodogastra*) species are all solitary, as are most *Lasioglossum* (*Lasioglossum*). They probably follow a life cycle similar to *Lasioglossum* (*Dialictus*) *tenax*, where male bees emerge in late spring, followed a few days later by females. A female drinks nectar for a few days, likely mates, and then begins to dig a nest about 5 inches deep. She builds nest cells for four or five eggs and provides for them. There appears to be just one generation per year, and females never share nests.

A male *Lasioglossum* on a sweet clover flower (*Melilotus*).

While species of *Lasioglossum* (*Dialictus*) build nest cells that are individual, separate, and sealed off from each other by earthen partitions, several species of *Lasioglossum* (*Evylaeus*) build subterranean chambers with combs containing their bee progeny. These chambers, with the combs, are supported by columns made of mud, and the combs are separated from each other by very thin, almost translucent, walls made of soil and a waxy substance secreted by the Dufour's gland. Several females work together to build this underground lair. In some cases females also share the responsibility of laying eggs, whereas in others one female lays the eggs and the others perform housekeeping tasks. Nests are just a few inches under the ground, which allows them to warm up quickly when the sun heats the earth.

A *Lasioglossum (Dialictus)* visiting a pussypaws flower (*Calyptridium*). While most *Lasioglossum (Dialictus)* have very little hair, this species, *Lasioglossum pilosum*, is covered with short blond hairs. PHOTO BY HARTMUT WISCH.

6.4 AUGOCHLORINI

Nocturnal bees

One genus of Augochlorini, *Megalopta*, visits flowers that bloom at night. Nocturnal bees are rare, but three of the groups that are night-flying are Augochlorini.

Augochlorini bees are generalists and can be seen on many types of flowers. This bee is visiting a Mexican hat flower (Rudbeckiinae).

An Augochlorini forages on a summer composite (Asteraceae) in the Midwest, where they are most common.

AUGOCHLORA, *AUGOCHLORELLA*, *AUGOCHLOROPSIS*, AND *PSEUDAUGOCHLORA*

Four genera in the Augochlorini tribe are found north of Mexico. Typical of the Halictidae bee family, several species exhibit various types of sociality. While there aren't many species of any of these genera in the United States or Canada, a few of them are widespread and common, more often seen in the east than in the west. Oddly, these four genera are most diverse in the tropics instead of in deserts, which is uncommon among bees.

Augochlora

Augochlora is Greek for "exceedingly green."

In the United States and Canada, these vibrant and variably colored bees can be found from Minnesota and Quebec south all the way to the Gulf of Mexico. While only four species are found north of the Mexican border, roughly 120 species occur in Central and South America. All are broad generalists, taking advantage of the pollen offered by a wide range of the flowers that might be in bloom. Most of them nest under bark and in dead rotting wood, though one species nests in the soil. The wood-nesting species are all solitary whereas the ground-nesting species is primitively eusocial. Unlike most bees that have just one generation per year, species of *Augochlora* can have two or even three generations in

RIGHT: Most Augochlorini genera nest in the earth. While some nests look like a hole in the ground, perhaps surrounded by a small pile of dirt, some species build a turret around the nest entrance. It is thought that the turret might help a bee locate its nest in a complex landscape, or it might deter predation by other insects. PHOTO BY DANA ATKINSON.

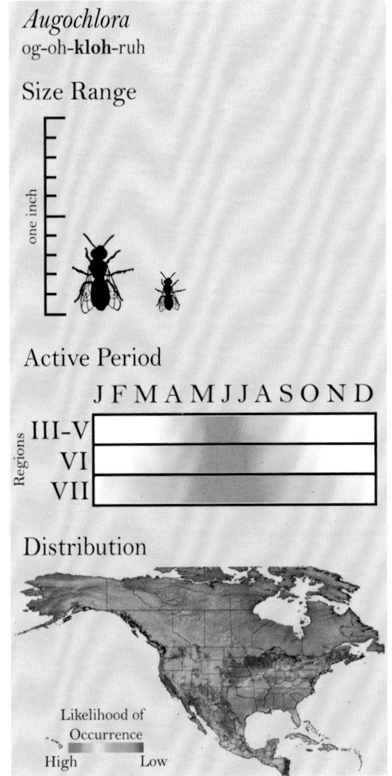

that time period. They can therefore be seen flying from April through September. In the fall, a fertilized female builds herself a hibernaculum, a small chamber in which to spend the winter. She crawls into a horizontally angled beetle burrow in a stump or rotting log and then chews her own tunnel straight down until she reaches the level of the soil. As she chews, she shoves the sawdust into the formed tunnel, filling it in behind her even as she moves forward and down. When she reaches ground level, she simply stops and spends the winter facing straight down, protected by a thick layer of sawdust in the otherwise open burrow behind her. She emerges around April and builds a new nest in a different rotten piece of wood. With *Augochlora*, most of the nest construction happens at night. When the sun first rises, the female begins foraging, provisioning new nest cells she constructed the night before. Her first round of offspring emerge around June. In turn, their offspring will include the females that overwinter, fertilized, in hibernacula.

While most Augochlorini nest in the ground, some *Augochlora*, like the one pictured here, nest in rotting wood. PHOTO BY STEVE SCOTT.

An *Augochlorella* resting on the petal of a blue-eyed grass flower (*Sisyrinchium*). PHOTO BY HARTMUT WISCH.

Augochlorella

Augochlorella means "little Augochlora" in Greek.

These are beautiful coppery to green bees found from Argentina all the way to Nova Scotia. North of Mexico, 7 species are found (2 in Canada), and approximately 20 worldwide. In the United States, most of them are found in the eastern states, with just two species west of the Great

A male *Augochlorella* on a rabbitbrush flower (*Ericameria*) in Nevada.

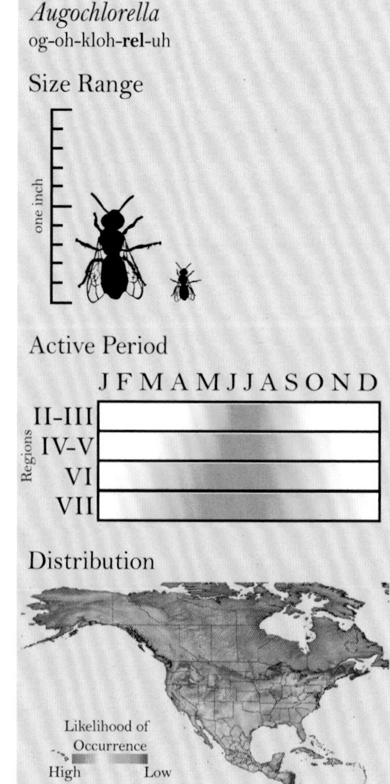

An *Augochlorella* forages on a desert marigold (*Baileya*) in the Mojave Desert. Note the dark eyes and the smooth green shading that stays consistent across each tergal segment (compare to *Agapostemon*).

Plains. *Augochlora* will visit a wide array of flowering plants. These bees are all primitively eusocial ground nesters. In the spring, after spending the winter in a hibernaculum, a fertilized female emerges and builds a new nest. Her nest is a tunnel straight down to a large chamber in which a cluster of cells is constructed. In addition to building a nest, she provides each cell inside with pollen and nectar, and lays eggs—mostly females. Her daughters, when they have grown to adulthood around June, take over the duties of gathering pollen and nectar. From then on the foundress (the original female) remains in the nest laying eggs—she is at this point a queen. In a few species, several females will work together to build the nest and lay the eggs in the spring, but after the first generation has matured, one of the original females will take over egg laying. Either way, all the workers cooperate to provide for one egg at a time, depositing enough provisions for the larva to mature once it hatches. The queen will lay one more generation of eggs, for a total of three generations per year. The last brood laid in the year develops by September, and the females from that brood become the foundresses of the following year. If the foundress or queen dies, one of her daughters will take over as egg layer. Many colonies may aggregate to nest in the same area.

Augochloropsis

Augochloropsis means "looking like Augochlora" in Greek.

Like their relatives, these are vibrant and shiny green bees. They are more common in South America and

Augochlorella pomoniella on a flower in the Sonoran Desert of Arizona.
Photo by Jillian H. Cowles.

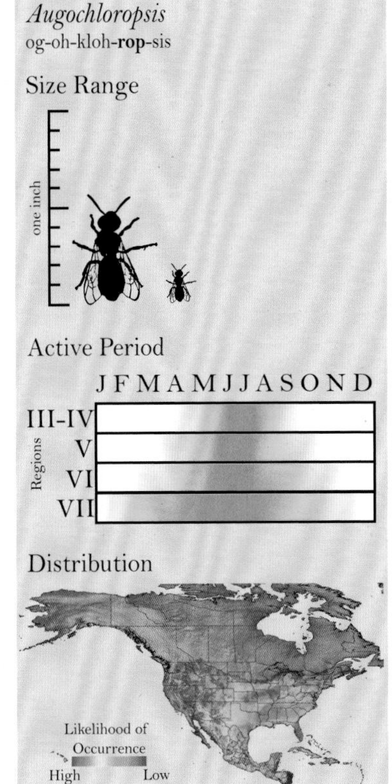

An *Augochloropsis* on a desert dandelion (*Malacothrix*). Photo by Hartmut Wisch.

A square meal

Augochloropsis are interesting in that the pollen masses they provide for their young are shaped into perfect cubes. Most bees form their pollen masses into oval or spherical shapes.

Theodore Dru Alison Cockerell (1866–1948)

Augochloropsis is just one of many bee genera named by Dr. Cockerell, a passionate taxonomist at the turn of the twentieth century. Working throughout the West, Cockerell described thousands upon thousands of organisms in his lifetime, the majority being bees. He began his taxonomic career studying mollusks. In one of his early bee publications, he mentions that having secured a job at what is now New Mexico State University he found there was only one slug species (a mollusk) in the whole state; he therefore took it upon himself to learn about bees, which were very diverse and mostly unknown. His goal was to use his knowledge of bees to teach biology undergraduates "to form just conceptions of the phenomena of life."

Central America, where there are about 140 species, but 3 species are found in the United States and Canada; these are restricted almost entirely to the Rocky Mountains and areas east. Two species can rarely be found in California, Arizona, Texas, and New Mexico. *Augochloropsis* can be found during the late spring and through the summer. These bees are all generalists. They nest in the soil and are communal—meaning that many bees will use the same nest entrance, but they build their own cells within.

Pseudaugochlora graminea

Pseudaugochlora means "false Augochlora," referring to the similarity in appearance of this genus to Augochlora.

This bee is bigger than the other augochlorines in the United States and Canada. Only seven *Pseudaugochlora* are known worldwide, and only one species in the United States. It has been collected only in Texas but ranges south from there to Argentina. As opposed to the other Augochlorini, it may be a specialist on *Cassia* and nightshades (*Solanum*). It nests in the soil—one long tunnel leading down to a chamber in which the female builds an assemblage of cells. These bees are primitively social, following a life history similar to the one described above for *Augochlorella*; however, the colony sizes for *Pseudaugochlora graminea* are very small and may consist of two or at most three females working together.

6.5 NOMIINAE

A *Nomia* resting on grass. Note the pearly colored bands on her abdomen. Photo by Jillian H. Cowles.

NOMIA AND *DIEUNOMIA*

Nomia and *Dieunomia* are the two North American genera in the bee tribe Nomiini. Among the *Nomia* are the alkali bee (*Nomia melanderi*), the most intensively managed ground-nesting bee in the world, and the pearly-banded bee (*Nomia tetrazonata*), a close relative of the alkali bee.

Nomia

Nomia was a nymph in Greek mythology. Nomo- is also the Greek word for "pastoral." Whether the bee was named for the Greek nymph or the habitat type is not known.

Description

Nomia are medium-sized bees with bright pearlescent bands running across their abdomens.

Distribution

Nomia is well represented in Africa, tropical Asia, and Australia. It is found across the United States and into Canada, but it is most commonly seen in the western United States. There are nine species of *Nomia* occurring north of Mexico.

Close quarters

Nomia nest in very large aggregations with many nest entrances per square yard. In fact, the largest concentrations of bees ever seen worldwide are for the alkali bee (*Nomia melanderi*). Scientists counted more than 700 bee nests per square yard in one nesting aggregation. These dense aggregations may be part of the reason this bee is manageable as a pollinator. Cubes of soil can be dug up from a nesting aggregation of *N. melanderi* and planted in new nesting habitats, near alfalfa fields that need pollination. The Columbia Basin (which covers parts of eastern Washington and Oregon) is the area with the largest managed populations of the alkali bee. At times in the past an estimated 16 million alkali bees have nested in the Touchet Valley near Walla Walla, Washington, with in excess of 5 million bees in one nesting aggregation of less than 4 acres. *N. melanderi* prefers to nest in moist soils that are somewhat salty or alkaline (hence the name alkali bee). Farmers in Washington have artificially created prime nesting habitat in order to encourage these bees to nest near their fields. These ingenious farmers first buried irrigation pipes deep underground in order to keep the soil moist, then spread loads of salt on the surface to create an alkaline environment. This artificially created nesting habitat successfully attracted alkali bees that now pollinate the farmers' crops more efficiently and less expensively than when they used to rely on European honey bees.

An alkali bee (*Nomia melanderi*) on an alfalfa flower. Photo by L. Saul-Gershenz © 2014 All rights reserved.

A *Nomia* visiting a legume flower (Fabaceae). Photo by Jillian H. Cowles.

BELOW LEFT: A male *Nomia* foraging for nectar on a creosote flower (*Larrea tridentata*). Photo by Jillian H. Cowles.

BELOW: *Nomia* nest in the ground, often in sunny open areas. The alkali bee (*Nomia melanderi*) in particular prefers to nest in moist, alkaline soils, hence the common name. Photo by L. Saul-Gershenz © 2014 All rights reserved.

Diet and Pollination Services

Nomia are summer bees, generally seen from June through September. They opportunistically visit a wide array of pollen sources in their habitat, and are thus generalists when it comes to providing for their offspring. Prior to the introduction of the alfalfa leaf-cutting bee (*Megachile rotundata*), the alkali bee (*Nomia melanderi*) was considered the most important pollinator of alfalfa in North America. It is still used to some extent. Because it is a broad generalist that can subsist on any flower, its success in alfalfa plantings is guaranteed by surrounding it with acres and acres of alfalfa, thus eliminating its propensity for visiting other pollen sources. Intriguingly, the alkali bee is perhaps the only managed ground-nesting bee in the world.

Other Biological Notes

In addition to nesting in large aggregations, *Nomia* are sometimes communal. Many female bees will share one entrance, even though within the nest they will continue to build their own nest cells, providing each cell with pollen and their own egg. Scientists discovered one nest with about 15 *Nomia tetrazonata* sharing a single entrance. Inside they found nearly 200 nest cells. Why these bees share a single nest entrance, but no other activities, is not known. One possibility is that these bees save energy by having to dig only one main tunnel. *Nomia* often nest in soils that have a hardpan surface layer with more pliable soils below. Having to dig through that hard layer just one time may be beneficial.

Dieunomia

Dieunomia means "second [or another] Eunomia," which was the name assigned to a genus of bees that has since been sunk to a subgenus of Dieunomia. Eunomia means "pleasant" or "good" Nomia.

Description

Dieunomia are larger bees with dusky wings, and often red abdomens.

Distribution

Found only in North America, they range as far south as Costa Rica and as far north as Manitoba, Canada. Nine species of *Dieunomia* can be found in the United States and Canada.

Diet and Pollination Services

In contrast with *Nomia*, *Dieunomia* are all specialists, limiting themselves to flowers in the sunflower family. The Helianthini plant tribe, which includes the annual sunflower (*Helianthus annuus*), the prairie sunflower (*Helianthus petiolaris*), and many other common roadside sunflowers, is popular with these bees. *Dieunomia* time their emergence with the peak bloom of sunflowers, which is typically from July until mid-September.

Other Biological Notes

Dieunomia appear to be solitary, with each female building her own nest and providing for the eggs she lays within.

The nests of *Dieunomia* are easy to find. Each nest entrance is surrounded by a giant mound of excavated soil. Oddly, the entrance is not in the center of this mound of soil, but at its base. *Dieunomia* build a free-standing tunnel running along the soil's surface for about an inch, using spit to harden the walls and ceiling so that they are like cement. The females then kick soil up and over the hardened tunnel, burying it under a mound so that just the entrances shows. Inside the nest, *Dieunomia* dig long shafts straight down, with horizontal branches reaching a depth of about a foot. The actual nest cells are dug into the floors of these lateral tunnels, so that they appear to hang below.

A few reports have been made of two females using the same nest entrance, but this appears to be rare and has not been studied. Though they are solitary, they do nest in very large aggregations and often nest in the same general area for decades. *Dieunomia triangulifera* is the best studied. Aggregations of more than 15,000 individuals have been found, often at densities of a dozen nests per square yard. They prefer to nest in bare fallow fields associated with farming; the nesting aggregation migrates annually, moving to whatever fields in the area are not being used in that

A female *Dieunomia nevadensis*. The crease in the front face of the first tergal segment is just visible as white light catches on it. Photo by Jillian H. Cowles.

A male *Dieunomia*. Note the dark wing tips; this is a common characteristic of *Dieunomia*. Several species have red abdomens, while others have black abdomens. PHOTO BY JILLIAN H. COWLES.

Wooing the ladies

In general, male bees don't engage in courtship or other premating rituals in order to encourage acceptance by females. Instead, they pounce on any female that flies within their field of view. Copulation appears forced, and often females try violently to escape their clutches. Interestingly, some species of *Nomia* and *Dieunomia* may be different. Scientists have observed male bees swarming over nesting sites. When a female emerges from a nest, the male mounts the female. He then gently strokes her legs, drapes his wings over her, and wiggles his antenna in front of her face in a predictable manner. These rituals may last for up to 4 minutes before mating actually begins. Males of some nomiines, including *Dieunomia heteropoda*, have acutely modified legs that are very adept at holding the legs and body of the female.

year. As with other bees that nest in large aggregations, *Dieunomia triangulifera* populations include "traditional" nesters, which dig a nest and provision it with pollen in the normal way, and "floaters," bees that look for abandoned nests they can take over. They still gather pollen and lay their own eggs, but they do not dig a nest. A third strategy has also been observed among these bees: intraspecific cleptoparasitism. These bees are parasitic on other members of their own species. Cleptoparasitic *Dieunomia* are slightly smaller than average. They not only don't dig their own nests, they also don't gather pollen, but instead lay an egg on the growing pollen mass that another female has been collecting. This behavior has been observed in only a few bees.

A large *Dieunomia heteropoda*, with its brown/black abdomen. *Dieunomia* all specialize on sunflowers (Asteraceae). PHOTO BY MICHAEL ORR.

6.6 ROPHITINAE

A *Dufourea, ready to* leave a suncup flower (*Camissoniopsis*). The slight metallic color typical of some Dufourea is just evident. Photo by Hartmut Wisch.

DUFOUREA, CONANTHALICTUS, MICRALICTOIDES, PROTODUFOUREA, SPHECODOSOMA, AND *XERALICTUS*

Rophitinae can be found worldwide; six genera within the Rophitinae are found in the United State, one of which is also found in Canada. Though clearly members of the family Halictidae, this subfamily is fairly distinct from other subfamilies. Most are rare, restricted to the southwestern United States, though one genus, *Dufourea,* is more widely distributed (and discussed at greater length below). They are all specialists.

Dufourea

Dufourea was named in 1841 by Amedee Louis Michel de Lepeletier, comte de Saint-Fargeau. It was named for his "learned friend," entomologist Leon Jean Marie Dufour, after whom the Dufour's gland is also named.

Description

Though at first glance these bees appear nondescript, their knobby antennae, fancy tufts of hair, and scrunched up faces make *Dufourea* stand out under a microscope. They are generally scrawny bees, ranging in color from black to lightly gunmetal blue.

Distribution

Dufourea are found in the Northern Hemisphere, including Europe, China, Russia, Japan, and North America. Around 140 species are found worldwide, and 70 occur in North America, ranging as far north as Yellowknife in the Northwest Territories of Canada, and as far south as Oaxaca, in Mexico. Even bee-depauperate northern states such as Maine harbor species of *Dufourea*. California has more species of *Dufourea* than any other state, though they are common throughout the west. East of the Mississippi River, only four species are found, and just eight occur across Canada.

A *Dufourea* resting on a composite flower.

A *Dufourea* gets covered with purple pollen as it forages on a desert gilia (*Gilia ochroleuca*). Photo by Hartmut Wisch.

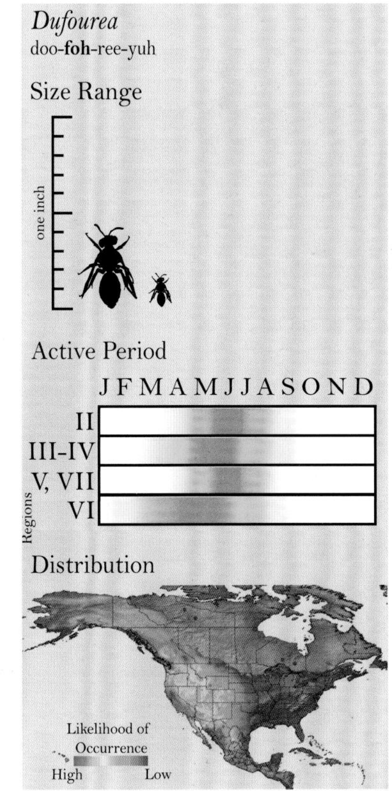

A *Dufourea* uses its petite legs to pry open the petals of a mustard flower (Brassicaceae), searching for nectar.

Diet and Pollination Services

Nearly all *Dufourea* in North America are thought to be specialists. Though they may visit a wide array of flowers for nectar, they restrict themselves to specific plant families or genera for collecting pollen. Across the *Dufourea* genus, the number of pollen hosts is amazingly diverse and includes the primrose (Onagraceae), sunflower (Asteraceae), mustard (Brassicaceae), cactus (Cactaceae), pea (Fabaceae), waterleaf (Hydrophyllaceae), lily (Liliaceae), mallow (Malvaceae), poppy (Papaveraceae), rose (Rosaceae), and figwort families (Scrophulariaceae). Each species specializes on just one of those listed.

One species (*Dufourea novaeangliae*) specializes on pickerel weed (*Pontederia*), an aquatic plant found in shallow bodies of water and muddy ditches. The tongue of this *Dufourea* has stiff bristles, and scientists think these are used in drawing pollen from the narrow flower tubes of those flowers.

Other Biological Notes

It appears that *Dufourea* are solitary, and none share the duties of laying eggs or provisioning nest cells. Like most bees in the family Halictidae, *Dufourea* nest in the ground. Though they may nest in loose aggregations, with half a dozen nests per square foot, they can be very hard to see. The bees are not big, making it difficult to observe them flying near their nests. While there may be a small mound of soil by each nest, the nests are often near leaf litter, other forest floor debris, or at the base of stones, making even the mound of soil difficult to see. If you do find the

Many *Dufourea*, like the one pictured here, specialize on scorpionweed (*Phacelia*).
Photo by Hartmut Wisch.

A *Dufourea* resting on a *Cryptantha* flower.
Photo by Hartmut Wisch.

A *Dufourea* male prepares to depart from a mustard flower (Brassicaceae).

mound of excavated soil, look closely and you will see a trail down its center. As *Dufourea* dig out their nests, they push the soil further and further back, creating a trough as they move the soil from the entrance. Underground, nests are shallow, only 5–8 inches below the surface. Each nest cell is lined with nectar, which is tamped into the soil; this hardens on the walls and provides some protection to the growing larvae inside. Pollen is formed into a perfect sphere, then coated with a thin layer secreted from the Dufour's gland, probably to reduce fungal growth.

Conanthalictus

Conanthalictus was named for the flower on which it was first collected, Conanthus hispidus, growing in a cow pasture in southern New Mexico.

Found along the Mexican border from southern California to Texas, *Conanthalictus* are small and rarely seen. All 13 species are specialists on plants in the waterleaf plant family (Hydrophyllaceae). Some species have two generations in a year (see voltinism, p.85), corresponding with the two periods of floral bloom typical of deserts that experience late summer monsoonal rain. While not all species have been studied, it appears that they nest in the ground, generally in areas where there is little vegetation, and often in small aggregations (i.e., a dozen nests in a square yard, but no more in the area). Several females have been seen using the same nest entrance, suggesting that some species are communal.

Micralictoides

Micralictoides is Greek for "tiny halictid-looking bee."

Throughout the southwestern United States, eight species of *Micralictoides* can be found, though they are concentrated in California. These bees are also very small and may be hard to see. They look very similar to petite *Dufourea* though some species have a red abdomen. The floral preferences of these species are not well known because they are so rare, but some individuals have been observed collecting pollen from poppies (*Eschscholzia*) and woolly sunflowers (*Eriophyllum*). Like their relatives, these bees nest in the ground.

Playing dead

Male *Dufourea* often patrol patches of flowers in search of females instead of patrolling nesting sites as other bees do. Females are apparently aware of this, and when a male attempts to mate with an uninterested female, she will drop instantly to the ground and hold very still. She is never mated while on the ground. The males do not appear to be aggressive toward each other in their search for females, however, and do not fight or establish territories near flowers.

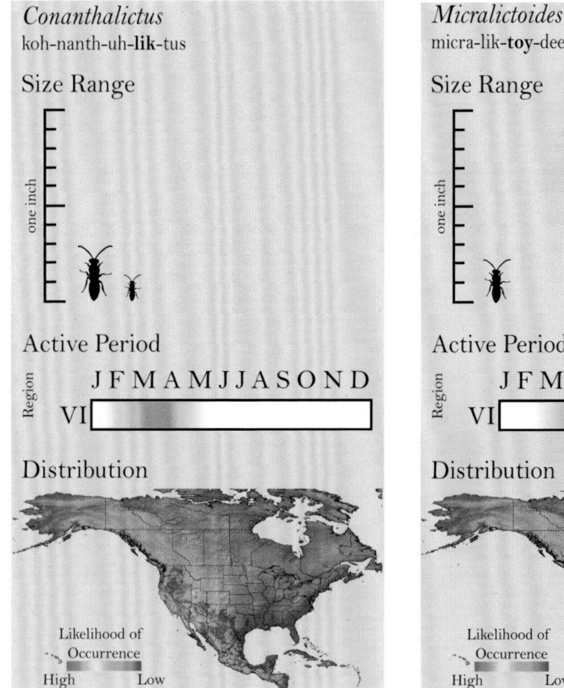

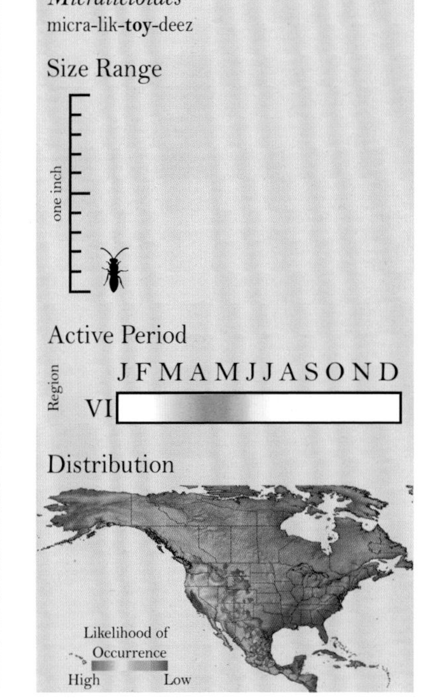

Protodufourea

Protodufourea means "early form of Dufourea."

These early spring bees are little and rare. The five species are found only in California and Arizona. Like other groups of Rophitinae, *Protodufourea* specialize on plants in the waterleaf family, specifically scorpionweed (*Phacelia*) and whispering bells (*Emmenanthe penduliflora*). While very few nests of *Protodufourea* have been found, it appears that they nest in the ground.

Sphecodosoma

Sphecodosoma translates from Greek to mean "body like Sphecodes," referring to the similar body form of this bee to the cleptoparasite Sphecodes (section 9.2).

Only three species of *Sphecodosoma* are known, and all live in the southwest. These tiny bees are specialists on fiddleleafs (*Nama*). *Sphecodosoma* nests that have been studied are in sandy areas with firm soil underneath. While seldom seen, when they are found, it is usually many nests in an area of a few square yards. Occasionally, several females may share the same nest entrance, though within the nest they each build their own nest cells.

Xeralictus

Xeralictus refers to the habitat of this bee and means "desert Halictus" in Greek.

These bees are among the biggest bees in the subfamily Rophitinae, at half an inch. Only two species are known in the entire world, limited to the deserts of California, Arizona, Nevada, and Mexico. Both species (*Xeralictus timberlakei* and *X. biscuspidariae*) are specialists on blazing stars (*Mentzelia*). The nesting habits of these bees are little known. Just one nest has been found; a female was seen flying into a large burrow made by a mammal. Inside, the bee made a long tunnel that meandered around stones and through cracks in the soil before ending with four nest cells.

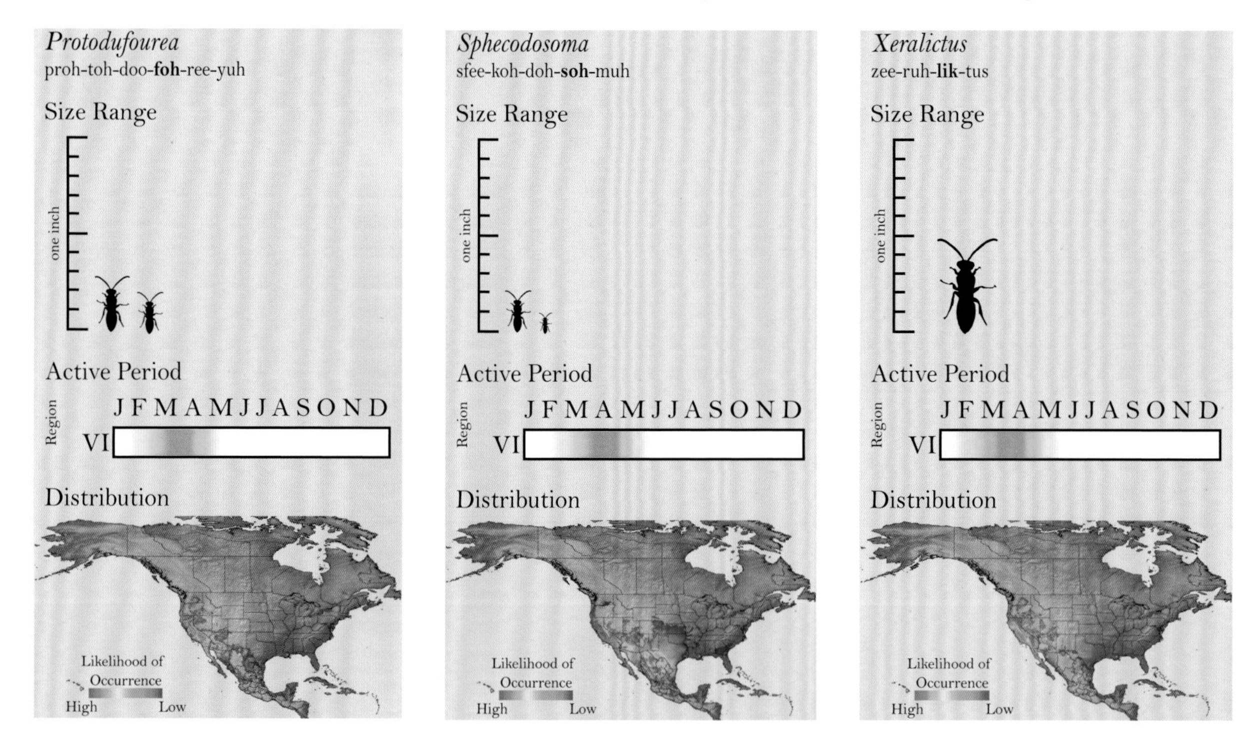

...Sting like a bee

Only female bees can sting, but that does not mean that all female stings feel the same! One courageous (some have said crazy) scientist, Justin Schmidt, created a pain scale to measure the potency of the stings of various hymenoptera (bees, ants, and wasps). Using his own body as a laboratory, he has provoked these insects to sting him, and then ranked the ensuing pain on a scale from 1 (unable to penetrate the skin) to 4 (beyond excruciating). To date, he has catalogued 80 stinging hymenoptera. It may be reassuring to know that honey bee stings only rank as a 2 on his scale, and that many of the smaller bees are not even noticeable. Sweat bee stings rank as a 1.0, while carpenter bees rank as a 3.0. Finally, another researcher has followed up on Schmidt's work and determined that the most painful areas to receive a sting are the nostril and the upper lip.

7

MEGACHILIDAE

Bees in the family Megachilidae are unique in that they carry pollen on the underside of their abdomen rather than on their legs. The diversity of colors and body shapes among the Megachilidae is stunning. Here, a collection of bees belonging to this family show some of the variety you can encounter, even around your house. The bees here are presented roughly to scale.

The bee family Megachilidae includes beautiful and recognizable bees, with a delightful assortment of species that are decorated in zebra-like stripes and spots. Other genera are brilliantly colored jewels tinged with purple, green, blue, or copper tones. Matte black beasts with horns and spines on their faces and other body parts are even known. Several members of this family have enormous mandibles (jaw) with heavy-duty teeth for chewing leaves, moving rocks, and gnawing at wood. The name Megachilidae, in fact, means "big-lipped family," a reference to these enormous mouthparts.

Megachilidae is a diverse bee family. At least 76 genera have been identified around the world, encompassing more than 4000 species. In the United States and Canada there are 18 genera and over 600 species. They are probably the most widespread bee family as well—with their propensity for nesting in wood, they are easily transported to the smallest of islands, and they exist in many areas where bees are otherwise few and far between. For the same reason, the majority of invasive bees are in this family.

A wide variety of nesting habits are demonstrated among the Megachilidae. Many do not construct their

A male *Hoplitis* resting on cinquefoil (*Potentilla*).

A *Dianthidium* on Russian sage (*Perovskia*).

Many bees in the family Megachilidae use plant material to line their nests. Here a leafcutter bee (*Megachile*) carries a piece of leaf back to her nest to construct nest cells.

own nests. They are able squatters and will use nearly any preconstructed "tunnel" as a nest. Species have been found nesting in pine cones and snail shells, between roof shingles, inside steel pipes, inside beetle burrows in dead wood, and within bee nests built in the ground (and abandoned) by other bees. In addition to the variety of structures used by Megachilidae for nests, these bees repurpose many objects from their natural surroundings for use in nest preparation. For example, many Megachilidae use their mandibles to snip pieces of leaves or petals off a plant, carrying small pieces back to their nests to use as wallpaper as they improve their nest cells. Some species gather balls of pine sap or plant resin to use as glue when creating walls between nest cells. Others choose not to reuse old tunnels, but instead gather dollops of mud and

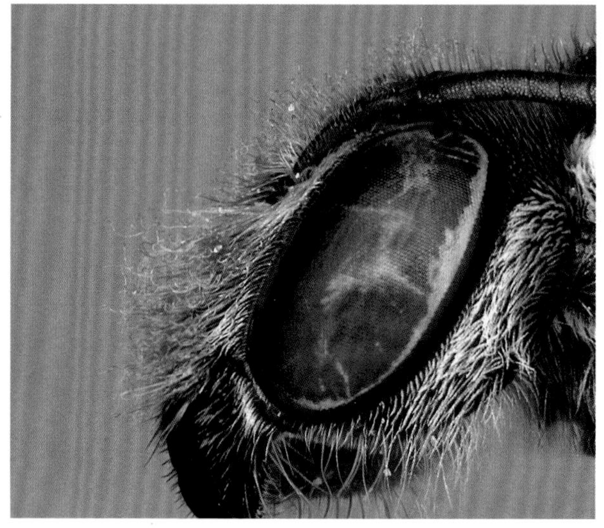

A close-up of the hooked and curled hairs on the face of a *Megachile*. These hairs help the bee collect pollen from small, tube-shaped flowers. Photo courtesy of USDA-ARS Bee Biology and Systematics Laboratory.

daub it, much as a brick mason would, to build their own nests as external dwellings on or between rocks. These few use mud and pebbles, which they carry to the nesting site with their strong jaws and legs. Scientists think that many of the amazing horns and other projections on their faces likely aid in carrying rocks or other items.

Many species of Megachilidae are very specialized when it comes to foraging for pollen. Some species have corkscrew or hooked hairs on their faces, specifically for reaping pollen from inside tube-shaped flowers that are too narrow for harvesting pollen with their legs or

A *Chelostoma* species. Note the pollen-collecting hairs (scopa) on the underside of the abdomen. This is a characteristic shared by all nonparasitic Megachilidae.

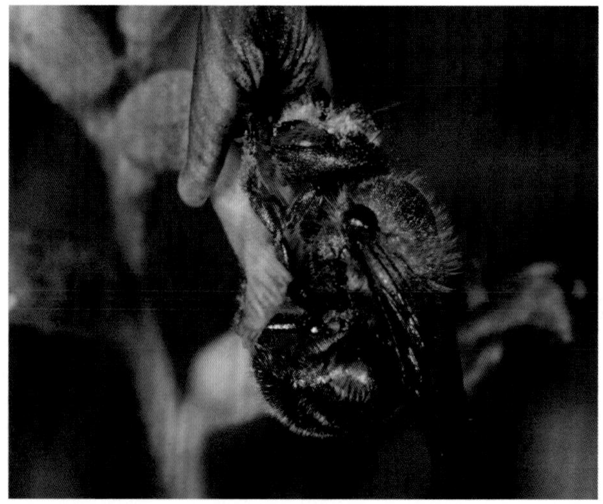

The family Megachilidae contains some of the most striking bees in North America, like this metallic green *Osmia*, here foraging on a legume flower (Fabaceae). Photo by Jillian H. Cowles.

abdomen. These adaptations make them capable pollinators, and many species are managed by growers to aid in the pollination of several high-value crops. The alfalfa leafcutter bee (*Megachile rotundata*) is essential to seed production for alfalfa in the United States. The blue orchard bee (*Osmia lignaria*) is used to pollinate almonds, cherries, and apples. The blueberry bee (*Osmia ribifloris*) has been domesticated for use in the pollination of both high and low bush blueberries.

A *Heriades* foraging on a desert wildflower. Typical of many Megachilidae, the abdomen is curled well under the body.

An *Anthidium* drinking nectar from the long tube of a mustard flower (Brassicaceae).

We have divided the Megachilidae into sections that roughly follow their relationships to each other. Four tribes and subfamilies have been identified in North America: Lithurginae, Osmiini, Anthidiini, and Megachilini. Rare or uncommon genera are grouped together by tribe, but we dedicate whole sections to common and widespread genera.

A male *Megachile* on sweet clover (*Melilotus*).

An *Osmia* foraging on sweet clover flowers (*Melilotus*). Note the metallic blue coloration on the abdomen of the bee. This is a common characteristic of *Osmia*.

IDENTIFICATION TIPS

The Megachilidae are relatively recognizable on the wing. Their bodies are robust and often cylindrical without the flattened look of many other kinds of bees. Several genera have yellow and black markings across their bodies. Females carry pollen on the underside of their abdomens, which no other bee family does. Any bee seen with a bushel of brightly colored pollen on its belly is in this family, giving you a leg up on identification.

IDENTIFYING FEATURES OF THE MEGACHILIDAE BEE FAMILY

- Look for two submarginal cells on the wing, with the second being noticeably longer.
- If you can see the labrum (the flap that hangs below the clypeus), it is big, hinged across its entire length, and longer than wide.
- Females of all genera except those that are parasitic have scopa on the underside of their abdomens (bellies) instead of on their legs.
- Megachilidae tend to look rounder, more cigar- or submarine-shaped than other bee families.

Examples of diagnostic features for different genera can be seen below.

LITHURGINAE (SECTION 7.1)

A large, dull black bee with pollen on her belly, visiting a cactus flower, is very likely a *Lithurgopsis*. Where in the United States the bee is seen can even aid with species identification. In the east, *Lithurgopsis gibbosus* is the most common species, ranging from North Carolina west to Texas and south to Florida. *Lithurgopsis littoralis* is a midwestern species that ranges as far east as Illinois but is most common in Texas. *Lithurgopsis apicalis* is abundant throughout the west, occurring as far north as South Dakota. All other species are restricted to the Mexico border, in Arizona, New Mexico, and California.

Identifying Features of Lithurginae

Male Lithurginae (left) have a pygidial plate—no other bee in the Megachilidae bee family has a pygidial plate. Females (right) have a "spine" instead of a pygidial plate.

The clypeus of a female *Lithurgopsis* is extremely bulbous, with a raised ridge at the top edge, giving them an unmistakable look. Photo courtesy of USDA-ARS Bee Biology and Systematics Laboratory.

On the hind leg, the tibia, especially of females, has lots of dull spines and bumps (called tubercles) on the outer surface. On a freshly dead specimen, you can actually feel these spines by running your finger or the edge of a pin lightly across the leg.

From the side, *Lithurgopsis* can look remarkably like *Megachile*. Note the subtle slope from the front facing to the upward-facing sections of the first tergal segment. In *Megachile* this is much more abrupt. The projections on the face of this female, which look like "horns," can just be seen behind the antenna.

Similar Species to the Lithurginae

Megachile (section 7.7)

OSMIA (SECTION 7.2)

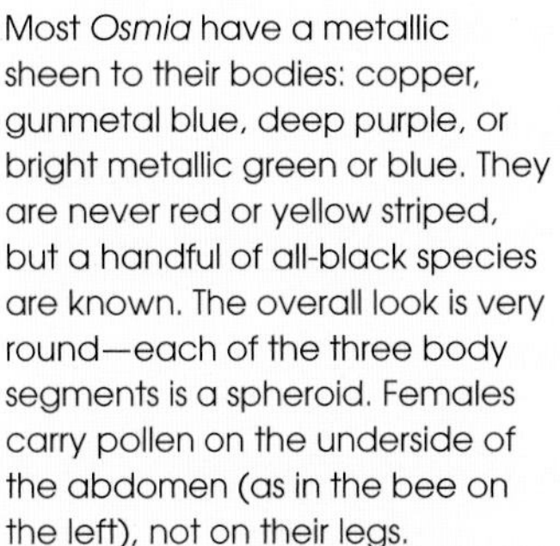

Most *Osmia* have a metallic sheen to their bodies: copper, gunmetal blue, deep purple, or bright metallic green or blue. They are never red or yellow striped, but a handful of all-black species are known. The overall look is very round—each of the three body segments is a spheroid. Females carry pollen on the underside of the abdomen (as in the bee on the left), not on their legs.

Identifying Features of *Osmia*

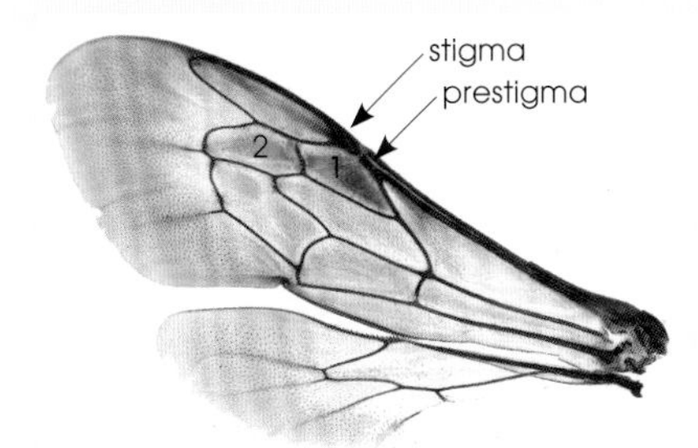

LEFT: *Osmia* wings have two submarginal cells (as with all Megachilidae). The stigma, which is big and easily visible, and prestigma are quite long, which is typical of the Osmiini bee tribe.

RIGHT: There is no pygidial plate at the end of the abdomen of male or female *Osmia*.

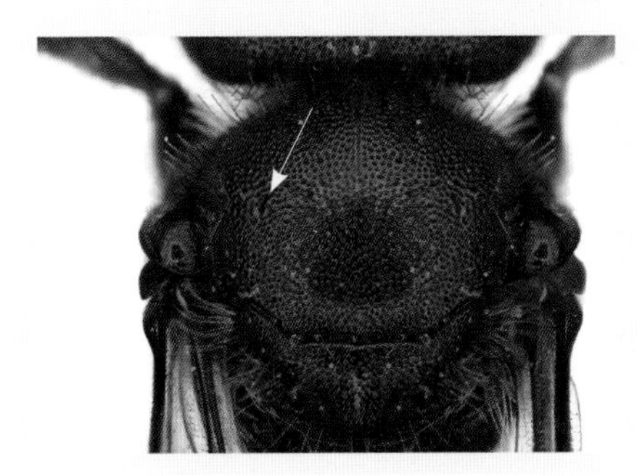

LEFT: A very important, though not easily seen, feature for *Osmia* are the parapsidal lines, which are little points or small round holes instead of actual long lines.

RIGHT: In *Osmia*, an arolium is located between the front claws. Just visible is the tiny cleft at the tip of the claws—a trait unique to *Osmia*, among the Osmiini.

continued overleaf

OSMIA (SECTION 7.2) *continued*

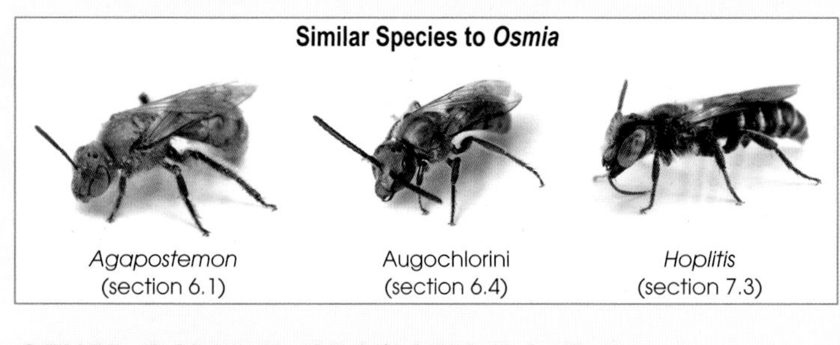

OTHER OSMIINI (SECTIONS 7.3–7.4)

In addition to *Osmia*, seven other genera are placed in the Osmiini tribe. They are easy to distinguish from *Osmia* (the characters are illustrated below), but often difficult to tell from each other. We list general characters for distinguishing Osmiini from bees outside the tribe, and a few characters for identifying some of the easier genera within the tribe.

Osmiini tend to have a long, cylindrical abdomen and thorax—a little like a cigar. Interestingly, it is thought that this long cylindrical body is in part an adaptation to nesting in narrow wood tunnels. There are often white hair bands running across the abdomen, and (though not always), these Osmiini are diminutive. Paying attention to the flower on which a bee was seen, the place, and the time of year can also help with identification. *CHELOSTOMA* PHOTO COURTESY OF USDA-ARS BEE BIOLOGY AND SYSTEMATICS LABORATORY.

Identifying Features of Other Osmiini

In all of the Osmiini except *Osmia* the parapsidal lines are truly lines—relatively long grooves running on each side of the thorax.

The Osmiini all have an arolium (a little pad between the tarsal claws), which can help distinguish them from small *Megachile*, which do not. The tarsal claws end in one distinct point.

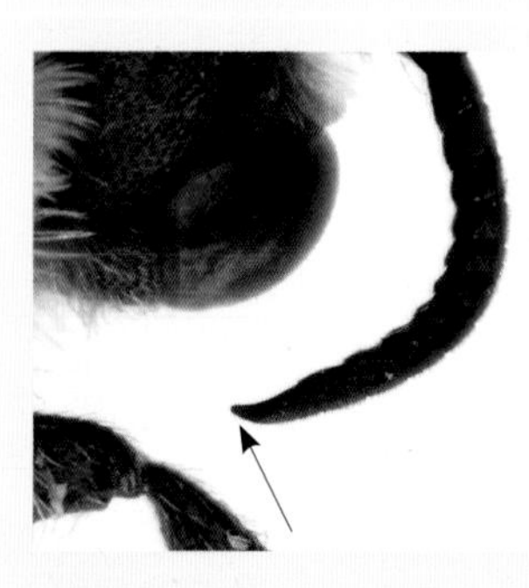

Distinctive to many species of *Hoplitis* (section 7.3), the antennae taper to thin hooks.

The abdomen of a male *Ashmeadiella* has four pointed teeth on the tip. This can be seen sometimes when the bee is rooting around inside flowers. *Ashmeadiella* are a relatively distinctive Osmiini, with bright white stripes on the abdomen, which is often red (seen in the compilation of all the Osmiini on previous page). A caveat: a handful of Ashmeadiella with red abdomen look very similar to Hoplitis.

Chelostoma are longer Osmiini, largely because of an elongated thorax. In fact, if you measure the length of the thorax from front to back, you will find that the ends of the tegula fall midway along the length of the thorax. In other Osmiini, the tegula fall much closer to the back half of the thorax. PHOTO COURTESY OF USDA-ARS BEE BIOLOGY AND SYSTEMATICS LABORATORY.

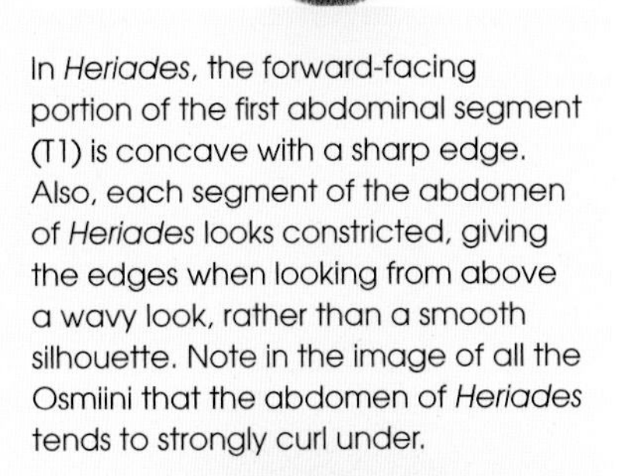

In *Heriades*, the forward-facing portion of the first abdominal segment (T1) is concave with a sharp edge. Also, each segment of the abdomen of *Heriades* looks constricted, giving the edges when looking from above a wavy look, rather than a smooth silhouette. Note in the image of all the Osmiini that the abdomen of *Heriades* tends to strongly curl under.

Though *Xeroheriades* are seldom seen, being restricted to the Mojave Desert, they stand out. The abdomen is rust red, on a rather long body. On the underside of the abdomen of females, the first sternal segment has a distinct projection, shaped a bit like the letter U.

The face of female *Protosmia* is one of the most distinctive among North American bees, though best seen with a microscope. These bees have a spatula-like projection extending from the clypeus' lower edge (shown here from two angles). PHOTOS COURTESY OF USDA-ARS BEE BIOLOGY AND SYSTEMATICS LABORATORY.

Similar Species to Osmiini

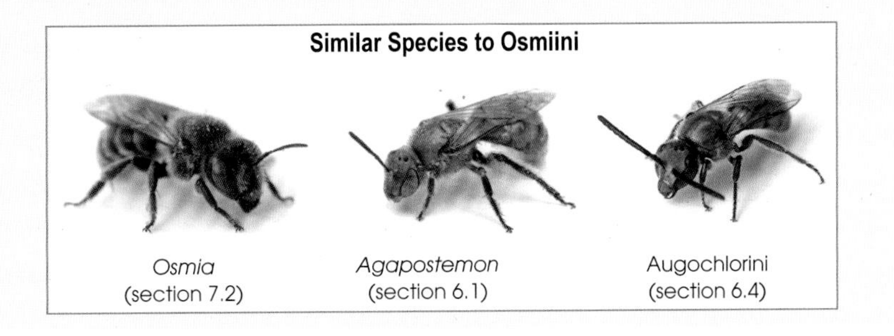

Osmia (section 7.2) | *Agapostemon* (section 6.1) | Augochlorini (section 6.4)

continued overleaf

ANTHIDIUM (SECTION 7.5)

Anthidium are burly bees whose bodies are some mix of yellow-and-black stripes. They are fast flying, with a tendency to hover near plants of interest. As with other Megachilidae, the scopa (pollen-collecting hairs) are on the underside of the abdomen.

Identifying Features of *Anthidium*

LEFT: There is no arolium between the front tarsal claws. This is important for distinguishing *Anthidium* from other Anthidiini.
ABOVE LEFT: The mandibles of an *Anthidium* have many teeth—generally between five and seven indentations. These can be seen from some distance and often without a hand lens or macrophoto.
ABOVE RIGHT: The subantennal sutures (only one on each side) are very straight on an *Anthidium*, with no curve as they run from the antenna to the top of the clypeus. PHOTO COURTESY OF USDA-ARS BEE BIOLOGY AND SYSTEMATICS LABORATORY.

Similar Species to *Anthidium*

Other Anthidiini (section 7.6)

Some wasps

Wasp mandibles (shown here looking up from below) tend to be pointed and pincer-like, while Anthidiini have well-developed mandibles with several teeth. Wasp females also lack pollen-collecting hairs (scopa) on the underside of the abdomen; female Anthidiini have them.

OTHER ANTHIDIINI (SECTION 7.6)

The Anthidiini included here are chunky bees with pollen collecting hairs (scopa) on the abdomen (belly) of females, and with striking yellow and black (and sometimes red) coloration on their otherwise black bodies. They tend to hover near flowers and dart about quickly—a little like a fly. "Other Anthidiini" is (obviously) not a true taxonomic grouping; *Anthidium* (section 7.5) is also in the Anthidiini bee tribe. Fortunately, there are a few characters that exclusively unite the bees in this chapter. These Anthidiini have fewer (up to four) teeth than *Anthidium*; the teeth are duller—sometimes so dull that they are hard to distinguish from each other. We highlight a few additional distinctive features of each genus. *Anthidiellum* photo courtesy of USDA-ARS Bee Biology and Systematics Laboratory.

Identifying Features of Other Anthidiini

There is an arolium between the front tarsal claws of all Anthidiini except *Anthidium*.

The mandibles of the Anthidiini bees tend to have fewer, less distinct teeth than *Anthidium*. The teeth of this *Anthidiellum* are almost a smooth surface with no sharp points at all.

On the female of *Paranthidium*, the mandible is strongly angled between the lower margin and the toothed edge, making the toothed edge extra long and giving the mandible a droopy look. Photo courtesy of USDA-ARS Bee Biology and Systematics Laboratory.

The pronotal lobes of *Dianthidium* individuals have edges that are paper thin and translucent. In other Anthidiini the edges may be thin to very thick, but they are never as thin as in *Dianthidium*. Photo courtesy of USDA-ARS Bee Biology and Systematics Laboratory.

In *Anthidiellum*, the subantennal suture is clearly curved out toward the eyes. In all other Anthidiini (including *Anthidium*) the subantennal suture is fairly straight. Photo courtesy of USDA-ARS Bee Biology and Systematics Laboratory.

Behind the simple eyes on top of the head of *Anthidiellum*, the angle between the top of the head and the back of the head is strong, shelflike even, and perhaps covering the beginning of the thorax (the middle body segment). Photo courtesy of USDA-ARS Bee Biology and Systematics Laboratory.

continued overleaf

Identifying Features of Other Anthidiini *continued*

On the thorax of *Anthidiellum*, the scutellum is swollen-looking and also shelflike. It very distinctly overhangs the propodeum. From the top it doesn't taper but suddenly ends in a more or less straight line. PHOTO COURTESY OF USDA-ARS BEE BIOLOGY AND SYSTEMATICS LABORATORY.

In *Trachusa*, the tibia of the middle leg is very wide, and the edges are rounded out.

Similar Species to Other Anthidiini

Anthidium (section 7.5)

Some wasps (see p.156)

MEGACHILE (SECTION 7.7)

Megachile are often the most frequently seen of all the Megachilidae bees. They are generally gray to matte black, often with a fair amount of hair and white stripes on the abdomen. These bees are best distinguished by their chunky demeanor, black (never metallic) bodies, and (with females) the ample scopa on their abdomens. That said, the diversity of form and body shape among the *Megachile* species is tremendous. Notable horns and other protuberances can often be seen on their faces. And many males have long manes of hair on their front legs.

Identifying Features of *Megachile*

LEFT: The rear end of males is often blunt and quickly turns down, rather than tapering to a gentle end. It gives the silhouette the look of a WWII helmet.

RIGHT: The switch from the forward-facing to the upward-facing surface of the first tergal segment (T1) is abrupt. In *Lithurgopsis* (section 7.1), which look similar at first glance, this transition is much more gradual.

Identifying Features of *Megachile* continued

LEFT: *Megachile* have no arolium between their front tarsal claws. The Osmiini do. You might also notice the tiny claws near the base of each tarsal claw. Osmiini have no such claws, though the tip of the *Osmia* claws has a very small notch.

Similar Species to *Megachile*

Osmia (section 7.2) | Osmiini (section 7.4) | Lithurginae (section 7.1)

7.1 LITHURGINAE

LITHURGOPSIS AND *LITHURGUS*

Lithurginae contains two genera in the United States and Canada: *Lithurgus* and *Lithurgopsis*. *Lithurgopsis*, while not as common as some bee genera, is much more common than its close relative *Lithurgus*. The *Lithurgus* in North America are introduced, and only locally abundant.

Lithurgopsis

Lithurgopsis means "resembling Lithurge," referring to the incredible similarity in the appearance of these two genera.

DESCRIPTION

Lithurgopsis are fairly large, burly, and matte black with white stripes on their abdomen. They are typically seen in cactus flowers.

You say po-tay-to, I say po-tah-to

Lithurgus, *Lithurgopsis*, *Lithurge* ... this group of bees has many names. *Lithurge* is the French scientific spelling of *Lithurgus*. The French entomologist Latrielle first identified this bee, but the description was corrected shortly afterward by the German entomologist Berthold, who used a Latin/Greek spelling (*Lithurgus*). Until recently *Lithurgus* and *Lithurgopsis* were both lumped together as *Lithurgus* within the tribe Lithurgini in the subfamily Megachilinae. Now they have their own subfamily, the Lithurginae. Tune in next time for another taxonomic about-face...

DISTRIBUTION

Worldwide, nine species of *Lithurgopsis* have been identified, ranging from the midwestern United States south through Argentina. *Lithurgopsis* are most common in dry areas with abundant cactus (i.e., the desert southwest), and seven species are known north of the Mexican border. They are uncommon in the northeastern United States and Canada.

DIET AND POLLINATION SERVICES

Lithurgopsis specialize on flowers in the cactus family (Cactaceae), and are important pollinators of these flowers. A scientist studying cactus flowers in Brazil found that when

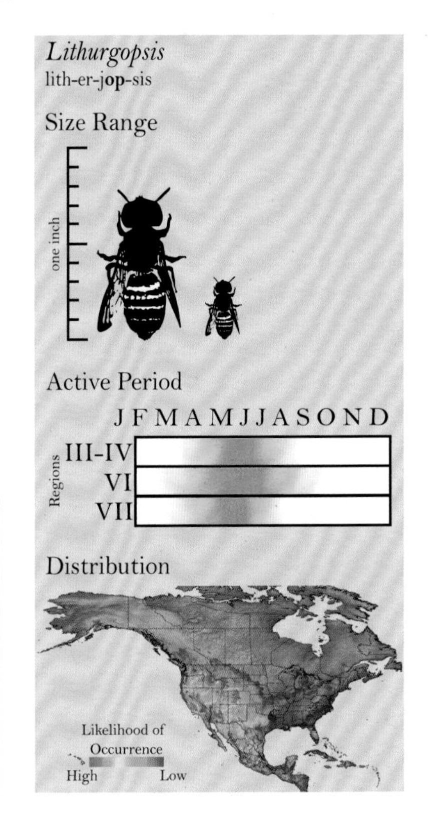

A *Lithurgopsis* visiting a cactus flower. PHOTO BY JILLIAN H. COWLES.

A *Lithurgopsis* foraging in a cactus flower (*Opuntia*).

ABOVE LEFT: All *Lithurgopsis* collect pollen only from cactus flowers. Notice the protrusions on her face, just below her antennae.

ABOVE: *Lithurgopsis*, like all Megachilidae, have their pollen-collecting hairs (scopa) on the underside of the abdomen. This *Lithurgopsis* has filled her scopa with the large pollen grains of the cactus flower.

LEFT: A female *Lithurgopsis* grooming herself, moving pollen from her head and thorax, onto the pollen-collecting hairs (scopa) under her abdomen. PHOTO BY JILLIAN H. COWLES.

a bee visits certain species of cactus, the cactus anthers close around the bee like a hand closing into a fist. Only *Lithurgopsis* and two other bee genera are savvy enough to force their way through the closed anthers to where most of the pollen is. As a result, they are among the best pollinators of these iconic desert plants.

OTHER BIOLOGICAL NOTES

All species of *Lithurgopsis* nest in wood, usually old dead pieces like rotting tree stumps or branches. Some species, however, nest in door frames and other man-made structures and may be destructive when many nest in one area. Whereas other bees in the family Megachilidae use preexisting holes, North American *Lithurgopsis* are unique in that they bore their own holes. They chew one main tunnel that branches, sometimes several times. Nest cells are separated from each other with loosely packed sawdust walls, though sometimes they skip building the walls. In these cases the pollen mass for each egg serves to separate the eggs, creating a sort of communal cavity in which the eggs hatch and the larvae grow. This arrangement is seldom seen among other bees.

Lithurgopsis can be somewhat destructive as they burrow into wood to build their nests. Here, a *Lithurgopsis* has chiseled her way into the wood siding of a house.

Lithurgus

Lithurgus means "rock worker" in Greek; in the broader sense, lith- can refer to any hard-packed material, which may explain its use in the name of a bee that chews into hard wood.

Worldwide, *Lithurgus* are especially abundant and common in Europe and South Africa. In the United States, only two species are known (none occur in Canada). They are rare and only occasionally seen. *Lithurgus chrysura* is an invasive species from the Mediterranean, and it can be quite destructive. This bee prefers to nest in man-made structures, including vinyl siding and wood. Because it nests in aggregations, a group of *Lithurgus chrysura* can cause extensive damage; however, they have been observed only twice in the United States, in New Jersey in the 1970s, and more recently in Pennsylvania in 2007. One other species of *Lithurgus* occurs in the United States: *L. scabrosus* is found only in Hawaii. It is hypothesized that *Lithurgus scabrosus* arrived there along with the Europeans about 200 years ago. *Lithurgus* specialize on flowers in the mallow and sunflower families (Malvaceae and Asteraceae). *Lithurgus scabrosus*, the colonist of Hawaii, visits *Hibiscus* flowers, and *L. chrysura,* the invasive New Jersey/Pennsylvania migrant, visits the invasive weed called starthistle (*Centaurea*). For additional nesting information see *Lithurgopsis* above; they nest very similarly.

Keystone species: more than meets the eye

In the ecological world, an organism whose importance to ecosystem function is inordinately greater than its abundance is referred to as a keystone species. As an example, consider cactus. In addition to supporting a number of specialist bees (like *Lithurgopsis*), they provide sustenance for many generalist bees as well. Those generalist bees go on to visit and pollinate other rare flowers that are not as well-visited. Thus, the cactus flower's presence is essential not only to a wide range of bees, but also to many uncommon flowers. In one area of southern Utah, one cactus species (*Opuntia polycantha*) was visited by 23 different bee genera and 53 different bee species. At least six of those species only visit cactus flowers for pollen.

A *Diadasia* visiting a cactus flower, its sole source of pollen.

7.2 *OSMIA*

The genus *Osmia* includes some of the most economically important bees in North America, responsible for pollinating many orchards and other commercial crops, and for picking up the slack when honey bee pollination falls short. Some species will build nests entirely from mud, thus earning them the common name "mason bees."

Description

Vibrant metallic green, deep blue, or inky black hues decorate the stout little bodies of bees in the genus *Osmia*.

Distribution

Roughly 500 species of *Osmia* have been identified in the world. These bees are most common in North America, Europe, and the Middle East, with some species found as far east as Japan. In the United States and Canada over 130 species are found, with close to 60 in Canada. More species are found in the west than in the east, and fewer than 30 occur east of the Mississippi River. *Osmia* thrive at a wide range of elevations, from below sea level in Death Valley to high in the alpine tundra on mountain peaks above 12,000 feet. At these higher elevations, *Osmia* are frequently seen sunning themselves on rocks, possibly to maintain their body temperature in the chilly alpine climates.

Osmia can be found from sea level to high above tree line. This *Osmia* was photographed above 12,000 feet in the Uintah Mountains of Utah.

LEFT: Many *Osmia* found in the United States and Canada are bright metallic green or blue.

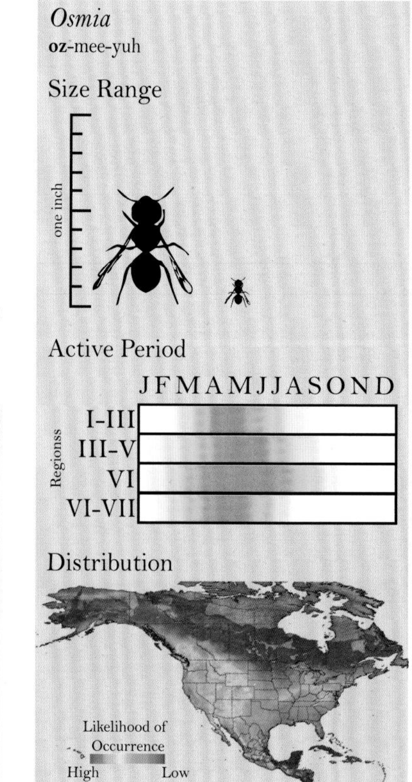

Home is where the scent is

Osmia means "odor" and refers to the somewhat lemony scent with which these bees mark their nest entrances, presumably so they can distinguish their nest entrance from those of other *Osmia* nesting nearby. Interestingly, each species and, in fact, each individual *Osmia* produces a unique scent so she can be sure she is going to the right nest. One researcher performed an elegant experiment to learn about this. She created a gridwork of nests in a block of wood. Straws were placed inside each nest entrance. Once the bees had started building inside the straws, she moved the straws to different positions in the grid. Rather than returning to the original position on the wood block (i.e., third row, fourth column), they returned to their own particular straw!

Busy as a bee

It takes numerous trips to gather enough pollen to provide for just one egg. For *Osmia lignaria*, it takes between 15 and 35 trips, with the bee visiting approximately 75 flowers each trip. That's a total of 1875 flower visits for just one egg.

A few species of *Osmia* have been introduced to North America. *Osmia cornifrons*, for example, is native to Japan, but it can be found in northern Utah and along the east coast, the result of two separate introductions.

Diet and Pollination Services

Many *Osmia* are generalists and will collect pollen from any flower, but some species are very specialized and focus only on a few types of plants. Some of these specialists focus on an entire family of plants while others focus on just a single plant genus. Frequently the plants favored by *Osmia* have flowers that are tube-shaped or are asymmetrical, such as beardtongues (*Penstemon*), mints (Lamiaceae), and plants in the pea family (Fabaceae). Some specialist *Osmia* species have unique structures on their bodies that aid them in collecting pollen from their host plants. For example, several species of *Osmia* in the United States have stiff or corkscrew-shaped hairs on their heads that may aid them in gathering pollen from the plants on which they specialize.

An *Osmia* foraging on a composite flower (Asteraceae).

An *Osmia* resting on the ground between visits to flowers. Many *Osmia*, like this one, have special hairs on their faces that help them collect pollen. Note the white pollen covering the face of this bee. Photo by Jillian H. Cowles.

Better than honey bees

Osmia lignaria is a more efficient pollinator of apples, almonds, plums, and cherries than the European honey bee. A pollination job that takes over 90,000 honey bees can be done with fewer than 300 *Osmia*.

Many species of *Osmia* specialize on flowers in the rose family (Rosaceae). Since most of the orchard crops grown in the United States and Canada are also in this family, the bees' preference for rose family flowers can be exploited to achieve high pollination rates in some crops. For example, when surrounded by a grove of apples, almonds, plums, or cherries (all of which are in the rose family), *Osmia lignaria,* which is also known as the blue orchard bee, is an excellent commercial pollinator. In fact, this little bee appears to be more efficient than the European honey bee at pollinating many important orchard crops. This higher efficiency can be explained partly by *Osmia*'s preference for these flowers, but it is also because *Osmia* visit more flowers per minute and transfer pollen more effectively between flowers than do honey bees. Considering that the honey bee is currently experiencing population declines across North America, the blue orchard bee may become increasingly important over the coming years. Scientists and growers are also exploring the use of other species of *Osmia* for pollinating kiwis, raspberries, blackberries, and blueberries.

Many *Osmia* are colored metallic blue or green, and people often mistake them for flies as a result.

Osmia are frequent visitors of tube-shaped flowers like those of this scorpionweed (*Phacelia*).

ABOVE: An *Osmia holding* onto a thistle (*Cirsium*) flower, tongue extended.

LEFT: An *Osmia* foraging on indigo bush (*Psorthamnus*). Note the metallic sheen to the skin and the full scopa under the abdomen, two clues that this is *Osmia*.

An *Osmia* hovering near a milkvetch flower (*Astragalus*). Many *Osmia*, like this brilliant green species, seem to favor flowers in the pea family (Fabaceae).

A male *Osmia* on a milkvetch (*Astragalus*). Note the black knobs on the antennae—little knobs like this are not uncommon in Osmia.

An *Osmia* manipulating the many petals of a legume flower (Fabaceae)

Many *Osmia* are a shiny metallic blue or green color. This *Osmia* is resting on a clover (*Trifolium*).

ABOVE: An *Osmia* alighting on a thistle flower (*Cirsium*). Note the two submarginal cells.
LEFT: An *Osmia* resting on a mustard plant (Brassicaceae).

A male *Osmia* on a composite flower (Asteraceae). Males generally look slimmer than females and they sometimes have more hair on the head and thorax.

Osmia males are often found on the ground near the plants on which they, and females, are foraging.

The United States isn't the only country to put *Osmia* to work. In Japan, the Japanese hornfaced bee (*Osmia cornifrons*) has been used to pollinate apple and cherry orchards for more than 50 years. It is now responsible for at least three-quarters of all the apples commercially produced in Japan. In Europe, the Spanish mason bee (*Osmia cornuta*) is similarly employed.

Other Biological Notes

Like many bees in the Megachilidae bee family, most *Osmia* use empty cavities rather than building their own nests. Some *Osmia* nest in old beetle burrows, others in the hollow stems of perennial plants, cracks in stones, galls on plants, or burrows in the ground dug by other bees. Even old snail shells and pine cones make fine homes for mason bees. Some cobble together a nest by sticking mud and pebbles inside a rock cavity or under the edge of a roof. Because most *Osmia* don't like to do any digging or constructing of their own, and also because their nests are often made in twigs or old stems, there is usually just one row of nest cells in a main tunnel; there are seldom branching tunnels. An *Osmia* female starts at the back of the nest and works her way forward. Eggs that will grow into female bees are at the back of the nest, and those that will become male bees are near the front. Each egg is separated from the others by a small wall, often made of mud. When a bee is ready to emerge, it will chew its way through whatever is blocking the way to the entrance, including mud walls and even its siblings that haven't already left the nest.

Some *Osmia* construct nests from mud (hence the common name mason bees). The nest pictured here was found on the underside of a rock. Some nest cells have been sealed off; one has been opened, exposing the yellow pollen mass inside.

Like the majority of North American bees, *Osmia* are completely solitary and do not share the task of gathering pollen or nectar for each other's offspring. A female *Osmia* waits a few days after she has emerged from the nest in which she developed before beginning to set up shop in a nest of her own. She will inspect several possible nests before finally settling on the perfect one. When she picks the nest that she likes best, the female will fly back and forth in front of the nest entrance numerous times. This zigzagging flight is thought to help the bee memorize the nest and the landmarks around it so she can find it again. Because they don't determine the length of the nest they use, *Osmia* may lay eggs in several different nests over her lifetime. When one is full, the bee moves on to another.

An *Osmia* female often gathers mud to build partitions between the cells that hold her offspring. She scoops mud together with her mandibles (jaws), shaping it into a transportable blob that she can carry back to her nest, balanced between her front legs and her jaw. Inside the nest, she sculpts the mud into a wall—it usually takes several trips to get enough mud for just one wall. She may also add chewed-up plant material to the mud mixture, or sap from pine trees. A few species take after their leaf-cutting cousins (section 7.7) and use leaves and petals to line their nests.

At the same time she is gathering pollen, the female *Osmia* gathers nectar in her crop (nectar stomach). In the nest she regurgitates the nectar and mixes it with pollen for her offspring to eat.

The ability of *Osmia* to nest in preexisting holes makes it easy to provide habitat for them. A block of wood or a bundle of straws will often encourage these bees to nest in your backyard. For more details on how to make bee nests, see section 2.2.

A female *Osmia* preparing to fly from a sweet clover (*Melilotus*).

7.3 HOPLITIS

Hoplitis were named after the Hoplites, Greek citizen soldiers who lived nearly 3000 years ago and were known for their distinctive shields and spears. Paradoxically, there is little to distinguish their namesake bees.

Description

Hoplitis are a morphologically diverse genus of bees that range from small and black to long and metallic green.

Distribution

Hoplitis is one of the most wide-ranging groups in the tribe Osmiini. The genus is found throughout Europe, Asia, Russia, the Middle East, and Africa. More than 300 species are known around the world. In North America they range as far south as northern Mexico. In the United States and Canada, almost 60 species have been found. Overall, bees in North America are most diverse in the deserts, but *Hoplitis* may be the exception. The majority prefer boreal habitats and cool mountains (though there are also desert species). There are more *Hoplitis* in western North America than in the east, though some species are found only in the east.

Most bee species are restricted to either the Eastern or Western Hemisphere. *Hoplitis robusta* is an interesting species in that it appears to occur naturally all across the Northern Hemisphere. One non-native species of *Hoplitis* occurs in North America. Found in the state of New York,

Specialist and pollinator aren't the same things

Just because some *Hoplitis* are specialists on specific flowers does not mean that these bees are their best pollinators. A scientist studying a rare beardtongue known as the blowout penstemon (*Penstemon haydeni*) found that *Hoplitis* were so good at hoarding the pollen they gathered from the flowers that very little was transferred to the flowers they subsequently visited.

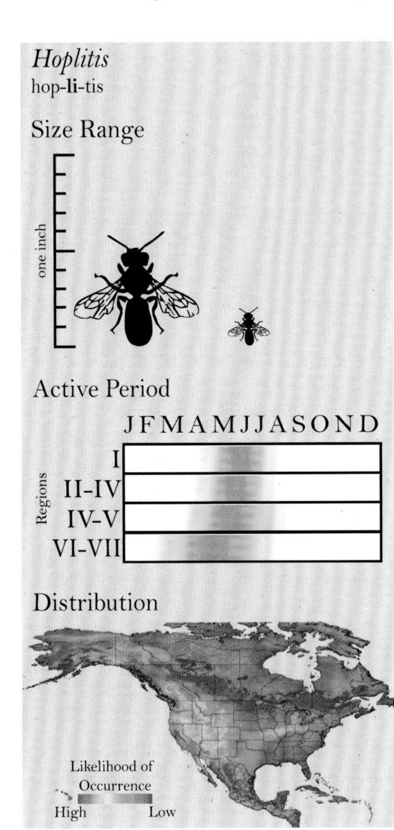

Hoplitis can range from more than half an inch long down to only slightly bigger than a mosquito. Pictured here is a small *Hoplitis* foraging on the equally-small flowers of woolly thyme (*Thymus pseudolanuginosus*).

A *Hoplitis* resting in a desert dandelion (*Malacothrix*). Notice that the ends of the antennae are needle-thin hooks, a common characteristic of *Hoplitis*.

A *Hoplitis* visiting a composite flower (Asteraceae).

Hoplitis anthocopoides was first noticed in the 1960s. It is native to Europe.

Diet and Pollination Services

A range of dietary preferences exist among *Hoplitis*. At one extreme are the generalists that will visit any plant in bloom. Others seem to prefer to forage on just a few plant families. For example, some species of *Hoplitis* visit plants in both the pea family (Fabaceae) and the borage family (Boraginaceae), but nothing else. Other species specialize on a single plant family, with many *Hoplitis* having a preference for the waterleaf family (Hydrophyllaceae), the pea family, or the borage family.

A few species visit just one plant genus. One group of *Hoplitis* found mainly in the southwest visit only cat's eyes, also known as popcorn flowers (*Cryptantha*). *Cryptantha* flowers have very narrow flower tubes, and the pollen and nectar inside are difficult for bees to harvest with their legs. *Hoplitis* species that specialize on *Cryptantha* have stout hooked hairs on their mandibles (jaws) and on their clypeus (just above the mandibles). When a *Hoplitis* sticks her head inside the tiny flower to drink nectar, the hairs on her face pull the pollen away from the anthers. The bee can then groom her face and move the pollen to her belly for transport back to her nest. Other plants that are host to specialist *Hoplitis* include scorpionweed (*Phacelia*), deervetches (*Acmispon* and *Hosakia*), creosote bush (*Larrea*), and beardtongues (*Penstemon*).

Other Biological Notes

All *Hoplitis* are solitary, and they do not share with other bees the work of laying eggs or raising young. The wide variety of nesting habits found among bees in the family Megachilidae is mirrored among species of *Hoplitis*. They nest in burrows in the soil, in cracks in rocks, in the stems of plants, and in preexisting tunnels made by other insects, such as old beetle burrows in the ground or

A *Hoplitis* foraging on a popcorn flower plant (*Cryptantha*). A few *Hoplitis* species visit only *Cryptantha*. The red abdomen can make a handful of *Hoplitis* look remarkably similar to *Ashmeadiella* (section 7.4).

A large *Hoplitis* forages on indigo bush (*Psorthamnus*), while another smaller bee flies by.

A mating pair of *Hoplitis*; the female is just visible below the smaller male on her back. PHOTO BY HARTMUT WISCH.

in wood. Some species nest in old ground nests built by other species in past years. Others nest in empty galls or in empty snail shells.

Hoplitis that nest in premade tunnels lay their eggs in a long series. They start at the back of the tunnel (which they often make a bit longer to suit their needs) and then work their way to the front as they provision, separating each egg's nest cell with a partition as they go. A few species, like the introduced *H. anthocopoides*, build exposed nests. They use "spit" and soil to make a mortar, cementing pebbles together until a cell is formed, large enough for one egg and its pollen provision. A female will build several (up to 30) of these cells each touching the next, and then will cover the whole nest with another layer of mortar. The resulting abode is so sturdy that it can be used a second year by any one of the females that emerge. It is not uncommon for these bees to nest in the same area, building on the same exposed surface, over the top and adjacent to their mother's nests, sometimes even modifying the very cells from which they emerged.

Hoplitis use an assortment of materials to build the partitions between each nest cell. Some *Hoplitis* use pebbles, cemented together with mud (one species uses resin). Others chew up leaves or sawdust. Some use whole leaves and flower petals to separate cells, folding the leaf around the egg and its provision like a tiny envelope.

Masters of remodeling

Many *Hoplitis* build nests in holes and burrows that are made by other insects. While they have little control over the size of these tunnels, the bees often remodel the existing burrow by making it longer or even by lining the walls with soil and chewed-up leaves to make the tunnel just the right size.

A *Hoplitis* foraging on a beardtongue (*Penstemon*).

7.4 OTHER OSMIINI

Ashmeadiella

Heriades

Atoposmia

Chelostoma

ASHMEADIELLA, *ATOPOSMIA*, *HERIADES*, *CHELOSTOMA*, *PROTOSMIA*, AND *XEROHERIADES*

These six genera of bees are rare and relatively uncommon in the United States and Canada. All are in the bee tribe Osmiini and are therefore closely related to *Osmia* and *Hoplitis*.

Ashmeadiella

Ashmeadiella is a group of small, burly bees, many with red-and-white striped abdomens. These bees are found only from Canada through Costa Rica, and nowhere else in the world. There are nearly 60 species, with 50 north of the Mexican border. *Ashmeadiella* prefer dry desert environments and only 2 species are found east of the Mississippi River. Though more common in the west, a few western species of *Ashmeadiella* are quite rare. For example, *Ashmeadiella sculleni* has been found in only three locations in Oregon and Nevada. There are both generalist and specialist *Ashmeadiella*. Among the various plant species known to host *Ashmeadiella* are

W. H. Ashmead (1855–1908)

Ashmeadiella is named for William Harris Ashmead, a biologist who became an entomologist later in life after working as the publisher of an agricultural magazine that devoted much of its page space to articles about injurious insects. He eventually became a curator at the U.S. National Museum. By the time of his death he had collected more than 60,000 specimens, many of which were new to scientists; in his lifetime he personally described over 3000 new species of bees, wasps, and ants.

An *Ashmeadiella* prepares to pry open a pea flower (Fabaceae).

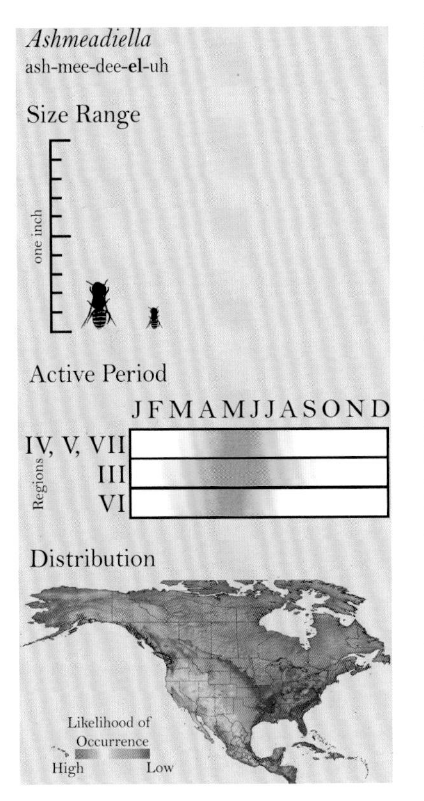

An *Ashmeadiella* foraging on a fleabane (Erigeron).

Many *Ashmeadiella* have a red abdomen, but others, like this one, are mostly black.

Ashmeadiella can be found on a variety of flowers. Here an *Ashmeadiella* forages on a desert marigold (*Baileya*). Note the tiny but stout body, with three round segments.

A female *Ashmeadiella* on a gumweed flower (*Grindelia*). When bees are foraging they often point their antennae down—absorbing the many scents associated with their flowers.

beardtongues (*Penstemon*), cactus (*Opuntia*, *Echinocactus*, etc.), mesquite (*Prosopis*), lupine (*Lupinus*), and *Salvia*. *Ashmeadiella* are solitary bees. They nest alone, and each female gathers pollen and nectar independently to provide for eggs that she lays. *Ashmeadiella* nest in a variety of locations, including holes in wood or stems, in abandoned beetle burrows, under rocks, or even in shallow holes in the ground. One *Ashmeadiella* nest was found inside a snail shell in Texas. Nest cells are separated from each other with chewed-up plant leaves (and occasionally petals), and the resulting mass is often sticky—likely because the bees mix the leaves with nectar or plant sap. Interestingly, even

A male *Ashmeadiella* flying by a cactus flower, likely searching for a mate. Note the miniscule spines at the end of the abdomen, a common characteristic of male *Ashmeadiella*.

An *Ashmeadiella* nesting in the end of a cactus (*Cylindropuntia*) stem. Photo by Laurelin Evanhoe and Matthew Haug.

An *Ashmeadiella* grooming his abdomen with his back legs, while resting on a flower in the pea family.

when these bees are nesting in the ground, their nests consist of one long tunnel that ends in a series of cells, just as if they were nesting in a hollow twig. In these ground nests, the female puts about four eggs, each in its own chamber, then leaves a long empty space before sealing the nest off at ground level.

Atoposmia

Atoposmia means "unusual Osmia"; Atoposmia look Osmia-like, but with some slight differences.

Atoposmia are medium-sized dark bees, found only in Canada, the United States, and Mexico. They are most diverse in the mountains of the western United States. Only a few species range as far east as Oklahoma and western Texas. There are nearly 30 species in the United States and Canada, seen most often in the late spring and early summer. Several species are specialists on beardtongues (*Penstemon*), and one species is a specialist on scorpionweed (*Phacelia*). Many others are generalists. These bees are all solitary, and there appears to be no social nesting. Nests are found inside plant stems, in the soil, or as clumps of individual cells built on the underside of rocks. These bees, like other Osmiini, chew up plant material to form a mortar for divisions between cells that are inside plant stems, or as the "walls" of cells built from sand and other soils on the underside of rock surfaces.

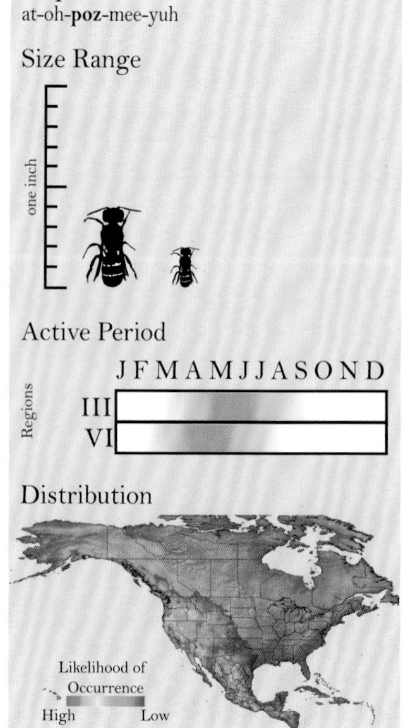

Many *Atoposmia*, like the one pictured here, specialize on *Penstemon* flowers.

Heriades

Heriades means "wool," and the name is probably a reference to the woolly patches of hair common on the abdomen of many species.

Fairly small, and black, *Heriades* are found around the world, including parts of Africa, Europe, Asia, India, Japan, and the Pacific Islands. In the Americas, they range from southern Canada (four species) all the way through Panama to the northern edge of Colombia. Close to 25 species occur in North and Central America, but east of the Rocky Mountains only 3 species are known. In the west, a dozen of them are common while others are rarely collected, and they have not yet been given names by scientists. *Heriades* are all generalists and visit a wide variety of flowering plants. These bees nest in small holes made by other insects, especially beetle holes in wood, but they may also use pine cones. Like *Hoplitis* and *Ashmeadiella,* they use resin to construct partitions between cells.

A *Heriades* foraging on a composite flower. Note how the tip of the abdomen is curled under, a common characteristic of *Heriades.*

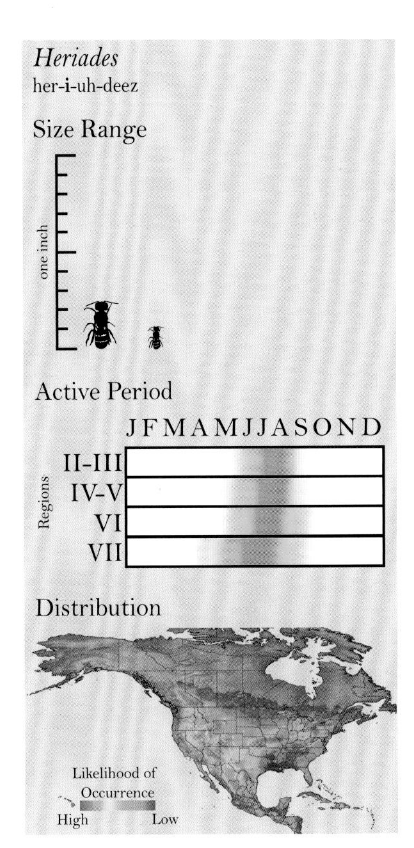

BELOW: Most *Heriades* in North America are long slender bees with dark body and some light stripes on the abdomen.

A *Heriades* foraging on a borage flower (Boraginaceae); each segment of the abdomen looks slightly inflated, so that the outline of the abdomen appears wavy, rather than smooth.

Chelostoma

Chelostoma means "clawed mouth," a reference to the shape of the mandibles.

Chelostoma are black, slender bees seen commonly throughout the Northern Hemisphere, although they do not occur in China and Japan. In North America, they are found in Canada and south to Mexico. There are 10 species north of the Mexican border, but only 3 of them occur east of the Mississippi River. All *Chelostoma* are specialists. Several specialize on scorpionweed (*Phacelia*), or yerba santa (*Eriodictyon*)—both in the waterleaf family (Hydrophyllaceae). One species (*Chelostoma philadelphi*), found in the eastern United States from New York to Georgia, is a specialist on mock orange flowers (*Philadelphus*). Two species (*Chelostoma rapunculi* and *C. campanularum*), both introduced from Europe, are found in New York and specialize on bellflower (*Campanula*). *Chelostoma* all nest in tunnels and cavities made by other insects in dead wood (tree stumps, fence posts, etc.). The walls between each cell are made of soil mixed with nectar or resin.

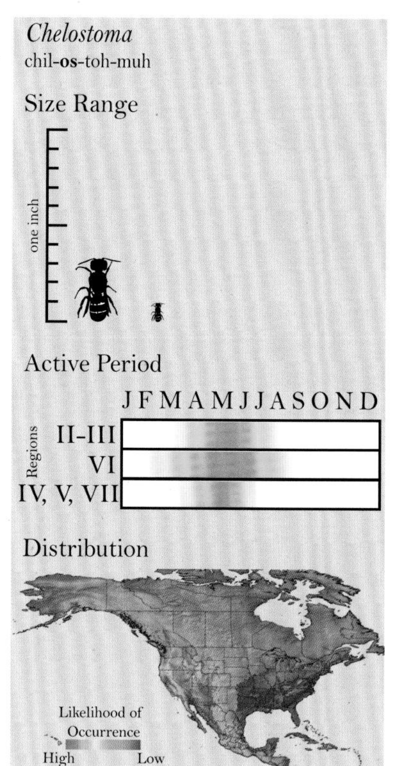

ABOVE: A female *Chelostoma* on a borage flower (Boraginaceae).

BELOW: A *Chelostoma* resting near a narrow flower, illustrating how well these tiny bees can fit inside.

What does it take to name a bee?

New species of bees are collected by scientists every year, even in the United States and Canada; however, officially naming a bee takes time and effort. First, a scientist (specifically a taxonomist) must establish the genus of the specimen, then look at all the species in that genus (and as many individuals of each species as possible) to be sure that the potentially new bee is truly different. The taxonomist must write up a description, complete with notes regarding each body part, and submit the description for review by scientific peers (including a naming committee: the International Commission on Zoological Nomenclature). He or she may add the bee to a preexisting key for that genus—or may rewrite an old key. Then, the taxonomist must look through collections to find any unidentified specimens that may also belong with the new species, so a record can be put together of all known locations for the new bee. Because this can take so much time, many new species are left unnamed in collections, dubbed simply, for example, "*Heriades* nr. *truncorum*," meaning that the bee looks similar to, or near, *Heriades truncorum*, but isn't quite the same. Sometimes scientists wait until they have five or six new species within one genus and name them all at once.

Protosmia rubifloris

The name Protosmia is a reference to this bee's similar appearance to Osmia. The name means "basic Osmia."

From a distance *Protosmia* are small, nondescript black bees with white bands on a dark abdomen. Up close, though, females of this species have a "nose" that would make Pinocchio envious—it is a very long projection that hangs from the clypeus. Only one species of *Protosmia* occurs in North America. Other species of *Protosmia* are found in Europe and northern Africa. *Protosmia rubifloris* is found throughout the west in drier areas. This bee appears to be a generalist and has been found on many types of desert flowers; however, it is not uncommon for scientists who study *Protosmia* nests to find cells with pollen masses made up entirely of one kind of pollen, indicating that this bee specializes with each foraging bout. *Protosmia* nest in pine cones and in pieces of wood. Nests are not lined with any material, but pine resin is used to separate each cell and to cap the nest once all eggs have been laid. Remarkably, one scientist found individuals spending the winter in their nest cells as adults—most bees spend the winter as prepupae.

Xeroheriades micheneri

Xeroheriades means "desert Heriades," in reference to the dry desert regions of the southwest where it is found.

Only one species of *Xeroheriades* is known in the world. It is found exclusively in the Mojave Desert in California. The bee is small, with a bright red abdomen. Because this bee is so rare, little has been discovered about its diet and nesting habits.

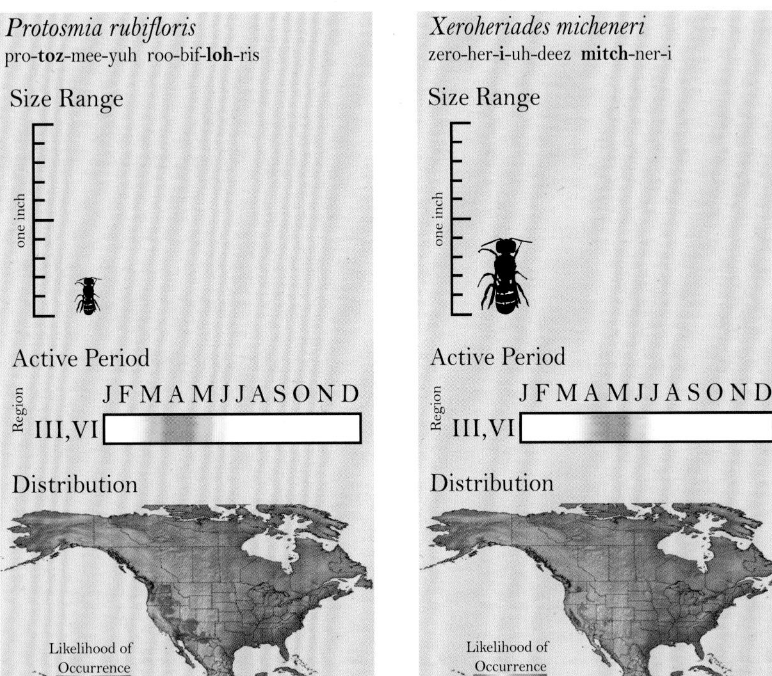

A *Protosmia* forages on a *Cryptantha* flower. Photo by Hartmut Wisch.

In honor of a bee scientist

Charles Duncan Michener is one of the most influential bee scientists of the last 100 years (and, it could be argued, ever). He published his first paper on bees in 1934, when he was only 16 years old. This was the beginning of a long career studying bees from around the world, culminating 66 years later in publishing of the "bee bible," *The Bees of the World*. This tome is an essential reference for any serious student of bees. As influential as his writings is the lasting impact he has made on hundreds of students over his long career, mentoring more than 80 masters' and PhD students, and countless other bee-lovers who have crossed paths with him. A large number of those students have gone on to describe new species and in appreciation have named many bees after him. At least six genera have been named after Dr. Michener, as well as dozens of species.

7.5 *ANTHIDIUM*

The name Anthidium means "little flower visitor."

Description

Bees in the genus *Anthidium* are compact, strikingly colored black and yellow bees found throughout North America. Like flies, these bees (especially the males) tend to hover around flowers, defending a bouquet. Female *Anthidium* shave the fuzz off plant leaves with their mandibles (jaws) and take it back to their nests where they use it to line their nest cells. This is why they are often called wool carder bees.

Distribution

Anthidium are found around the world, though they are absent from Australia and rare south of the equator. More than 160 species are known worldwide, but only 36 occur in the United States, with 5 of those in Canada. Species of *Anthidium* appear to have distinct habitat preferences. For example, two *Anthidium* species are found only on sand dunes in the western United States. Many species seem to prefer dry habitats and deserts and are therefore more common in the western United States. Some researchers have recently suggested that *Anthidium*'s greatest diversity occurs in the Great Basin and on the Colorado Plateau. Only four species of *Anthidium* occur east of the Mississippi River, two of which are recent introductions (see below). Similarly, two species are unique to the Great Plains

The two recent immigrant *Anthidium* species in North America are *Anthidium oblongatum* and *A. manicatum. Anthidium oblongatum* is native to Europe and the Middle East but is now found from Maryland to New York. *Anthidium manicatum*, the European wool carder bee, was first seen in the United States in the 1960s on the east coast but has spread rapidly across the country and is now commonly seen in most states. This bee has been introduced elsewhere in the world as well and can be found in New Zealand and the Canary Islands.

Seasonal bees

Most species of *Anthidium* come out in the spring, but one introduced species, *Anthidium manicatum,* is often out in the hottest part of the summer. If you find an *Anthidium* in your yard in the late summer, chances are it is *Anthidium manicatum*.

ABOVE RIGHT: Many *Anthidium* are confused for wasps because of their yellow-and-black color patterns and because they often fly very quickly around flowers. Here, a female *Anthidium*, with scopa full of pollen, is resting on a sweet clover flower (*Melilotus*).

RIGHT: An *Anthidium* preparing to land on a *Penstemon* flower, antennae forward as it "smells" the flower's fragrance, tongue extended in anticipation.

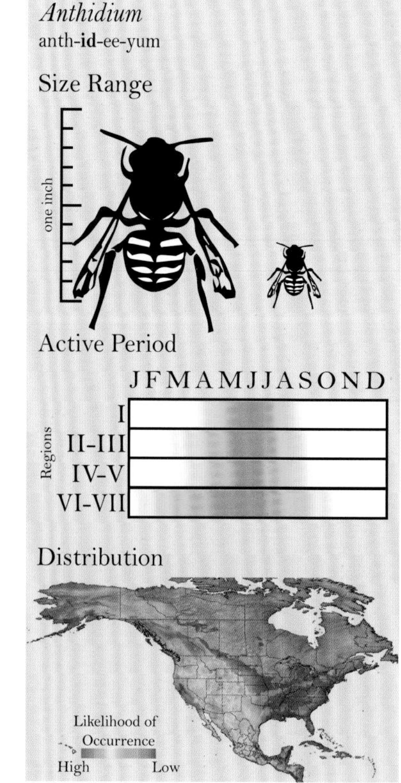

Diet and Pollination Services

The dietary preferences of most *Anthidium* are not well known. A few species are clearly generalists and have been seen visiting a wide array of plants. Other species, while collected on many plants, show strong preferences for one or a few plant families. These preferred plants are often in the waterleaf family (*Hydrophyllaceae*), the pea family (*Fabaceae*), or the sunflower family (*Asteraceae*). A few species appear to be specialists on just one plant genus. As examples, some species prefer milkvetch (*Astragalus*) and other species favor scorpionweed (*Phacelia*). Several *Anthidium* have interesting wiry or hooked hairs on their faces and mandibles (jaws). Other bees with hairs like these have been found to be specialists on flowers with very narrow flower tubes that have anthers hidden deep inside, making them hard to reach with bee legs. The bees use the hooked hairs on their faces to pull pollen from the flower tube when they insert their heads to drink nectar. It may be that these unique hairs indicate that they are specialists on narrow flowers, but this isn't known for certain.

Other Biological Notes

All *Anthidium* species are solitary and do not share the responsibilities of building a nest or providing for offspring.

Unlike nearly all other bees, male *Anthidium* are significantly bigger than females. These males are among the most territorial bees in North America, and they will guard a patch of flowers from any intruders—including humans, dogs, and other insects. This guarding behavior secures their food sources (which, in desert regions, may be rare) as well as their mating turf. Males don't have stings, however, and are not at all dangerous to humans who stop to observe.

An *Anthidium* on scorpionweed (*Phacelia*). Some *Anthidium* species will collect pollen only from scorpionweed flowers.

A male *Anthidium* resting on a *Penstemon* flower. The spines on the end of the abdomen of this bee give away its identity: the introduced species *Anthidium manicatum*.

A male *Anthidium manicatum* patrols his territory around a Russian sage plant (*Perovskia*).

An *Anthidium* on a snakeweed flower (*Gutierrezia*).

An *Anthidium* on a sweet clover (*Melilotus*); the straight subantennal sutures are obvious even from this distance.

Nearly all *Anthidium* in the United States and Canada make their nests in preexisting tunnels, either in wood, inside thick plant stems, in the ground, or in man-made walls. Any tunnel of the right diameter will do, and they have been known to take over abandoned beetle, wasp, and other bee burrows. *Anthidium* line the inside of their nests with plant fuzz (trichomes) that they scrape off hairy leaves and stems. They use this material not only to line the outside walls of their nests, but also to create the walls between nest cells. Once she has prepared all the cells, provided food, and laid eggs, the female packs the nest entrance with more plant fuzz and also pebbles, bits of wood, chewed-up plant material, or (as one scientist found) lizard feces!

An *Anthidium* carries a load of plant hairs into her nest (the white cottony material is just visible below the head and front legs). *Anthidium* cushion their nest cells with these plant fibers.

Just a few species of *Anthidium* actually dig their own nests. Two of them are found exclusively on sand dunes. The females have a long brush of hairs on their foretibia (forearms) that aid them in digging up soft sand. Even these tunnels dug in sand are lined with plant fuzz.

In most bees, males develop from the eggs laid near the front of the nest. *Anthidium* are different in this regard because the male eggs are laid at the back of the nest. This is likely because male *Anthidium,* being bigger than females, take longer to develop. Because they are laid first in the nest, they have a little extra time to grow compared with their sisters, which are in the nest cells between them and the exit.

Many *Anthidium* prefer flowers in the pea family (Fabaceae); the lack of an arolium between the front tarsal claws is clear.

An *Anthidium* balancing on the lip of a *Penstemon* flower.

Anthidium are known as wool carder bees because they scrape the hairs from fuzzy plant leaves to line their nest cells.

Many *Anthidium* have yellow-and-black stripes on their abdomen, but others, like this bee, have a checkerboard pattern.

7.6 OTHER ANTHIDIINI

Dianthidium

ANTHIDIELLUM, DIANTHIDIUM, PARANTHIDIUM, AND *TRACHUSA*

In addition to the *Anthidium* of the previous section, there are four other groups in the Anthidiini. They are less common in the United States and Canada than *Anthidium*. These other *Anthidiini* all use plant resin to construct their nest cells, and some call them collectively the resin bees.

Anthidiellum

Anthidiellum means "tiny Anthidium" and refers to the fact that they look like miniature Anthidium.

Description

Anthidiellum are small compact bees with distinctive yellow and black zebra-like markings.

Distribution

In North America, *Anthidiellum* can be found from British Colombia to Quebec, south through Central America and into Brazil. In the United States and Canada there are three species that are common, one of which occurs commonly both east and west of the Mississippi River (*Anthidium notatum*). In the east is *Anthidiellum perplexum*, and in the west is *A. notatum*. An additional few species are occasionally seen along the shared border of Mexico, Arizona, and New Mexico.

Diet and Pollination Services

There appear to be no specialists among the six North American *Anthidiellum,* though a few species may have a slight preference for flowers in the sunflower family (Asteraceae).

Other Biological Notes

All *Anthidiellum* are solitary and do not share the tasks of rearing young or building nests. Many species build external nests, meaning that instead of building inside preexisting structures (i.e., wood or in the ground), they build their nests on the outside of branches and twigs. While not all species have been fully studied, from those that have been it appears that nests are made almost entirely of plant resin. Each resiny domicile contains just egg.

RIGHT: An *Anthidiellum* foraging on a composite flower. PHOTO BY JILLIAN H. COWLES.

BELOW: The compact *Anthidiellum notatum* resting on a flower head. PHOTO BY JILLIAN H. COWLES.

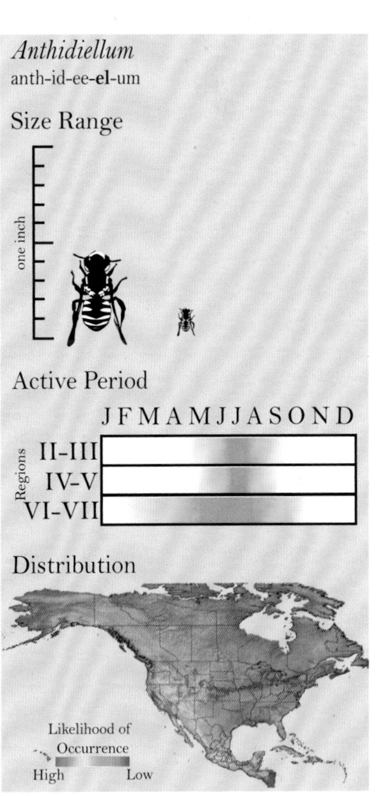

Dianthidium

Dianthidium means "second" or "another" Anthidium—a reference to the similarities between this bee and Anthidium.

Description

Small and black, but with strong yellow spots on all parts of the body, *Dianthidium* are stocky and hard to miss. Like the other Anthidiini, *Dianthidium* are often mistaken for wasps.

Distribution

Dianthidium are found across North America and south into Mexico. In the United States, 21 species are found. Almost all of them are in the west. Just two species are common along the east coast: *Dianthidium curvatum*, and *D. simile*. One species, *Dianthidium floridense,* occurs only in Florida. Three species have been collected in Canada: *Dianthidium subparvum* in western Canada, *D. ulkei* in western and central Canada, and *D. simile* near Ontario.

Diet and Pollination Services

Some *Dianthidium* species are generalists, but many have a preference for plants in the sunflower family (Asteraceae). They collect not only pollen and nectar from these flowers, but also resin for use in nest building.

Other Biological Notes

Dianthidium are solitary bees and do not share the duties of nest building or egg laying. These bees nest in riverbanks, sand dunes (in stabilized sands at the base of grasses growing in these dunes), in holes in pieces of wood and plant stems, or in nests made by other insects, including other bees. Also, many species build exposed nests; instead of taking advantage of the protection afforded by preestablished burrows in the ground or in wood, these bees build their own fortresses. Using pebbles, plant material (leaves and plant hairs), resin, and soil, they construct nest cells stuck to the surface of rocks or twigs. Carefully searching out pebbles of the right size, a female carries each back to her nest site with her mandibles (jaw), and glues them together with resin that she also gathers. Exposed nests of *Dianthidium* are disguised on the outside with mud and pebbles to look like debris stuck to a rough surface. Several nest cells may be stuck together, making one large pebbly conglomeration with many growing bees inside. If you are lucky enough to find one of these, search other rocks or twigs nearby, as other nesting *Dianthidium* are likely in the vicinity. One interesting *Dianthidium*, *D. heterulkei*, nests only inside natural cavities in volcanic rocks.

Dianthidium are unique in that females do not lose their attractiveness to males after they have mated. In other

Back and forth

One scientist estimated that because *Dianthidium* nests require so many different materials (soil, pebbles, resin, etc.), it may take up to 1000 trips for a bee to build, provision, and conceal one nest.

An exposed *Dianthidium* nest, constructed of mud and pebbles that have been glued together with plant resin.

A *Dianthidium* foraging on a gumweed flower (*Grindelia*).

bees, once a female has mated, males avoid her (likely because she smells different). In cases where a female bee mates multiple times, scientists hypothesize that it is important to be the last male to mate with a female before she lays an egg, to increase the likelihood that his sperm will fertilize the eggs when they are laid. As a result, males of *Dianthidium* hover near nest building sites in hopes of maximizing mating success.

RIGHT: A *Dianthidium* drinking nectar from a gumweed flower (*Grindelia*). While most of the Anthidiine bees are yellow and black, some *Dianthidium*, like the one pictured here, can also have red color patterns. Unfortunately, the color is variable and not indicative of any particular species.

BELOW: A small male *Dianthidium* resting on a Russian sage flower (*Perovskia*). Observe the beautiful tufts of hair on the back side of the front legs.

BELOW RIGHT: Many *Dianthidium*, like this one perched on a Russian sage flower (*Perovskia*), fold their wings beside, rather than across their bodies, making them look wasplike.

ABOVE: A female *Dianthidium* resting on a Russian sage flower (*Perovskia*).

RIGHT: A mating pair of *Dianthidium*. Note that the male is larger than the female; while size differences between sexes are not uncommon, it is rare, outside of the Anthidiini, for males to be larger than females. PHOTO BY JILLIAN H. COWLES.

Paranthidium

Paranthidium means "equal to Anthidium," another reference to how similar the Anthidiini look to each other.

Description

Looking almost hornet-like, these medium-sized yellow and black bees are more rare than the other anthidiines, but no less distinctive.

Distribution

About half a dozen species of *Paranthidium* are known worldwide. All occur in the Western Hemisphere, but in the United States there is just one species. It does not range as far north as Canada. *Paranthidium jugatorium* is found from the east coast to the west. Four subspecies can, to some extent, be separated by the area of the country in which they occur. In the east, *Paranthidium jugatorium jugatorium* ranges as far north as New York, as far west as Nebraska, and as far south as Indiana. *Paranthidium jugatorium lepidum* ranges north to Kentucky and Virginia, and south to Georgia. In the west, *Paranthidium jugatorium butleri* occurs only in Arizona, and *Paranthidium jugatorium perpictum* is found in Colorado, New Mexico, and Arizona.

A *Paranthidium* visiting a composite flower (Asteraceae). Photo by Jillian H. Cowles.

Diet and Pollination Services

Paranthidium are generalists, but they are often found on flowers in the sunflower family (Asteraceae).

Other Biological Notes

Paranthidium nest in old holes, burrows, and tunnels made by other insects. One scientist, for example, found that ground nests made by the bee *Melitta* had been taken over by *Paranthidium*. It did not appear that they had stolen active nests, but just set up housekeeping in those that had been abandoned. Only a few cells with eggs and pollen provisions are left in each burrow, and they are separated from each other by walls of plant resin. Two resin walls are made between each cell, with tiny pebbles filling the space between.

Paranthidium are generalists, though they can often be found on plants in the sunflower family (Asteraceae). Photo by Jillian H. Cowles.

Paranthidium jugatorium
pair-anth-**id**-ee-yum joo-guh-**toh**-ree-yum

Size Range

one inch

Active Period

J F M A M J J A S O N D

Regions

III

VI

V, VII

Distribution

Likelihood of Occurrence

High Low

Trachusa

The name Trachusa means "roughened," likely referring to the coarse pock marks called "punctations" covering the thorax (back) of this bee.

Description

Trachusa are on average the biggest of the Anthidiini. While many species have the characteristic yellow and black markings that distinguish this tribe, there are also species that are almost entirely black and hairy.

Distribution

Trachusa are found throughout the Northern Hemisphere and into northern Africa. They are common in both North and South America. In the United States and Canada, 19 species are found, mostly in the United States (just a few make it into Canada), and mostly in the west. Four species are found east of the Mississippi River.

Diet and Pollination Services

The dietary preferences of many *Trachusa* are not known. A handful appear to be specialists. *Trachusa larrea* visits only creosote bush (*Larrea tridentata*), where males can often be seen hovering and guarding flowers. Female *Trachusa larrea* collect not only pollen, but also creosote resin. They can be seen chewing on the green branches of the creosote bush to get at the gummy resin inside. These bees then mold the resin into a sphere, which they carry back to the nest with their mandibles (jaws) and legs.

A *Trachusa* visiting a flower of creosote bush (*Larrea tridentata*). Photo by Jillian H. Cowles.

Other Biological Notes

Trachusa are all solitary and there is no sharing of the responsibilities associated with building nests and providing for eggs. All *Trachusa* nest in the ground, and unlike many Anthidiini bees, they dig their own nests. *Trachusa* may nest individually, or many individuals may nest in close proximity to each other (one researcher found between 12 and 14 nest entrances per square yard extending for a distance equal to the length of a football field). Nests are generally shallow, long tunnels. They may end in a row of stacked nest cells, or there may be many offshoots, each containing a nest cell. Cells are lined with resin and fine gravel, or with leafy material, depending on the species.

Trachusa, like the one pictured here, are some of the largest of the Anthidiine bees. Some species visit only creosote bush (*Larrea tridentata*) both for pollen and for resin used to line their nests. Photo by Jillian H. Cowles.

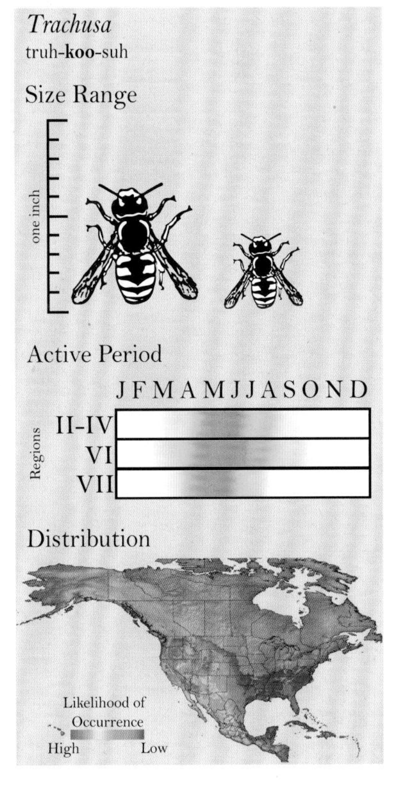

7.7 *MEGACHILE*

The name Megachile means "large lipped" and refers to the enormous mandibles (jaw) characteristic of this group.

Megachile are common bees found worldwide. Like some species of *Osmia*, a few *Megachile* are highly managed in the United States, where their pollinating services are employed to fertilize important crops. The genus includes the giant resin bee (*Megachile sculpturalis*), recently introduced to the United States and Canada from the Far East and almost an inch long. It also includes *Megachile pluto*; at 1.5 inches long it is the largest bee species in the world.

Hemisphere (e.g., *Megachile centuncularis*). Although *Megachile* are widespread in general, some species are very rare and may be known from only one specimen or one sex. A new species (*Megachile chomskyi*) was discovered in 2005 in Texas.

Because these bees nest so readily in wood and other preexisting cavities, they are easily moved between countries and have benefitted notably from globalization, the slave trade, and the trade routes of the last 300 years. As a consequence, at least seven non-native species can now be found in the United States. *Megachile concinna* is thought to have come to the United States from the West Indies sometime after World War II, and it may have been brought to the West Indies from Africa sometime in the 1800s. *Megachile rotundata*, the alfalfa leafcutter bee, was accidentally introduced into the United States sometime prior to 1940, but it is well established and now one of our most important pollinators of alfalfa. *Megachile lanata* has been collected in Florida, but because collection of this species is rare, scientists are unsure whether it has established itself in the state. It is, however, found in Cuba, where it likely arrived from India, 300 years ago during the slave trade. Similarly, *Megachile ericetorum* was found in Ontario in 2003, and it may or may not have become established there. *Megachile sculpturalis*, the

Description

Megachile species sport an amazing array of bumps, horns, and hooks on their stocky black bodies. These small to large bees have a propensity for snipping pieces of leaves from plants and carting them back to their nests as wallpaper. This practice has led to the common name of leafcutter bees.

Distribution

More than 1400 species of *Megachile* are found globally. They are most diverse and abundant in the Eastern Hemisphere, but there are many species in North America too. Nearly 140 species of *Megachile* can be found in the United States, including Alaska. Over two dozen occur in Canada, where they range as far north as Newfoundland. While most species are restricted to either Eurasia or the Americas, there are a few species that naturally span the entire Northern

A *Megachile* hovering near a Russian sage plant (*Perovskia*).

A *Megachile* drinking nectar from a gumweed flower (*Grindelia*).

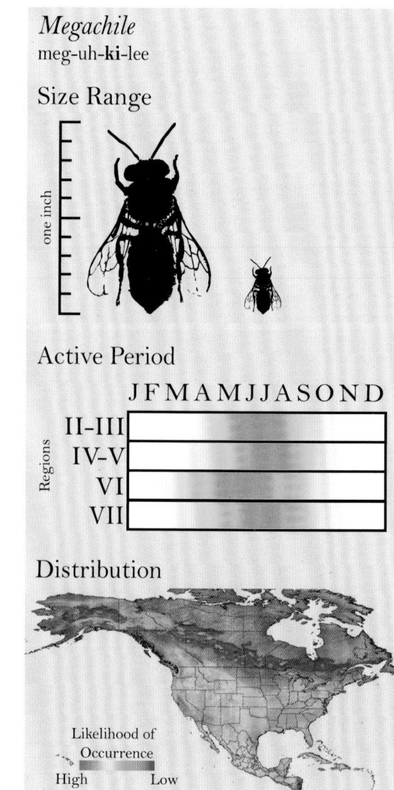

Unwelcome guests

There are more non-native bee species in the United States and Canada than you might imagine. These hail mainly from Europe, but a number of invaders have come from Asia, and even one from Africa. Perhaps the most notable is the honey bee, which didn't occur on this continent until the Pilgrims brought it with them over 400 years ago. Most non-native species here don't nest in the ground. Bees that nest in wood are the most easily transported to a new place. At least two dozen bee species in the United States and Canada originated in foreign lands.

A male *Megachile* on a blanket flower (*Gaillardia*). Males often have long hairs and spines on their front legs.

giant resin bee, was very recently (1994) found for the first time in North Carolina, but it has already made its way at least as far west as Kansas. While not all of these newly introduced species are harmful to native bees, some evidence suggests that they increase competition for limited resources like nesting holes and pollen.

Diet and Pollination Services

With so many species of *Megachile*, it isn't surprising that this group's preferences run the gamut from specialist to very broad generalist. Most *Megachile* are generalists; *Megachile texana*, for example, visits in excess of 80 different types of plants. At the other extreme, there is a group of *Megachile* that will visit only evening primrose (*Oenothera*). *Megachile davidsoni* appears to specialize on golden eardrops (*Ehrendorferia chrysantha*), a plant found in California that requires fire in order to germinate.

A male *Megachile* on a lupine flower (*Lupinus*).

Surviving the heat

Fire is a natural ecological process important in many ecosystems around the world. How do tiny bees survive the heat of a blaze? It appears that most wildfires move so quickly across the ground that the temperature just a few inches down does not become lethal. Mortality rates in ground nests exposed to surface fire are usually less than 10%. Twig-nesting bees are not so fortunate, but many nest in reeds and other materials typically associated with moister areas, where fire is more sporadic. Also, many twig nesters live in desert environments with sparse vegetation and an infrequent fire regime; presumably the rarity of fires here makes this behavior acceptable. Though fire in some cases is extremely devastating, too hot, and likely decimates all bee nests in its path, it also provides sunlit openings in canopies for wildflowers to bloom in years following. This boon in pollen sources may offset the destructive nature of the fire and allow for rapid recolonization.

A pollen-covered *Megachile* resting on a prairie clover flower (*Dalea*).

A female *Megachile* flying near a cluster of composite flowers. You can just see the bright yellow pollen under the abdomen of this bee. *Megachile*, like all Megachilidae, carry their pollen under the abdomen rather than on their legs.

A male *Megachile* hovering near some flowers. Males are generally longer and skinnier than females, and they lack pollen-collecting hairs on the underside of the abdomen.

Scientists do not fully understand how this bee specializes on a flower that follows fires, which are an unpredictable occurrence. Many other species specialize on the sunflower family, with some species preferring specific groups within the family (for example, *Megachile fortis* prefers *Helianthus* over all other sunflowers).

Some specialist *Megachile* are important crop pollinators. The alfalfa leafcutter bee (*Megachile rotundata*) is an important pollinator of this common crop, and populations are managed by farmers in order to maximize seed production each season. Prior to management of the alfalfa leafcutter bee, honey bees were used as the major pollinator of alfalfa, which is used to feed dairy cattle and other livestock. In the presence of honey bees, alfalfa seed production increased fivefold compared with fields with no introduced bees. When the honey bee was replaced by the alfalfa leafcutter bee, seed production increased fifteenfold. Alfalfa leafcutter bees are currently responsible for two-thirds of the world's alfalfa production. In terms of economic value, only the honey bee is a more important pollinator of crops.

There are wildflowers that also benefit from the presence of *Megachile*. Milkweeds (*Asclepias*), for example, have been found to set more seed when *Megachile* are around to pollinate them.

The dietary preferences of these bees do not always benefit land managers, however. The introduced bee *Megachile apicalis* is a common pollinator of yellow starthistle (*Centaurea solstitialis*), an introduced weed that conservationists are actively trying to remove from many wild areas.

Other Biological Notes

Like most bees in North America, *Megachile* do not nest in hives. *Megachile* are not social—each female builds and provisions her own nest. There are no known cases of *Megachile* sharing nest entrances either. They are gregarious, however, and several females may build nests in cavities in close proximity to each other. Males often hover near these nest entrances, creating a small swarm, as they wait for females with which to mate.

Females will often drag their bellies as they enter their nests, secreting a substance from a gland on their abdomens as they go. The smell of this secretion is unique to each female and can help her find her own nest entrance among the many that may be close together.

Megachile build nests in a variety of places. Most of them nest above the ground, often in preexisting holes like dead plant stalks and insect burrows, between rocks, under dried cow patties, or in decomposing wood. Some species aren't picky and will nest in any of these places, whereas others are more particular. *Megachile gentilis* will nest only inside twigs. *Megachile inimica* prefers mesquite trees. There are

No leaves, no *Megachile*

Obviously, for bees that line their nests with leaves, leaf material is an important natural resource that must exist in the vicinity of the bee's nest. Scientists have found that prairies with few broad-leaved plants are particularly lacking in *Megachile*.

Pictured here is a *Megachile* carrying a cut piece of leaf back to the nest she is preparing in the crack of a log. Most *Megachile* nest in premade cavities like this crack in a log, in old beetle burrows, in crevices in rocks, or in many other natural or man-made cavities.

also species that are ground nesters. *Megachile texana* nests in preexisting holes in the ground. All members in the subgenus *Xeromegachile* nest in sandy areas, and they often dig their own burrows. *Megachile* are opportunistic and will nest in many man-made objects, including copper tubing, folds on bags of fertilizer or seeds, and holes in concrete.

Almost all *Megachile* line their nests with leaf or petal material. While resting on the narrow edge of a piece of leaf, and holding it with her claws, a female will chew the leaf around herself (imagine the cartoon character that saws a hole in the floor around herself). When the leaf or petal breaks free, the bee and her leaf free-fall until she, with her prize, is able to fly away.

In the nest, she chews the edges of the leaf material until they become gummy. She then presses leaf edges together, creating long leafy sheets that she pushes against her nest walls. Once she has created a cup-shaped envelope out of leaf or petal material, she fills it with pollen and nectar, and lays an egg. She finishes the cell by closing the egg and its food resource in more vegetative material. She folds and weaves the ends of the leaves together, closing her offspring into its leafy capsule. The process of gathering leaves and weaving them into a suitable chamber is time consuming and it may take anywhere from an hour and a half to three hours to create one nest cell.

Old habits die hard

It appears that leafcutting bees have been snipping off pieces of leaves for more than 30 million years, because many fossilized leaves with notches characteristic of *Megachile* bites have been found.

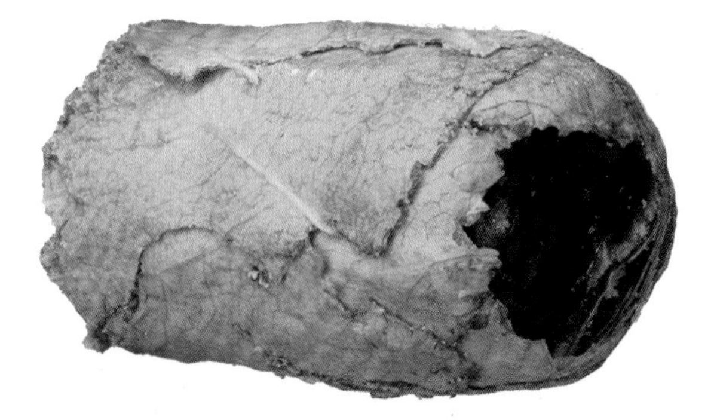

ABOVE: Nest cells of *Megachile* are lined with the pieces of leaves that the female bee cuts with her mandibles (jaw) and brings back to the nest. Pictured here is the nest cell of a *Megachile relativa* that was removed from the nest after the newly emerged bee chewed its way out of one end. PHOTO COURTESY OF USDA-ARS BEE BIOLOGY AND SYSTEMATICS LABORATORY.

LEFT: While many *Megachile* use leaves to line their nests, some species use flower petals. Pictured here on a globe mallow (*Sphaeralcea*) is a *Megachile* cutting off pieces that she will use to line her nest. PHOTO BY JILLIAN H. COWLES.

A *Megachile* on a mint flower (Lamiaceae). The substantial mandibles are hard to miss.

Megachile are known as leafcutter bees because they cut pieces of leaves that they bring back to their nests to line nest cells. PHOTO BY JILLIAN H. COWLES.

Some bees use a variety of leaves and petals, but others stick to specific types of leaves. *Megachile montivaga*, for example, is very particular about what kinds of material it uses to line its nest, preferring the petals of farewell-to-spring (*Clarkia*) over all other materials. The bees that specialize on evening primrose also cut off pieces of those petals for cell lining. *Megachile subparallela* uses whole leaves, picked from ticktrefoil (*Desmodium*) to line her nest cells. Many *Megachile* seem to prefer rose or lilac leaves, and you will probably see some evidence of leafcutter bees if you examine these plants.

As with all characteristic behaviors we describe, there are exceptions. Some species of *Megachile* forgo the leaf-cutting activities and build little walls between each cell. These partitions can be made of sand, small pebbles, mud, or sap. Among all *Megachile*, though, once a tunnel has been entirely filled with eggs and pollen, the mother seals the opening with debris of her choice (more soil or pebbles, resin, mud, or leaf material).

A close-knit family

Almost all bees begin their lives alone, growing up and pupating in isolation; however, one species (*Megachile policaris*) is different. Rather than sealing each little egg into its own individual cell, the mother prepares a gargantuan mass of pollen and nectar in one giant cell (a brood chamber). The mother lays her eggs randomly in little pockets on the pollen, and all the bee larvae hatch and develop together. Most bee larvae that come in contact with their brothers and sisters prior to emergence kill and eat them, but these siblings grow together peacefully.

A male *Megachile* sleeping inside a mariposa lily (*Calochortus*). As with some *Osmia*, some male *Megachile* species have knobs at the ends of their antennae.

8

APIDAE

The family Apidae includes some of the most recognizable bee genera in North America. From the big fuzzy bumble bee to the ubiquitous honey bee, this family contains most of the bees used in commercial crop production. Here, a collection of bees belonging to this family shows the amazing diversity of colors and body shapes, including some of the biggest and smallest bees in North America. The bees here are presented roughly to scale.

Characterized by fluffy hairs and hasty dashing from one flower to another, the Apidea bee family is large, with more than 5700 species in over 200 genera around the world. They are probably the most well-recognized bee family worldwide. Comprising many famous bees, the Apidae include hard workers like the honey bee (*Apis*), the well-known bumble bee (*Bombus*), and the oft-seen carpenter bee (*Xylocopa*). Though less renowned, other Apidae bees are equally interesting and beautiful. Male orchid bees (*Euglossa*), which just make it into the United States collect scent compounds from flowers and use them to attract females. Eucerini have antennae that stretch the entire length of their backs. The majority of the parasitic bees found in North America are in the Apidae family. Another interesting Apidae is *Ceratina,* some species of which reproduce entirely without males. Though the biggest bee in the world is in the Megachilidae bee family, the average size of North American Apidae is by far greater than the average size of any of the other families. From polished black to vibrant green, and bristling white to downy red, the appearance of the Apidae is varied and beautiful.

Largely because of the honey bee, the Apidae can be considered the most economically important bee pollinators in the world. The honey bee is a machine, responsible for

Xylocopa virginica forages on a sunflower. Species of *Xylocopa* may be all black or may have yellow to red hair on the thorax.

A *Ceratina* prepares to land in a cactus flower (*Opuntia*). *Ceratina* are generalists and will visit many different kinds of flowers.

Exomalopsis, already burdened with a huge load of pollen, stops for a drink of nectar and perhaps some more pollen collecting. These bees can carry more than their body weight in pollen on their legs.

Though all *Diadasia* (and in fact the entire tribe to which it belongs) are specialists, they will visit a variety of nonhost flowers when drinking nectar. This *Diadasia* is a globe mallow specialist (*Sphaeralcea*), but is visiting a daisy for nectar.

nearly the entire almond crop of North America, not to mention countless other fruits and vegetables (see chapter 2). Though wildflowers are not of commercial importance, the fuzzy bodies of Apidae bees make them important pollinators of the many wildflowers they visit as well. A large number of Apidae are strict specialists, limiting themselves to a narrow subset of available floral resources for their pollen collecting and nest-cell provisioning.

How honey bees cast their votes

Recently, scientists discovered that honey bees have the amazing ability to vote when making decisions about where to move their hives. Several scout bees will investigate potential new nest-sites, and fly back to the nest to tell the other worker bees about their findings. Each scout describes what she has found to the other watching bees, using a unique series of wiggles, waggles, and figure eights ("left past the tulip patch, straight ahead about 10,000 wing beats, and then right at the top of that crack in the wall"). If the other bees like what they see, they begin imitating the bee dance, and when a critical mass of bees are all dancing the same steps, a decision has been reached. Interestingly, the better the nesting site, the more enthusiastically the original scout will dance, repeating her steps over and over—maybe even doing the same moves several hundred times. If the site is only so-so, she may repeat her dance just a few times before giving up. Like any mass media stunt, the bees that dance the longest will come in contact with the most other bees, and each bee that sees her movements repeats them (and then passes the dance on to other bees, etc.). The bee that dances the longest is ground-zero for the New Nest Line Dance.

The majority of Apidae nest in the ground, with some species building long-running aggregations of thousands of individuals. The group also includes some very interesting twig nesters; carpenter bees in the southwest can be found inside dead yucca stalks, for example. And honey bees and bumble bees modify an existing crack or burrow to make it their own highly modified home. These two bee genera are the most social of all of the bees in the United States and Canada, with honey bees having highly structured social hierarchies established across colonies of several thousand individuals. Honey bees can even communicate complex pieces of information with each other and may vote democratically on new nest sites.

The Apidae include 33 tribes in 3 subfamilies worldwide. Excluding the many parasitic genera, 9 tribes are known within the United States and Canada: Xylocopini, Exomalopsini, Emphorini, Eucerini, Anthophorini, Centridini, Bombini, Apini, and Euglossini. We have divided this section to reflect these natural taxonomic divisions within the Apidae, though more commonly seen genera within these tribes receive their own section (i.e., *Xylocopa* and *Ceratina*). The Eucerini are split across three sections.

A jet-black *Anthophora* foraging on sage (*Salvia*). Note the stark contrast in the white color of the pollen-collecting hairs (scopa) on the legs.

Melissodes and *Eucera* are in the tribe known as Eucerini. This is a large group and the females (on the left) can be difficult to distinguish from other Apidae. The males (like the one on the right), however, are hard to miss with their inordinately long antennae.

Svastra, also in the Eucerini tribe, are large bees with a predilection for sunflower pollen.

Martinapis are Eucerini bees with males that have striking lemon-yellow antennae.

An *Anthophora* struggling with the flower of a milkvetch (*Astragalus*). Large *Anthophora* are one of the earlier bees to fly each spring.

A *Centris* male holding still on indigo bush (*Psorthamnus*). These bees are large and more frequently seen hovering than resting. Many collect floral oils in addition to pollen and nectar from their host flowers.

LEFT: Most people are familiar with bumble bees (*Bombus*). Large, ubiquitous, and with beautiful yellow, black, red, or even white markings on their abdomens, they are hard to miss. This one is foraging on beardtongue flowers (*Penstemon*). Note the bright orange pollen on the back legs. These bees carry their pollen as a wet mass, glued to their hind legs with nectar.

RIGHT: A honey bee (*Apis*) forages on Russian sage (*Perovskia*). Though they are perhaps the best known of all the bees, they should be considered the black sheep of the bee world because their life history is so very different.

Two genera of bees are closely associated with squash flowers: *Peponapis* (left), and *Xenoglossa* (right). These bees rely solely on squash flowers for pollen, and in the process of collecting it, they enable most of the fruit set in these plants.

Eucerini males commonly aggregate at night, sleeping while snuggled up in tight balls, presumably to stay just a little warmer. PHOTO BY B. SETH TOPHAM.

IDENTIFICATION TIPS

Most Apidae are easy to recognize: relatively large, extremely hairy, and fast flying. Figuring out what *kind* of Apidae is trickier. A few characteristics can lead you to the right tribe, but distinguishing genera may require more detail than we provide here.

IDENTIFYING FEATURES OF THE APIDAE BEE FAMILY

- All Apidae have long tongues (see section 1.8, p.28).
- Though not always true, many of the genera are very hairy, and females often have dense oversized scopal hairs that, when full, make the bee look as though it were wearing chaps.

Examples of diagnostic features for different genera can be seen below.

Similar Species to *Xylocopa*

Bombus (section 8.10)

Anthophorini (section 8.8)

XYLOCOPA (SECTION 8.1)

Because of their large size and the loud buzz they make when flying, it is hard to mistake a *Xylocopa* (carpenter bee) for much else. Smaller species may resemble some Anthophorini (section 8.8), and those with yellow on the thorax may superficially resemble bumble bees (*Bombus*, section 8.10). Note the very dark wings. The wings of Anthophorini are not this dark. Knowing about a few other key features can help.

Identifying Features of *Xylocopa*

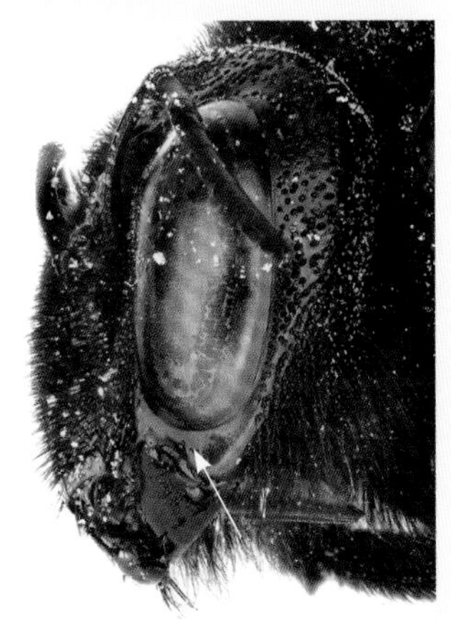

The area between the eye and the start of the mandible (the malar space) on *Xylocopa* is short. This can help distinguish between *Xylocopa* and *Bombus*, in which this area is much longer.

The pollen-collecting hairs of *Xylocopa* are stiff bristles, widely spaced so that many grains of pollen can fit between. *Bombus*, which might be confused with *Xylocopa*, carry pollen as a wet mass instead.

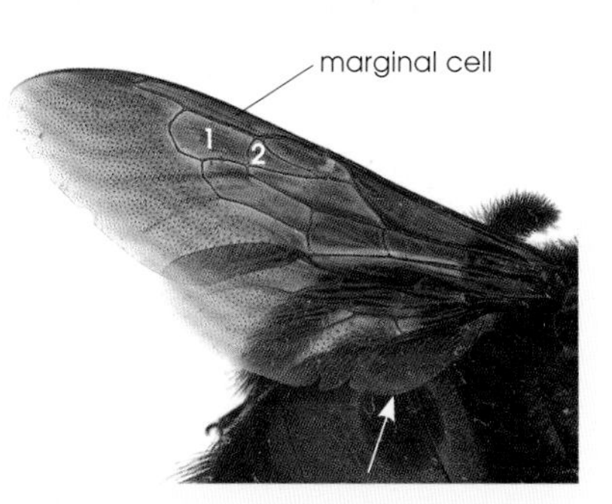

Though *Xylocopa* are difficult to confuse with other bees because of their large size, if you are unsure, the wing has several useful features. First, the hind wing has both a jugal and a vannal lobe (compare to the *Bombus* wing). Second, the marginal cell is extremely long and thin (compare with *Centris*, Eucerini, and even Anthophorini). Finally, the second submarginal cell has a unique shape, tapering to an extreme point at the edge nearest the body.

continued overleaf

CERATINA (SECTION 8.2)

Identifying Features of *Ceratina*

Ceratina are small, shiny black bees with few hairs. They often have yellow markings centered on the clypeus.

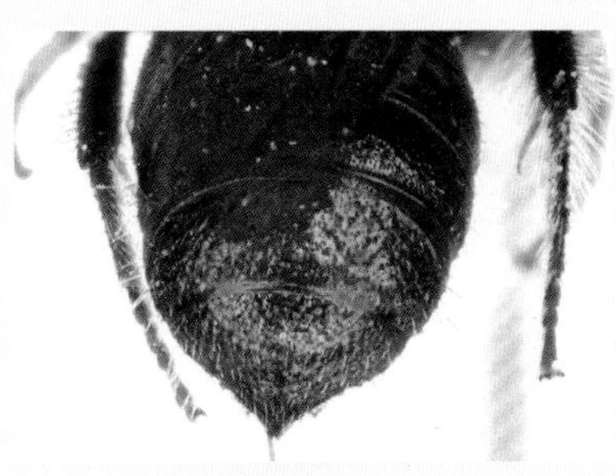

Ceratina females do not have pygidial plates at the end of their abdomens. The shape of the abdomen is unique; the sides stay more or less parallel before tapering quickly to a point.

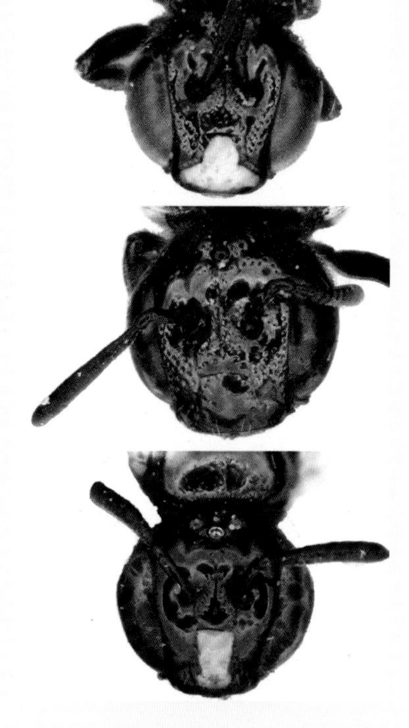

ABOVE: The facial patterns on *Ceratina* are variable between, and even within, species; however, the yellow mark on the clypeus is usually present and quite distinctive.

Similar Species to *Ceratina*

Hylaeus (section 4.2)

Protandrenini and Panurgini (section 3.2)

EXOMALOPSINI (SECTION 8.3)

Exomalopsis

Anthophorula

LEFT: Exomalopsini are the smallest of the furry Apidae. They have pale hair bands running across their abdomens that are sometimes red. The clypeus and mandibles may occasionally have yellow on them. The thorax of the Exomalopsini tends to be extremely shiny and often lacks much hair.

Identifying Features of Exomalopsini

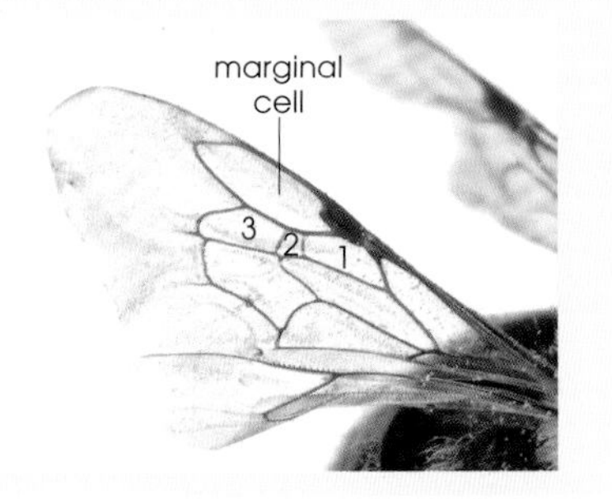

On the wing of Exomalopsini bees, the marginal cell is shortened, looking as though snipped at an angle away from the edge of the wing. There might be two or three submarginal cells.

Identifying Features of Exomalopsini *continued*

Though hard to see unless looking at the face from just the right angle, a unique row of long, widely spaced, stiff hairs line the inner margin of the complex eye (showing as yellow in the image). These hairs are not found in other Apidae bee tribes.

Exomalopsis and *Anthophorula* have very polished exoskeletons, especially on the top of the thorax. Often there is no hair here, and the mirror-like shine can even be seen as the bees forage on flowers. Also visible in this photo, the head of the Exomalopsini tends to be about as wide as the thorax, which can help distinguish it from the Emphorini (section 8.4).

Similar Species to the Exomalopsini

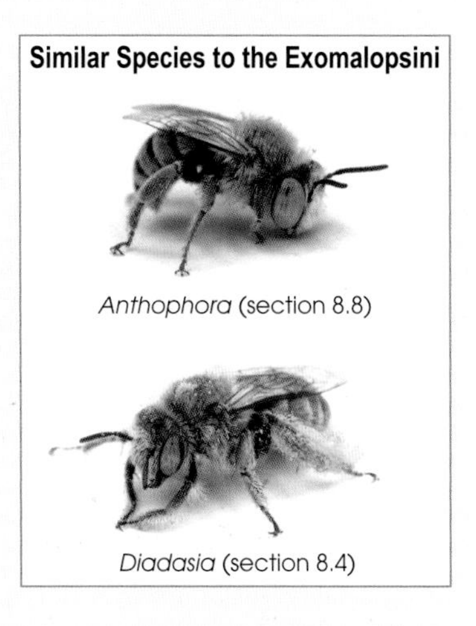

Anthophora (section 8.8)

Diadasia (section 8.4)

EMPHORINI (SECTION 8.4)

Ancyloscelis *Melitoma* *Diadasia* *Ptilothrix*

Emphorini are very hairy bees, often with pale bands of hair stretching across the abdomen. The flower on which the bee was collected can be very telling here, as all of the Emphorini are pollen specialists.

Identifying Features of Emphorini

Both male and female Emphorini have short antennae (unlike some other Apidae tribes). Looking straight at the face of the bee, the face is smoothly oval-shaped, with the simple eyes on top of the head falling below the highest point, not disrupting the head's outline. Generally the head is small, not as wide as the thorax.

continued overleaf

Identifying Features of Emphorini *continued*

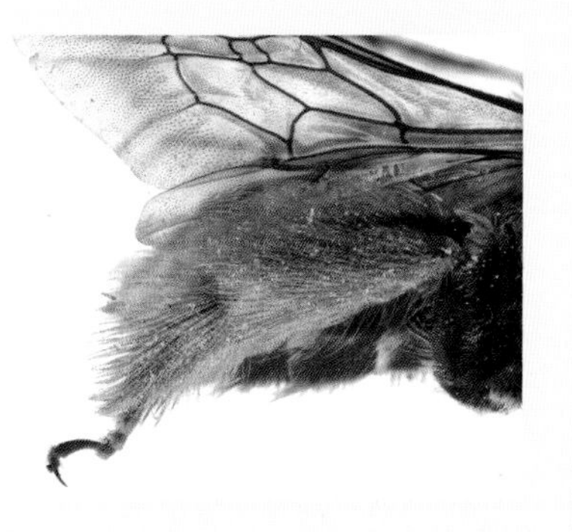

Because the Emphorini bees often collect pollen from plants with extra-large pollen grains, their scopal hairs are extraordinarily stiff and more widely spaced from each other than those of many other species of Apidae. Note that the hairs are feathered, with many little bristles to help hold in the pollen grains.

Similar Species to the Emphorini

Eucerini (sections 8.5–8.7) Exomalopsini (section 8.3)

EUCERINI (SECTIONS 8.5–8.7)

Eucera, *Melissodes*, *Melissoptila*, *Syntrichalonia*, *Martinapis*, *Cemolobus*, *Tetraloniella*, *Svastra*, *Gaesischia*

Identifying a bee as belonging to the Eucerini bee tribe is fairly straightforward if it is a male, because all the males have extremely long antennae. Females are trickier. Once you are sure a bee is a Eucerini, figuring out which of the 14 genera it belongs to only gets harder. Identifying the genus may require shaving sections of hair from the bee, studying mouthparts, and taking measurements of various rather hidden body parts. We do not delve into the finer points of Eucerini identification and below provide only pointers on Eucerini genera that are easier to identify. If you are interested in more information, we refer you to Charles D. Michener, *The Bees of the World*. *Eucera* and *Melissodes* are the two you're most likely to see.

Identifying Features of Eucerini

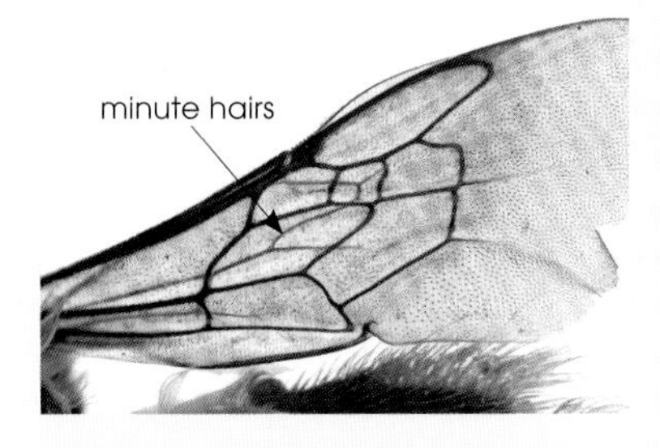

Eucerini may have two or three submarginal cells on their wing. These cells, as well as the other wing cells of Eucerini, are covered with minute hairs—small dots in the accompanying image. Anthophorini, which look similar to Eucerini, have no hairs on the cells inside their wings. These hairs can sometimes be seen without a microscope if you look across the plane of the bee's wing.

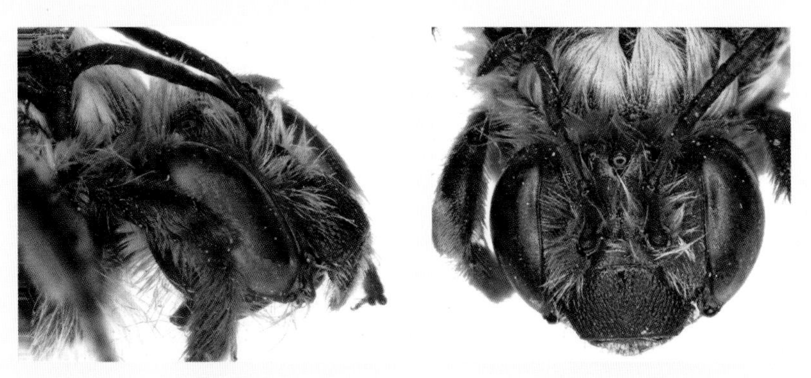

The clypeus of a Eucerini bee tends to protrude notably, like a long stout nose. While it sticks out further in some species than in others, none of them have a flat face. Beside the inner margin of the compound eye, there is no row of stiff hairs sticking straight up; this distinguishes them from the Exomalopsini. And looking straight on, the top of the head is rather flat, or even slightly concave (sagging in), so that the simple eyes sometimes seem to bulge out. This separates them from the Emphorini, whose heads are more evenly rounded. The head is the same width or wider than the thorax.

Among the Eucerini, *Peponapis* and *Xenoglossa* (section 8.6) are perhaps the easiest to identify to genus. Found almost exclusively in squash flowers, they are typically seen only in the early parts of the day. *Xenoglossa* typically have strong yellow or ivory markings on their faces and mandibles, and *Peponapis* (shown here) have strong hair bands running across the tergal segments. Also, the clypeus sticks out noticeably from the face.

Cemolobus (section 8.7) is a rare bee, medium to large, generally seen in the morning glory flowers (*Ipomoea*) on which it specializes. It is found only east of the Great Plains. If you observe a bee on this flower, and notice that the bottom margin of the clypeus is wavy, with three distinct "bumps," you can be sure that it is *Cemolobus*.

Florilegus condignus is uncommonly seen; however, it is a specialist on pickerel weed (*Pontederia*). If you see a Eucerini foraging on this plant, there is a good chance it is *F. condignus*. The bands of hair that run across the abdomen are distinctive; they look flattened compared with the hair bands of many other Eucerini.

FAR LEFT AND LEFT: *Eucera* (section 8.5) are the most common Eucerini in the early springtime. Males (far left) can be identified by their long antennae, but females are more difficult. The clypeus sticks out noticeably, like a "Roman nose" when viewed in profile. The stripes across the abdomen tend to be continuous, with no breaks in the middle.

continued overleaf

Identifying Features of Eucerini *continued*

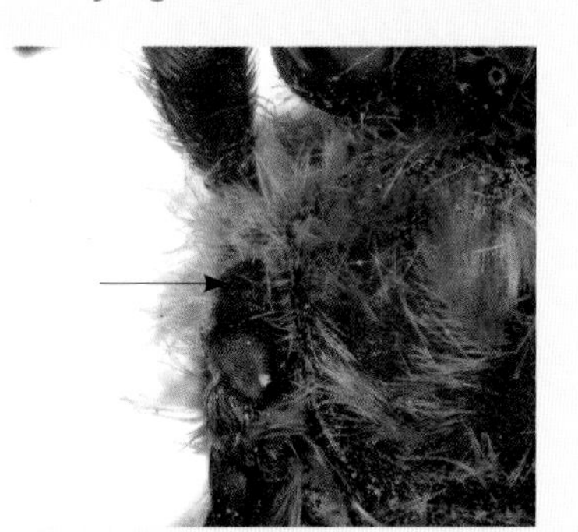

If your *Eucera* specimen is old and worn, you may be able to see the tegula covering the wing bases—it is perfectly oval, with no tapering point at the front end. This can separate it from the closely related, and also common, *Melissodes*.

The other very common Eucerini is *Melissodes*. They can be difficult to distinguish from *Eucera*, but typically *Eucera* are seen in the spring, while *Melissodes* are late summer and fall bees. The stripes that run across the abdomen are often broken in the middle, less continuous than those of *Eucera*.

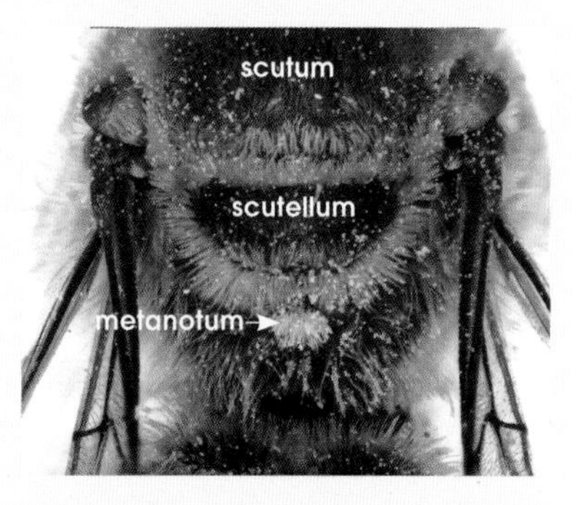

The tegula of *Melissodes*, if you can see it, is teardrop-shaped, tapering to a point at the end near the head. The pointy tip is often hidden beneath hair, however, and you might need to gently scrape that away with a pin to see it.

Svastra are also out by midsummer, flying through early fall. Like *Melissodes*, they frequently visit sunflowers, and the two can therefore be difficult to distinguish. Typically, they are larger, though without two specimens to compare, size can be hard to assess. The antennae of males are shorter than for most eucerines (though still long compared with other bee tribes).

With *Svastra*, the metanotum (the thorax segment located just before the thorax folds over toward the abdomen) often has a long tuft of hair on it. Though not clear in this image, the tegula is perfectly oval and not teardrop-shaped, as it is in *Melissodes*.

Similar Species to the Eucerini

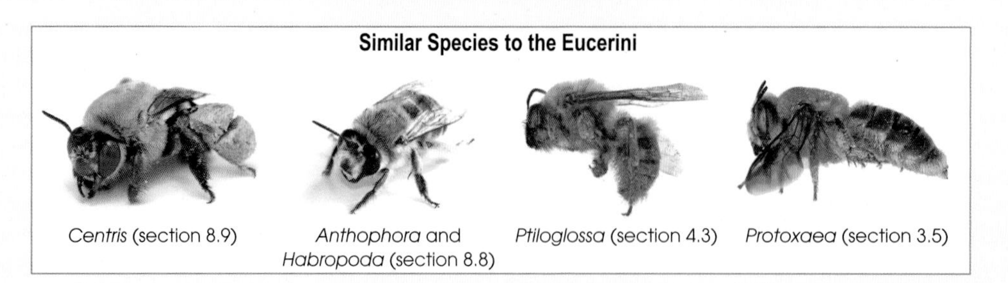

Centris (section 8.9) — *Anthophora* and *Habropoda* (section 8.8) — *Ptiloglossa* (section 4.3) — *Protoxaea* (section 3.5)

ANTHOPHORA AND *HABROPODA* (SECTION 8.8)

Anthophora *Anthophora* *Habropoda*

Anthophora and *Habropoda* are two ubiquitous bee genera seen flying most of the flowering season. The head is wide, as wide as the thorax (compare with Emphorini). These bees can look superficially similar to Eucerini, but wing characters can aid in distinction. The two genera can be difficult to distinguish from each other. In general the abdomen and the first tergal segment of *Habropoda* are the same color, which may or may not be the case among *Anthophora*.

Identifying Features of *Anthophora* and *Habropoda*

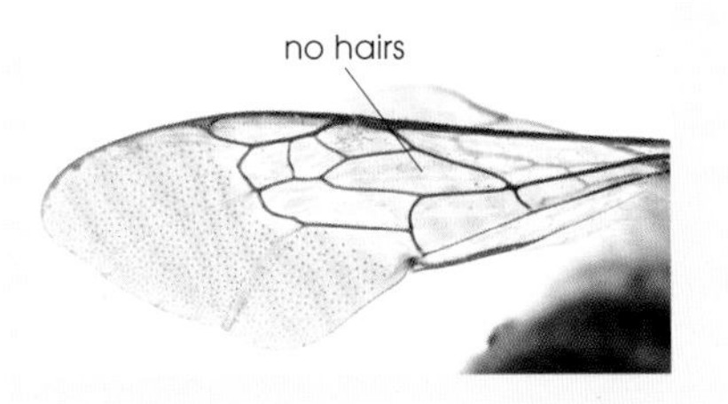

RIGHT: Anthophorini have an arolium, which is absent from *Centris*, a bee they may superficially resemble.

ABOVE: The cells making up the inner portions of the forewing are mostly hairless—there may be small bumps, but the numerous hairs seen on the wings of most other Apidae genera (especially Eucerini) are distinctly missing. This is easy to see but may require looking across the plane of the wing, rather than straight down on it. Note that the marginal cell is not sharply angled away from the wing margin, distinguishing these bees from Exomalopsini bees.

Similar Species to *Anthophora* and *Habropoda*

Eucerini (sections 8.5–8.7, especially 8.5) *Bombus* (section 8.10)

CENTRIS (SECTION 8.9)

Centris are among the larger Apidae. They are often seen hovering conspicuously in front of flowers, perhaps shifting their hind legs back and forth and rubbing their forelegs together. These bees often collect floral oils, and the leg-shuffling helps them readjust their harvest for transport back to the nest. Males have large eyes, and both males and females may also have colorful eyes (red or green); with death the color fades, however. The scopal hairs on *Centris* are huge, like overfilled saddle bags, and can give the bees an unmistakable appearance.

continued overleaf

Identifying Features of *Centris*

LEFT: Unlike some of the other Apidae that might otherwise look similar to *Centris*, these bees have no arolium between their front claws.
MIDDLE: The scopa of *Centris* females is extraordinarily large and dense.
RIGHT: The stigma of *Centris* is smaller than that of either *Anthophora* or *Eucera*. Compared with *Anthophora*, the base of the wing is slightly hairier, though not as hairy as *Eucera*. *Centris* may also look similar to some large bees in other families. Note that there is no "hook" in the marginal cell, as is seen in *Caupolicana* and *Ptiloglossa* (section 4.3), and that the first recurrent vein does not meet the vein between the first and second submarginal cells, as it does in the oxaeids (section 3.5).

Similar Species to *Centris*

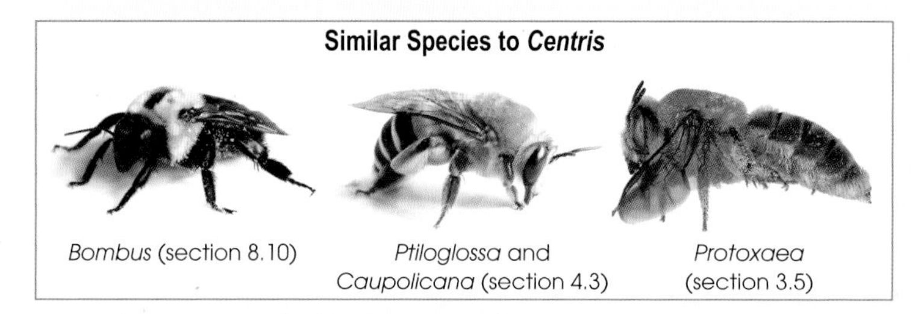

Bombus (section 8.10) | *Ptiloglossa* and *Caupolicana* (section 4.3) | *Protoxaea* (section 3.5)

BOMBUS (SECTION 8.10)

Bombus are one of the most recognizable of North America's bees, with bright yellow hair on the head, thorax, and abdomen. The abdomen can also sport stripes of red, orange, black, or white. While they are generally hard to confuse with anything else, there are still telling features unique to this group.

Identifying Features of *Bombus*

The abdominal hair patterns of *Bombus* can often be used to tell species apart. Several online guides and print books provide keys.

LEFT: The face of a bumble bee often appears very long, owing in part to the extended malar space, the area between the eye and the top of the mandible. This area is generally not as long in other North American bees.
RIGHT: *Bombus* are distinct in that they have no jugal lobe, the last segment of the hind wing. In a pinch, when other characters fail, this can be a useful feature for identification.

Identifying Features of *Bombus* continued

LEFT: *Bombus* females (the ones you are most likely to see) collect pollen not in scopal hairs, but rather in a corbicula—a flattened area on the hind tibia where scopal hairs are usually found. This area is often shiny and easy to see from a distance, and it has stout hairs that wrap around the sides. If pollen is in the corbicula, the pollen usually looks more like play dough than like dry pollen, because the bees mix the pollen with nectar to make it stick to the corbicula.

Similar Species to *Bombus*

Xylocopa (section 8.1) Anthophorini (section 8.8) Eucerini females (sections 8.5–8.6)

Mesoxaea (section 3.5)

Ptiloglossa (section 4.3)

APIS (SECTION 8.11)

Though one of the best-known bees, honey bees (*Apis*) may be difficult to identify on the wing. As they are medium-sized and fairly slow flying, watch for the hind legs, which often hang behind them as they fly. They are only moderately hairy, but the integument (skin) can be colored light orange to almost black. While they may look superficially like many other medium-sized bees from a distance, under a microscope they are very distinctive.

Identifying Features of *Apis*

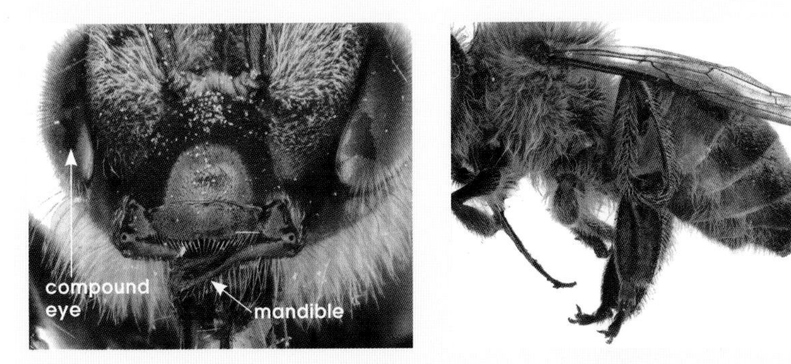

LEFT: Unlike any of the bees that might otherwise appear similar to *Apis*, these bees have a corbicula for carrying pollen back to the nest, instead of stiff scopal hairs. The corbicula is the hind tibia and femur, broad and slightly concave, with a banner of hairs curling in over the top. *Apis* mix pollen and nectar to create a "dough" that they press into the corbicula for transportation purposes.

ABOVE LEFT: *Apis* are among the few bees with hair on the compound eyes. The mandibles are very distinctive as well, with no teeth and a twisted-in appearance. Each mandible looks like a broad spatula rotated 90 degrees from that of other bee mandibles.

continued overleaf

Identifying Features of *Apis* continued

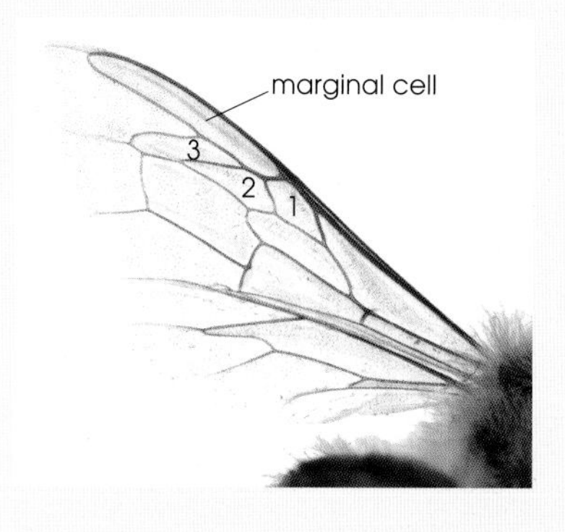

The wings of *Apis* look nothing like the wings of other bees in the United States and Canada. The marginal cell is ridiculously long, extending almost to the end of the wing. The stigma is very small, and the three submarginal cells are uncommonly shaped.

Similar Species to *Apis*

Andrena (section 3.1) — *Colletes* (section 4.1)

EUGLOSSA (SECTION 8.12)

Euglossa are the most outstanding of the Apidae in the United States and Canada. The brilliant green color adorning their large bodies and the substantial, inflated-looking hind legs are a dead giveaway as to this bee's identity. Beautiful as they are, *Euglossa* are found only in Florida.

Identifying Features of *Euglossa*

Euglossa have two exceptional features. One is the length of the tongue, which is easily as long as the body. The other is the highly modified hind legs, which are used to store floral oils.

Similar Species to *Euglossa*

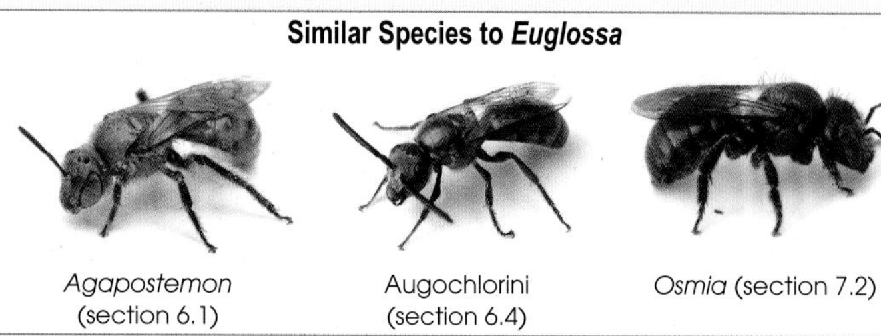

Agapostemon (section 6.1) — Augochlorini (section 6.4) — *Osmia* (section 7.2)

8.1 *XYLOCOPA*

Xylocopa is Greek for "wood-worker," referring to this bee's ability to make a home in a piece of wood.

Description

Xylocopa are massive, dark-colored bees. They often nest in wood, where they chew large holes for their nests (thus the common name carpenter bees).

Distribution

Around the world more than 500 *Xylocopa* species are known, most of them in tropical and subtropical areas. Thus, though around 50 species occur in Brazil, only 10 species are found in the United States and Canada. None are found across the entire United States; a few are restricted to areas east of the Great Plains, three species are common in the west, and the rest are vagrants in the southwestern United States, barely crossing the Mexican border at the northern edge of their range.

Diet and Pollination Services

Xylocopa are generalists and will take nectar and pollen from almost any bloom (in excess of 50 different flowering plants in some areas). Even though they are generalists, carpenter bees are often consistent in the flowers they visit, meaning that they will visit the same types of flowers in succession.

Because of their large body sizes, *Xylcopoa* can carry enormous bushels of pollen. Their constancy and body size make them good pollinators of many flowers. In the deserts, *Xylocopa* are important pollinators of many typically bat-pollinated flowers; when bats are few, these giant bees can pick up the slack.

Xylocopa are also important pollinators of many crops valued by humans. In Brazil, they are pollinators of passion fruit (*Passiflora*) and Brazil nuts (*Bertholettia*). In Australia, *Xylocopa* are reared in greenhouses for tomato pollination—tomatoes pollinated by carpenter bees have heavier fruits than those pollinated by other insects. In the Middle East, *Xylocopa* are important pollinators of cotton.

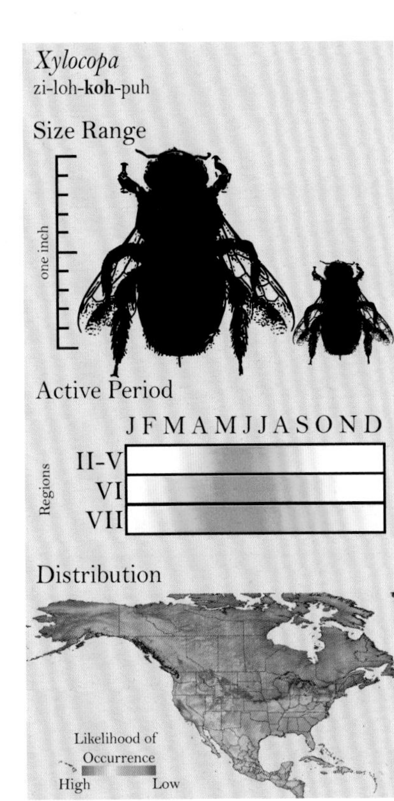

A *Xylocopa* foraging on a century plant (*Agave*). Photos by B. Seth Topham.

Flower constancy vs. flower specialty

Some bees, like *Xylocopa*, will visit the same type of flower over and over again as they prepare their nests. Such behavior, when a bee visits a specific type of flower throughout the nest preparation, is called flower constancy. The type of flower a bee visits is often based on what flowers are abundant in a given area, or what flowers are particularly rewarding. It also depends on the bee—scientists hypothesize that it is much easier for a bee to memorize the details of just one flower type at a time, and that a bee may become temporarily constant on that type in order to save time. Flower specialty, on the other hand, is a hard-wired tendency of a bee to visit only a specific type of flower, regardless of its abundance. Floral specialists will never switch from their favored floral type.

A male *Xylocopa* visiting a composite flower (Asteraceae). While most of the female *Xylocopa* in North America are shiny and black, the males of several species have yellow hair on their head and thorax. Note the compound eyes that almost touch on top (typical among male *Xylocopa*).

An eastern carpenter bee (*Xylocopa virginica*) visiting a plant in the sunflower family (Asteraceae). The eastern carpenter bee differs from many of the other North American *Xylocopa* species in that the female has blond hair on its thorax (back).

A *Xylocopa* caught in the act of nectar robbing (p.54). This bee is biting a hole in the base of a flower so it can drink the nectar without having to squeeze into the narrow confines of the flower tube.

Even though *Xylocopa* can be excellent pollinators of some flowers, they are a hindrance to the pollination of others. Their large size prevents them from getting into many tubular flowers, including beardtongues (*Penstemon*), ocotillo (*Fouqueria*), and *Salvia*; however, their mandibles and mouthparts are specially modified so they can rob nectar from these tubular flowers. They cut a hole in the base of the flower petals and then lap out the nectar without ever entering the flower or coming in contact with any pollen (see p.54). This can have a significant negative effect on flowers like blueberries, where scientists have found that *Xylocopa* steal enough nectar to make the flowers unattractive to other bees that could actually pollinate them.

Other Biological Notes

Around the world *Xylocopa* nest in plant material: dead or dying wood or plant stems (reeds, bamboo, etc.). The resulting nest is often called a "gallery." There is one exception—a group that lives in Europe, the Middle East, and Asia nests in the ground.

Male *Xylocopa* often establish territories near the nest entrances of females building close together. They will hover just a few feet from nest entrances and attempt to keep at bay any object that gets too close, including humans and their pets, hummingbirds, other insects, and (most importantly) other male carpenter bees. Associated with mating in this species, you may see male and female *Xylocopa* rising straight up in the air until they are out of sight.

In the east all *Xylocopa* nest in solid wood, and not plant stems. *Xylocopa virginica* often nests in human-made structures, including telephone poles, porch and deck wood, and lawn furniture. Using their strong mandibles, *Xylocopa* chew a round hole into the wood about a body-length deep, and then make a 90-degree turn; often this turn is made so they can chew with the grain of the wood. They do not eat the wood they chew up when making their tunnels, but shove it out of the nest, or compact it to serve as walls (similar to particle board) between nest cells. *Xylocopa* in the deserts of the southwest often nest in the dry flower stalks of yucca plants, agave plants, sotol plants, palm fronds, and bear grass plants.

Xylocopa females lay one brood per year in most of the United States and Canada. In wamer southern climates they may occasionally lay two. Mother *Xylocopa* meet their offspring, and will even guard them in their nest and feed them as they grow. Their offspring are fully developed by fall and may emerge to forage for pollen and nectar for a short period. They return to the nest their mother built them to overwinter, together, in a huddled group. The following spring, females will depart and find new nesting sites to raise their own brood.

A *Xylocopa* nest in the stalk of an *Agave* plant.

ABOVE: A *Xylocopa* near a century plant (*Agave*). If you look closely near the top right of this photograph, you can also see a very tiny *Lasioglossum* (section 6.3). PHOTO BY B. SETH TOPHAM.

RIGHT: A *Xylocopa* visiting an indigo bush flower (*Psorthamnus*).

RIGHT: The male of the western species *Xylocopa varipunctata* is not black like many of its relatives; instead it is covered with orange-blond hair. PHOTO BY HARTMUT WISCH.

BELOW: The female of *Xylocopa varipunctata*, on the other hand, is completely black. PHOTO BY LON BREHMER AND ENRIQUETA FLORES-GUEVARA.

8.2 *CERATINA*

Ceratina is Greek for "horned." The scientist who first saw this bee in 1802 thought the short antennae on his male specimen looked like a pair of horns. He had originally named this bee Clavicera but thought better of it the following year, when he realized Clavicera combined Greek and Latin roots, which he found unacceptable.

Share and share alike

It is not uncommon to find five or more *Ceratina* species nesting in the same vicinity. They are all roughly the same size, nest in twigs, and share the same diet—how is it that one species doesn't outcompete the others? Scientists have found that, even though it seems they are sharing the same resources, each species has carved out its own niche, allowing them to coexist. For example, in the east, *Ceratina dupla* and *C. mikmaqi* both nest in the same kinds of twigs, but *C. dupla* nests early in the season and has two generations. In contrast, *Ceratina mikmaqi* nests later in the season and has just one generation. In other words, *Ceratina dupla* nests around the season of *C. mikmaqi,* and this may limit their need to compete for nesting sites or floral resources. Similarly, *Ceratina sequoiae* may specialize on *Clarkia* in order to limit competition for resources with other *Ceratina* with which it commonly co-occurs.

A *Ceratina* departing from a composite flower (Asteraceae).

Description

Ceratina are tiny to medium-sized, shiny and nearly hairless, black or light metallic (green or blue) bees found commonly throughout North America.

Distribution

There are more than 350 *Ceratina* species around the world and they are found on every continent except Antarctica. North of Mexico, 22 species are known. Some species are widespread, but none occur across the entire continent. One species (*Ceratina dallatorreana*) has been introduced to the United States from the Mediterranean, probably first appearing in California sometime around 1950. Two species have been introduced to Hawaii, *Ceratina arizonensis* (originally from the mainland) and *C. smaragdula* (originally from Asia).

A *Ceratina* feeding on a desert marigold (*Baileya*). Note the characteristic shiny body and white markings on the face.

A small *Ceratina* on a scorpionweed flower (*Phacelia*), telltale white face clearly visible. This *Ceratina* is only 0.1 inch long.

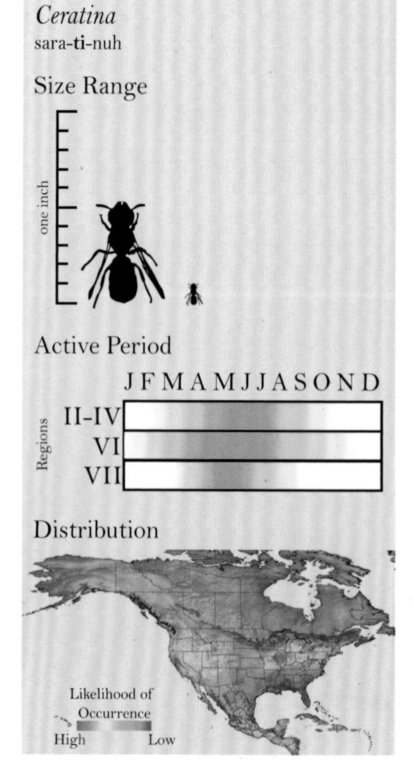

Diet and Pollination Services

Almost all *Ceratina* species are generalists, visiting many different plant species when gathering pollen, and just as many for gathering nectar. Just one species, found mostly in

A *Ceratina* assessing a wild rose flower (*Rosa*).

BELOW: A *Ceratina* feeding on a Russian sage flower (*Perovskia*) in a flower garden. Russian sage is a popular plant with many bee genera.

California (*Ceratina sequoiae*), is thought to specialize on farewell-to-spring flowers (*Clarkia*), though it visits other plants for nectar.

Other Biological Notes

All *Ceratina* nest in old wood, usually inside the stems of dead, burnt, or broken twigs. It is a requirement that the stem or twig be broken or the tip somehow damaged; these bees won't nest in the tips of intact branches, because they can't chew their way through the harder outer material. Species in the eastern United States are partial to roses (*Rosa*), raspberry (*Rubus*), sumac (*Rhus*), plumegrass (*Saccharum*), and teasel (*Dipsacus*). In the west, many *Ceratina* prefer sagebrush (*Artemesia*), buckwheat (*Eriogonum*), false willow (*Baccharis*), blackberries and raspberries (*Rubus*), and elderberry (*Sambucus*). *Ceratina*

A *Ceratina* resting on a golden aster (*Heterotheca*).

A *Ceratina* foraging on a desert marigold (*Baileya*) while another *Ceratina* hovers nearby.

Ceratina, like all bees in the family Apidae, have very long tongues.

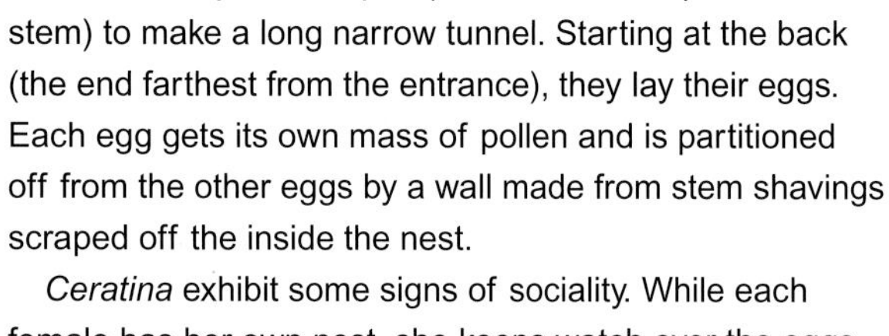

chew their way into the pith (the softer inside parts of a stem) to make a long narrow tunnel. Starting at the back (the end farthest from the entrance), they lay their eggs. Each egg gets its own mass of pollen and is partitioned off from the other eggs by a wall made from stem shavings scraped off the inside the nest.

Ceratina exhibit some signs of sociality. While each female has her own nest, she keeps watch over the eggs she has laid until they develop into adults. After laying eggs in a twig and sealing off each nest cell, she sits at the nest entrance (her hibernaculum), guarding her young until they are completely developed—sometimes for a whole year. She will even open up nest cells she has previously sealed in order to check in on the developing young.

Going it alone

Ceratina are unique among bees in that some species are parthenogenic, meaning they can reproduce without ever mating. While all bees are capable of producing male offspring without sex (see haplodiploidy in section 1.3), most bees can't produce female offspring without sex. *Ceratina dallatoreana* is an exception. Though both males and females are found in the Mediterranean region where they are native, almost all specimens collected in the United States are females, amounting to several thousand individuals across a wide geographic area over the last 50 years. It seems unlikely that males have simply missed being netted. The bee *Ceratina acantha* is also parthenogenic, but only in some areas of its range in southern California. Where it occurs in other areas of the west, males and females are equally common.

A *Ceratina* on a flower in the sunflower family (Asteraceae).

Many *Ceratina* species have distinctive white markings on their faces.

8.3 EXOMALOPSINI

EXOMALOPSIS AND *ANTHOPHORULA*

There are four genera in the tribe Exomalopsini. Two (*Tetragnatha* and *Eremapis*) are rare bees found only in South America, each represented by one species. And two (*Exomalopsis* and *Anthophorula*) are relatively common throughout South American and into the United States.

Anthophorula

Anthophorula is Greek for "little Anthophora."

Description

Anthophorula are tiny furry bees with oversized pollen-collecting hairs (scopa) on their miniscule back legs; they look as though they are wearing cargo pants with full pockets.

Distribution

All *Anthophorula* are found in the Western Hemisphere: North, Central, and South America, as well as the Bahamas and the West Indies, Jamaica, and Puerto Rico. Most of the 63 species are found north of the Mexican border; 40 species occur in the United States, but none make it as far north as Canada. Concentrated in the southwest, they range east as far as New York.

Diet and Pollination Services

Most *Anthophorula* are generalists and will visit the entire bouquet of flowers that may be in bloom at a given time when gathering pollen. A few species are specialists, though. One species appears to specialize on plants in the mallow family, one species has been collected only on *Tidestromia* (sometimes known as honeysweet), another likely specializes on gumweed (*Grindelia*), and one more seems to prefer flowers of leafy spurge (Euphorbiaceae). It also appears that some species have a preference for plants in the sunflower family (Asteraceae) and buckwheat family (Polygonaceae).

Scientists have noticed that *Anthophorula* (and also *Exomalopsis*) may be good pollinators of flowers in the plant family Solanaceae. This family includes important crops like potatoes, chilies, tomatoes, and eggplants. Exomalopsini bees might be particularly good at pollinating these crops because of their ability to buzz pollinate (p.21).

An *Anthophorula* feeding on a flower in the sunflower family (Asteraceae). Photo by Jillian H. Cowles.

An *Anthophorula* drinking nectar from a composite flower (Asteraceae). Photo by Jillian H. Cowles.

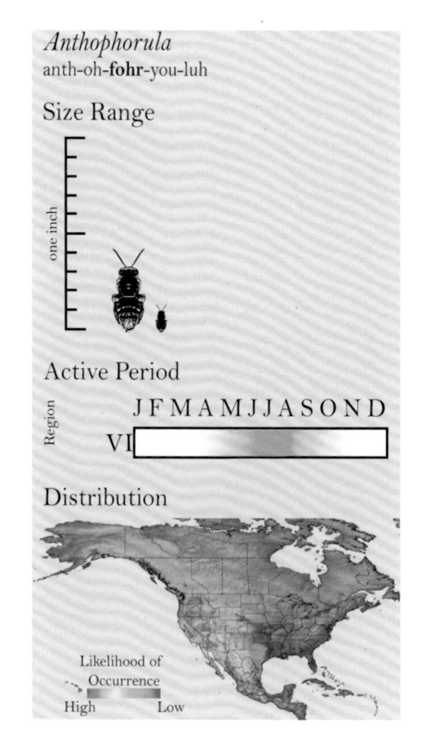

Group nesting

One scientist found a nest more than 15 feet deep, with more than 800 female *Anthophorula* nesting together. It is likely that in cases like this the nest is used for many years, with each successive generation adding on and modifying the nest left by the ones before.

An *Anthophorula* visiting buckwheat flowers (*Eriogonum*). Photo by Hartmut Wisch.

Other Biological Notes

Anthophorula nest in the ground. Though they are solitary, they do frequently nest communally, with several females sharing the same nest entrance. In a few species, hundreds of females may share one nest entrance. It is possible that in some of these communal nests the females work together to provision nest cells—if so, they could be considered quasisocial (a rare social condition in bees, see p.20). In the nest, cells are lined with a waxy material. Scientists have tried to melt the wax, and have found it resistant to temperatures in excess of 700 °F. Interestingly, *Anthophorula* mothers place their acquired pollen masses on a "foot," a little pedestal that keeps the pollen off the ground. It is thought that this keeps the pollen from growing mold, since it stays drier this way.

Exomalopsis

Exomalopsis means "without a bad appearance" (i.e., good looking).

Like *Anthophorula, Exomalopsis* occur only in the Americas. The 88 species are mostly found south of the Mexican border, but 10 species occur to the north. These little bees do not occur widely outside the southwestern deserts and are not found in Canada. About half the *Exomalopsis* in the United States are rare, just crossing into the United States from Mexico. *Exomalopsis* includes both generalists and specialists. The pollen preferences of *Exomalopsis* are similar to *Anthophorula* (see above for more details). They also nest in a similar manner to each other, though *Exomalopsis* do not place their pollen masses on a "foot."

An *Anthophorula* visiting a flower in the family Euphorbiaceae. These bees are very tiny, but the red abdomen is visible from a distance.

An *Exomalopsis* visiting a composite flower (Asteraceae).

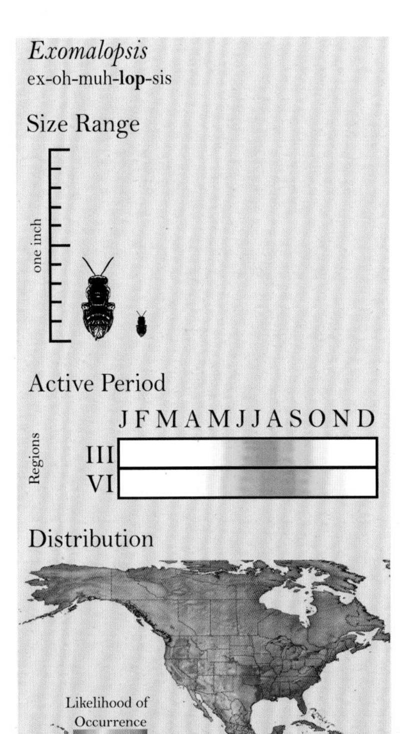

An *Exomalopsis* resting in an open flower. Note the shiny back and large pollen-collecting hairs (scopa) on the back legs. Photo by Jillian H. Cowles.

An *Exomalopsis* preparing to forage from a small flower. Photo by Jillian H. Cowles.

An *Exomalopsis* drinking nectar from a beardtongue (*Penstemon*) with a full load of pollen on its legs. Photo by Jillian H. Cowles.

8.4 EMPHORINI

DIADASIA, PTILOGLOSSA, MELITOMA, AND ANCYLOSCELIS

The Emphorini bees are all limited to the Western Hemisphere. Several additional genera occur in South America, where they mirror the specializations found in North American species. *Diadasia* is the most common of the Emphorini bees found north of Mexico, with *Ptilothrix*, *Ancyloscelis*, and *Melitoma* being less frequently seen.

Diadasia

The name Diadasia means "completely thick and hairy."

DESCRIPTION

Diadasia are medium to large, furry blond to tawny brown. The pollen-collecting hairs on their hind legs are often big and widely spaced, likely as an adaptation to the giant-sized pollen grains typical of the plants on which they specialize.

DISTRIBUTION

Bees in the genus *Diadasia* are found exclusively west of the Mississippi River, usually in desert or grassland habitats. Roughly 45 species are known, all found in the Western Hemisphere, with 25 of them occurring in the United States and southern Canada. They are most common in the southwest deserts, where cactus and globe mallow (*Sphaeralcea*), the most common of their host plants, abound. The one species that specializes on sunflowers is common across the Midwest. A few are restricted to California.

A nest of a *Diadasia* with a curved turret. *Diadasia* prefer to nest in very hard-packed dirt like that found on dirt roads or well-hiked trails. The turrets are fairly unique among bees.

Amphitropical distributions

Many bee groups like *Diadasia* live in the arid parts of North and South America but are absent from the tropics of Central America and northern South America. This geographic distribution is referred to as amphitropical (*amphi-* means "both sides of" in Greek) and is strangely common to many groups of bees. While not always the case, it seems that changing climate patterns over the last 15 million years provided avenues for northward expansion of many originally South American natives. As the Northern Hemisphere warmed and many plants found new habitat farther north, suites of bees followed. Later, the climate changed again and the tropical rainforests expanded across the equator, effectively cutting off North and South American populations.

DIET AND POLLINATION SERVICES

Some *Diadasia* species collect pollen only from cactus species (Cactaceae). Other *Diadasia* species collect pollen only from mallow flowers (Malvaceae). Among the mallow specialists, one species visits only checkerblooms (*Sidalcea*), another visits only bush mallows (*Malacothamnus*), yet another visits just wild hollyhocks (*Iliamna*), and many will visit only globe mallow (*Sphaeralcea*). Three *Diadasia* species specialize on plants other than mallows and cactus. *Diadasia enavata* will collect pollen only from sunflowers (*Helianthus*). A second species

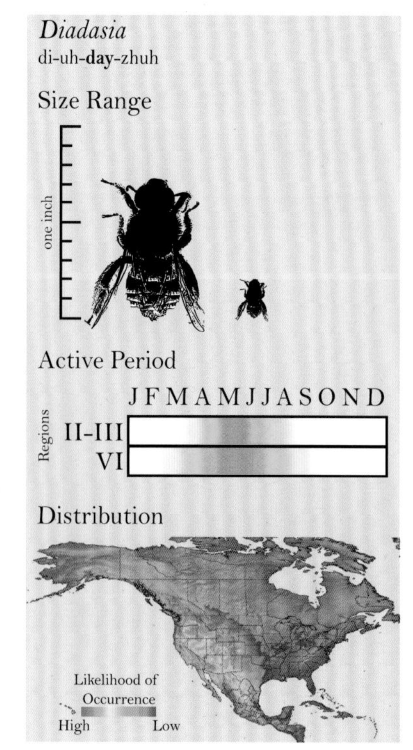

Some *Diadasia* nesting aggregations can have from several dozen to several thousand nests.

Diadasia often nest in large groups or aggregations. Here a female *Diadasia* peeks out of the turret surrounding her nest entrance. You can see several other nests with turrets of varying size nearby.

A male *Diadasia* taking off from a daisy (Asteraceae). All *Diadasia* are specialists with regard to the pollen they gather, but they will sip nectar from any number of flowering plants.

A male *Diadasia* exits a cactus flower (*Opuntia*) after spending the night inside.

A male *Diadasia* perching on the stigma of a cactus flower (*Opuntia*), preparing to take off.

will gather pollen only from farewell-to-spring flowers (*Clarkia*). Finally, one *Diadasia* species will collect pollen only from morning glory flowers (*Calystegia*, also known by the common name bindweed).

A *Diadasia* resting inside a bush mallow flower (*Malacothamnus*). Many *Diadasia* species specialize on flowers in the family Malvaceae.

Strange decorations

One species, *Diadasia afflicta*, gathers pollen only from poppy mallows (*Callirhoe*). The pollen and resulting feces are pure white. Interestingly, when females digest this pollen, which they eat as adults, they use it to line the very rim of their chimney-like nest entrance, creating a ring of white bumps around the edge of the nest entrance. Why this bee utilizes its fecal matter in this way is entirely unknown.

Other Biological Notes
Diadasia nest in the ground. A female *Diadasia* begins building her nests by first rubbing her head on the hard ground, then chewing at the soil until it is loose, and finally softening the loose soil with saliva. Each female digs a tunnel straight down and fashions 4 to 12 nest cells near the bottom of this main shaft, which is usually about 12 inches deep. In each cell she lays an egg and prepares a ball of pollen and nectar. Once the cell is completed, she seals it off with a mud cap and moves on to the next cell. When she has gathered enough pollen and nectar to fill all the nest cells she's made (it may take a week), she seals off the whole nest and digs another one.

It is not uncommon to find their unique nest entrances in the middle of dirt roads and trails. Because people don't expect bees to be nesting on trails and roads, nest entrances are often destroyed without anyone ever noticing them. While each bee constructs its own nest, many species nest in aggregations (section 1.5), ranging from a few closely spaced nests to thousands of nests in one area.

Interestingly, *Diadasia* build little walls (turrets) around the entrance of their nests using soil they remove as they dig the nest. Different species build differently shaped turrets; some are straight, while others have a bend in the end, like a periscope. No one knows why these bees build turrets over their nests, but scientists have come up with several possible reasons. Some researchers have suggested that perhaps the turret keeps rain or loose soil from falling into the nest. Others think it might keep enemies out of the nest; for example, velvet ants (Mutillidae) often attack bee nests, and the turrets could limit the number of nests a velvet ant can get into. Some fly species (e.g., bee flies) also attack

A *Diadasia* foraging on a globe mallow flower (*Sphaeralcea*).

A female *Diadasia* simultaneously drinks nectar and gathers pollen from her floral host, globe mallow (*Sphaeralcea*).

A *Diadasia* constructing a turret over her nest entrance. Turrets may be built to make it less likely that parasites will attack her offspring.

This large male Diadasia was patrolling cactus flowers, waiting for a female with which to mate.

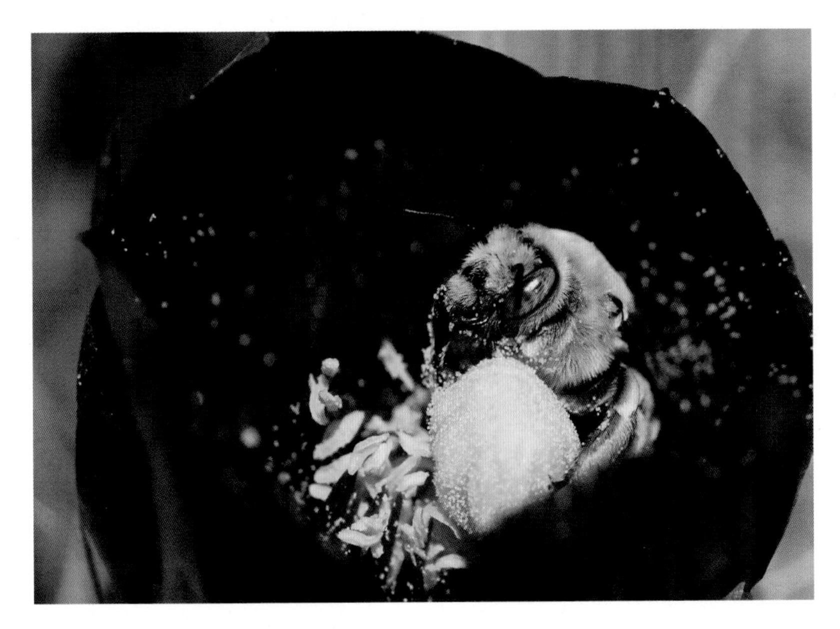

ABOVE: A male *Diadasia* resting on a fleabane flower (*Erigeron*).

LEFT: A female Diadasia resting inside a cactus flower (*Cylindropuntia*), the sole source of pollen for some species.

bee nests. A female fly will hover over a nest entrance and "shoot" an egg down the tunnel using her long egg-laying tool (ovipositor). Turrets may make it harder for flies to get their eggs all the way down the tunnels. Other scientists hypothesize that turrets aid in nest recognition in large aggregations, and some think they might help the bees warm up in the morning before their first flight—bees will wait in the chimney as it heats in the sun for 15 or 20 minutes before flying; bees of the same species whose nests don't have chimneys take longer to warm up before their first flight.

Ptilothrix

Ptilothrix means "feathered hair." To quote Frederick Smith, who originally described this genus while examining a specimen of Ptilothrix plumata: "This beautiful species is remarkable for having the entire pubescence plumose, each individual hair being pectinate; that on the metathorax, viewed under a pocket lens of good power, resembles fine down; even the short pubescence which forms the fascia on the abdomen is equally beautiful."

Ptilothrix are large bees with dark abdomens and light brown, fuzzy thoraxes. There are two *Ptilothrix* species in the United States. *Ptilothrix bombiformes* occurs throughout the eastern United States, mostly east of the Mississippi River. It is most abundant in the southern states, becoming rarer to the north. Rarely, in Arizona and New Mexico, *Ptilothrix sumichrasti* can be seen. *Ptilothrix* will collect pollen only from hibiscus flowers (*Hibiscus*) or morning glories (*Ipomoea*). Interestingly, *Ptilothrix sumichrasti* collects pollen only from *Ipomoea* flowers while this plant is in bloom, but if it stops flowering while the bees are still active and foraging for pollen for their offspring, they will begin collecting pollen from other flowers.

Ptilothrix species have a special adaptation for nesting in hard-packed soil. These bees are able to land on water, much like a water strider. With feet spread wide apart, a female sucks up water, flies it to where she's building a nest, and drops the water to soften the soil into mud. As she digs a tunnel, she forms the mud into round pellets, then rolls the pellets away from the developing nest. As the nest grows, she will bring the water further and further into the tunnel to soften the dirt, continuing to roll the mud pellets up and out of the nest. You can recognize a *Ptilothrix* nest by the little round pellets of mud all around the nest entrance.

Floaters and owners

Many bees that nest in groups (aggregations) can be divided into two types: the "floaters" and the "owners." Floaters are female bees that don't dig their own nest but stake out a nest that another bee has built. They'll wait in the nest for a while, to see if the owner is planning to return. If she doesn't, the floater will take the nest as her own. If the owner does return, she'll fly on and try another nest. This may be beneficial to the owner, since while the floater is waiting in the nest, no other predatory insects will likely try to enter.

A *Ptilothrix* foraging on a hibiscus flower (*Hibiscus*). Photo by Robert Klips.

A *Ptilothrix* nest. The small balls of mud surrounding the turret are distinctive among bee nests. As they excavate their nests, *Ptilothrix* will roll up the dirt and mud that they carry out. Photo by Thom Wilson.

Ptilothrix have an interesting habit of landing on the surface of the water (like a water strider) so they can collect water to bring back to their nest-site. They use the water to soften the ground as they dig their nest. Photo by Thom Wilson.

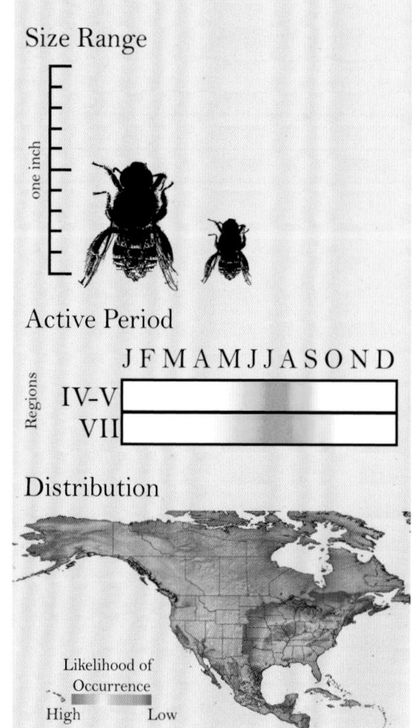

TOP: A *Ptilothrix* coming out of her nest. Note the small turret surrounding the nest entrance. Photo by Robert Klips.

ABOVE: A *Ptilothrix* entering her nest with a full load of hibiscus pollen. *Ptilothrix* are specialists on hibiscus flowers (*Hibiscus*). Photo by Robert Klips.

Melitoma

Melitoma means "honey parts." This bee was described from a specimen collected in Brazil in the 1820s. Little was known of its life habits, but it was assumed that the long tongue was used to collect honey and nectar.

These bees are medium-sized and fairly dark, with white stripes running across each segment of the abdomen. Ten species of *Melitoma* occur in the world, but just two of them occur in the United States; neither makes it as far north as Canada, nor west of the Rocky Mountains. *Melitoma* visit morning glories (*Ipomoea*) almost exclusively, and rely on the flowers of these plants for pollen. Nests are usually found within the vicinity of *Ipomoea* flowers, and near water they use (in a similar manner to *Ptilothrix*) to soften the clay soils in which they nest. Before human activity was as prevalent as it is now in the New World, the majority of these bee nests were likely made in river cuts through clay soils. Humans have greatly increased the number of potential nesting sites with road cuts and trail cuts. In addition, the widespread use of adobe bricks in the southwest has provided ample nesting habitat for these little bees; the bricks appear to be the perfect texture for *Melitoma* nests.

A *Melitoma* foraging in her host flower, a wild morning glory (*Ipomoea*). PHOTO BY SONNY MENCHER.

Ancyloscelis

Ancyloscelis means "crooked leg" or "curved leg" in Greek.

Ancyloscelis, with only three species in North America, can be found in the southwestern deserts of the United States, and south through South America. *Ancyloscelis* is highly specialized, visiting only morning glories (*Ipomoea*) for pollen. These little bees nest in the ground, or in vertical embankments. In fact, *Ancyloscelis* have been repeatedly observed nesting in the adobe walls common in the American southwest. Often consecutive generations will use the same wall, riddling it with holes and interlacing mazes of tunnels. They chew out a tunnel (or use old tunnels made by other insects) of the appropriate diameter, spackling a very thin layer of mud on the walls as they go. Tunnels are not straight but wind around obstacles in the adobe, and even through old burrows from previous years. Once nest cells have been provisioned, they are capped, but the tunnel leading to them is not filled back in.

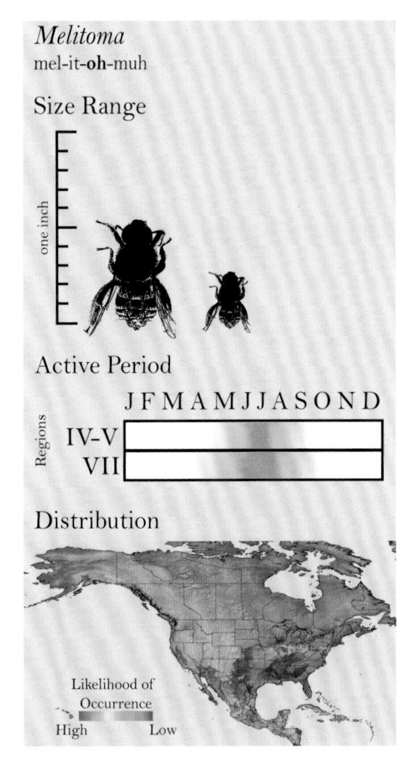

An *Ancyloscelis* visiting a morning glory flower (*Ipomoea*). *Ancyloscelis* collect pollen only from these flowers. PHOTO BY JILLIAN H. COWLES.

8.5 COMMON EUCERINI

EUCERA AND *MELISSODES*

The tribe Eucerini consists of 14 genera in the United States and Canada. The males have inordinately long antennae, thus the common name "long-horned bees." Long-horned bees can be very difficult to tell apart, and they have been grouped and regrouped by scientists many times over. We have included the two most common genera in this section, the squash bees (*Peponapis* and *Xenoglossa*) in the next section (8.6), and the other 10 rarer genera in a final section (8.7).

Eucera

The name Eucera is Greek for "well-horned," a reference to the long antennae of the males.

Description

Abundant throughout the United States and Canada, these fast-flying bees are hairy and generally large. Males have extraordinarily long antenna.

Distribution

More than 300 species of *Eucera* are found in Europe, Central Asia, and south through India. Only one subgenus (*Synhalonia*) occurs in the United States (55 species) and Canada (7 species). Ranging from British Colombia east to Massachusetts and south through Mexico, these bees are common, especially in the west (about 12 species occur in the east).

Diet and Pollination Services

Eucera includes both specialist and generalist bees. Specialists often limit themselves to flowers in the pea family (Fabaceae), including milkvetch (*Astragalus*), prairie clover (*Dalea*), and vetches (*Vicia*). *Eucera* can also contribute to the pollination of alfalfa (*Medicago*). Interestingly, they are seldom seen on flowers in the sunflower family (Asteraceae), though these are popular with many other bees in the Eucerini tribe.

In the Mediterranean, *Eucera* frequently visit wild orchids (*Ophrys* and *Orchis*). In these cases it is the males that are the best pollinators. Some orchid species produce the sex pheromones of female *Eucera,* deceptively luring males. When males attempt to mate with the flower, they become covered in pollen, which is then transferred to the next sexually scented orchid flower the male visits.

Other Biological Notes

Of the Eucerini, *Eucera* are among the first to emerge in the spring and are rare by August. Like other Eucerini, they nest in the ground and are solitary, though a few species

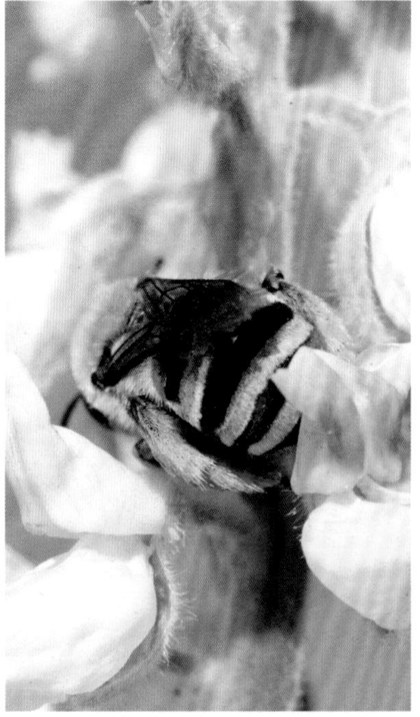

A female *Eucera* foraging on a lupine flower (*Lupinus*); note the continual bands of hair across the abdomen.

A male *Eucera* resting on the ground.

A male *Eucera* resting on a fleabane flower (*Erigeron*).

A male *Eucera* inserting its tongue into the tiny flowers that make up a composite flower (Asteraceae).

A male *Eucera* in the early spring, sipping nectar from a scorpionweed flower (*Phacelia*).

A female *Eucera* on a flower in the sunflower family (Asteraceae).

A male *Eucera* on a thistle (*Cirsium*).

Beauty is in the eye of the bee

Locating a female appears to be possible for males via a combination of visual and scent cues, with the relative importance of each differing between species. For example, male *Augochlora* completely ignored dead female specimens presented to them by scientists. They were then exposed to the scent from a live female on a piece of filter paper, and this was also ignored. But when the scientists wiped the live scent onto the dead specimen, the males attempted to mate with it. In contrast, males of *Andrena flavipes* appear to be attracted to the bright red hairs on the legs of females. When these are shaved off the females, males do not attempt copulation.

A male *Eucera*, head buried in a composite flower (Asteraceae).

appear to nest communally. Even in those cases, each female provides for her own young. *Eucera* prefer sand or clay soils, and species nest in flat areas or shallow embankments. Nest entrances have a mound of excavated soil heaped symmetrically around them.

When females emerge for the very first time from the nest in which they developed, they are often greeted by several males at once. Males hover in swarms over nesting sites in order to be the first to mate with virgin females. Several males will try simultaneously to copulate with a female; fighting is aggressive and often deadly. Once a female has been mated, however, her scent changes, and males leave her alone.

A male *Eucera* rests on a flower early in the morning. Many male long-horned bees, like *Eucera*, spend the night sleeping on flower stems or on twigs.

A female *Eucera* on a scorpionweed flower (*Phacelia*). Photo by Hartmut Wisch.

Melissodes

Melissodes means "bee-like." Melissa means "bee" and -odes means "looks like" or "resembles."

Description

Looking remarkably similar to *Eucera*, *Melissodes* are fall fliers, seen typically on sunflowers.

Distribution

Melissodes are found only in North and South America, concentrated in Mexico and areas north. There are at least 125 species altogether. Nearly 100 species are found in the United States and about 20 occur in Canada.

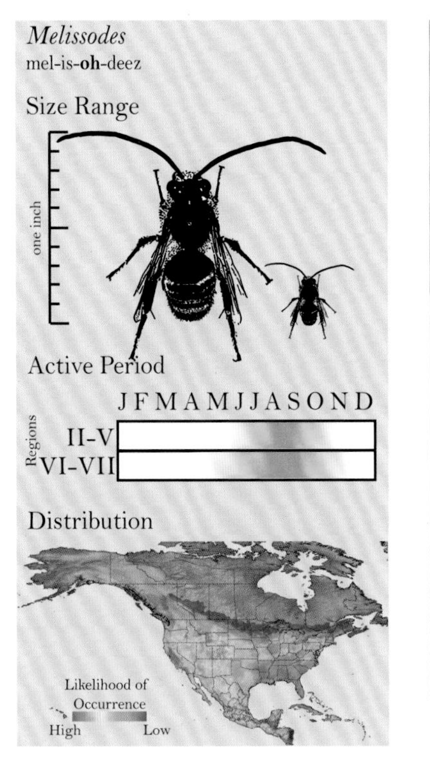

A female *Melissodes* on a composite flower.

Diet and Pollination Services

A handful of generalist *Melissodes* occur, but the majority are specialists. Many specialize on plants in the sunflower family (Asteraceae), including the entire subgenus *Eumelissodes* (comprising more than 60 species). Sunflowers may be popular hosts in part because *Melissodes* emerge so very late in the year, when flowers in the sunflower family are especially common and abundant but many other blooms are dwindling. While most of the sunflower specialists go to any plant genus in the family, a few will visit only specific genera. For instance, the two species of *Melissodes* in the subgenus *Heliomelissodes* visit only thistles (*Cirsium*) for pollen. There are also specialists on flowers in plant families other than Asteraceae. Popular hosts include poppy mallows (*Callirhoe*), farewell-to-spring flowers (*Clarkia*), prickly poppies (*Argemone*), and evening primrose (*Oenothera*). One species appears to prefer pollen from plants in the pea family (Fabaceae).

A male *Melissodes* on a gumweed flower (*Grindelia*).

A male *Melissodes* feeding on a gumweed flower (*Grindelia*).

A female *Melissodes* resting on a leaf.

Fashionably late

Melissodes druriella emerges in October, when most flowers are finished blooming and early morning frosts are common. It is thought that by emerging so late in the season, they may be able to avoid competition with other bees. Also, emerging so late may help them avoid parasitism by *Triepeolus* (section 9.1), a common parasitic bee associated with other *Melissodes* species, which does not fly that late in the season.

Other Biological Notes

Melissodes are found mostly in the late summer and fall, with some in the Midwest flying extremely late (i.e., October and November).

Melissodes are all solitary ground nesters, and there is no sharing of the duties associated with laying eggs or providing for them. While these bees are not well studied, scientists have observed several females using the same nest entrance. Therefore some species may be opportunistically communal; those same species also nest alone when other bees are scarce. Aggregations are not uncommon, with nests as close as just a few inches away from each other. Some species of *Melissodes* are particular about the soils in which they nest, seeming to prefer sand or sandy loam soils.

In nests of some *Melissodes* species the first part of the nest is filled with dirt—even while it is being used. The female has to dig through loose dirt every time she enters her nest tunnel. Using her front feet and mandibles, she digs a tunnel a few inches into the ground, piling the soil outside the nest. Then, she makes an abrupt 90-degree turn, which extends for a short distance before turning downward again for another 2 to 5 inches. The two 90-degree angles create a stopper; all soil excavated below the bends gets piled above them, at the tunnel entrance. In the hollowed-out tunnel below, the female builds offshoots, anywhere from half an inch to 5 inches long, with nest cells dug vertically at the end. Each nest cell is dug slightly bigger than it needs to be and is then backfilled with lightly tamped soil. A nest cell is lined with a thin layer of a waxy substance secreted by the Dufour's gland of the female bee. She fills the nest with pollen and nectar, and lays an egg on top. Finally, she seals the nest cell with soil and then fills in the side tunnel with tightly packed earth. From the few species that have been studied, it appears that a female can create and provision one nest cell per day. At the end of the day, after the nest cell has been taken care of, at least some species of *Melissodes* make one final trip and gather pollen and nectar for themselves. Even adults need nutrition.

Bee roost

While *Melissodes* females sleep in their nests at night, 10 to 20 males will huddle together on flower heads, or on plant stems. When they sleep on flower stems, they grab on tightly with their mandibles, use their legs to groom themselves completely, and then hang, perfectly still, from the stem until the next morning. One scientist marked males while they were sleeping and found that the same males returned to the same sleeping sites each night. Not all males sleep in this way, though; some species huddle together in groups of 6 to 12 in cracks or rocky crevices near nesting sites instead of on plants or stems.

A sleeping male *Melissodes*. Photo by B. Seth Topham.

A group of sleeping male *Melissodes*; note the two sizes, probably representing two different species. Photo by B. Seth Topham.

ABOVE: A male *Melissodes* feeds on a sunflower. This photo was taken in August, when *Melissodes* are much more common than *Eucera*.

RIGHT: A female *Melissodes* foraging on a sunflower (*Helianthus*).

BELOW: A female *Melissodes* on a gumweed flower (*Grindelia*); the hair bands running across the abdomen are not continuous, typical of many *Melissodes* species.

A male *Melissodes* resting on a leaf.

A male *Melissodes* in a desert marigold (*Baileya*).

8.6 *PEPONAPIS* AND *XENOGLOSSA*

Peponapis

Xenoglossa

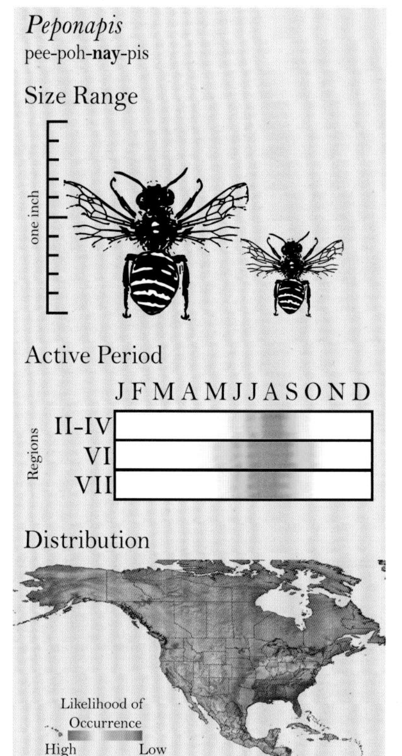

The squash bees, *Peponapis* and *Xenoglossa*, appear to be closely related, and some scientists think they should be included in one genus.

Peponapis

The name Peponapis means pumpkin bee; pepon is Greek for "pumpkin," and apis means "bee."

Description

Peponapis are medium to large fast-flying bees generally associated with squash patches. The thorax is coated in soft hair. The abdomen has white stripes of hair, and this is the part most commonly seen on the bees, as they are usually nose-down inside a squash flower.

How special are specialists?

It is questionable whether specialist bees are good pollinators of the plants from which they collect pollen. Because they are so efficient at hoarding the pollen grains they collect, they are not always helpful in transferring pollen from one flower to another. In the case of squash bees, however, it appears that they are in fact excellent pollinators of their host plants. Scientists have found them to be equivalent to, and sometimes better pollinators than, both bumble bees and honey bees. For one, they are bigger and hairier than honey bees, with more surface area to catch stray pollen grains. For another they fly fast, so they visit more flowers than honey bees in the same amount of time, and finally, they begin flying much earlier in the day, when most squash flowers are open. Though honey bees are often imported to agricultural squash and pumpkin fields to aid in pollination, it is possible that their services, when squash bees are present, are entirely unnecessary. Interestingly, it has been found that male squash bees, which are also big and fuzzy, fast flying, numerous, and *not* in the business of hoarding pollen, are also very good pollinators, and perhaps even better pollinators of squash plants than female squash bees.

Distribution

The 13 species of *Peponapis* range as far south as northern Argentina, though most occur in Mexico. In the United States and Canada, 6 species occur, and they can be found from southern California east to Georgia, north all the way to Maine, and into Ontario, Canada. Two of the U.S. species are found only in the deserts of southern California. The eastern cucurbit bee (*Peponapis pruinosa*) is the most widespread squash bee species and occurs coast to coast.

Diet and Pollination Services

As their name implies, squash bees collect pollen only from the flowers of squash plants, both wild varieties and cultivars. They usually fly very early in the morning when squash flowers open, often before full daylight, presumably using their sense of smell to find flowers in the near-dark. By midmorning, when squash flowers have wilted, females are busy in their burrows, and males are sleeping inside closed flowers. It is not uncommon to peel open a wilted squash flower and find several males nestled at its base.

Other Biological Notes

All squash bees are solitary, though many nest in large aggregations with several hundred bees in the same area. *Peponapis* and *Xenoglossa* nest in the ground, often very near their host plants. In fact, because they may nest very

A male *Peponapis* inside a squash flower. *Peponapis* visit flowers only in the squash family (Cucurbitaceae).

Multiple male *Peponapis* can often be found in the same flower; in fact, abdomen-up is the most common way that these squash bees are seen.

A female *Peponapis* in a pumpkin flower; notice the strongly protruding clypeus, typical of these bees.

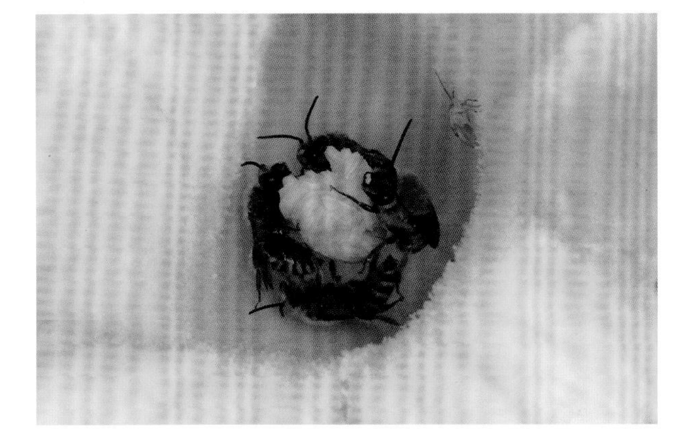

Several *Peponapis* resting in a pumpkin flower. *Peponapis* are most active early in the morning, often right at dawn. As the morning turns to afternoon, the female *Peponapis* will take shelter in their nests while the males often find a nice squash flower to stay in until the next morning.

Male *Peponapis*, like this one, often transfer more pollen between flowers than the females do. Here, you can see how pollen readily sticks to the thorax of the bee.

close to their host plants, they can be negatively impacted by tillage practices in farm fields—while tilling does not wipe out all bee nest cells, since many are quite deep, scientists have found that no-till farms can have triple the number of squash bees compared with farms that till.

Upon emerging from the nest her mother built, a new female spends two to three weeks gathering nectar before initiating a new nest. During that time the females, like males, sleep in flowers. Eventually they mate and begin nest construction. They dig a tunnel straight down a foot and a half or more into the soil. They'll make four or five cells in a nest in side-chutes off the main tunnel. After gathering pollen for those cells, they'll seal off the nest and start a new one. The larvae spend the winter as prepupae and metamorphose into adults in the early summer, just before squash begin to bloom.

A *Peponapis* inside a gourd flower. PHOTO BY JILLIAN H. COWLES.

Xenoglossa

The name Xenoglossa translates to "strange tongue."

Description

Like *Peponapis*, these bees are usually seen in the morning before squash flowers have wilted. The faces of males have strong yellow markings, and the bodies are a rust red, which distinguishes them from *Peponapis.*

Distribution

Xenoglossa is found only in the Western Hemisphere. It ranges as far south as Nicaragua, and north almost into Canada. In the United States it can be found from coast to coast. There are seven species all together, five of which can be found in the United States. Two are restricted to southern California, but the other three are fairly common and widespread.

Diet and Pollination Services

As with *Peponapis, Xenoglossa* is extremely specialized, gathering pollen only from squash plants (*Cucurbita*), including wild species as well as cultivars (i.e., zucchini, pumpkin, and butternut squashes). *Xenoglossa* have been observed foraging for pollen as early as 4:00 in the morning. By 2:00 in the afternoon they are often in their nests, finished with foraging until the next day. Once females retire to their nests for the day, males will curl up in wilted squash flowers for a long nap.

Other Biological Notes

The nesting habits of Peponapis and *Xenoglossa* are very similar—see *Peponapis* above for details.

An older *Xenoglossa* on a squash plant. *Xenoglossa,* while closely related to *Peponapis,* are generally bigger and have reddish abdomens.

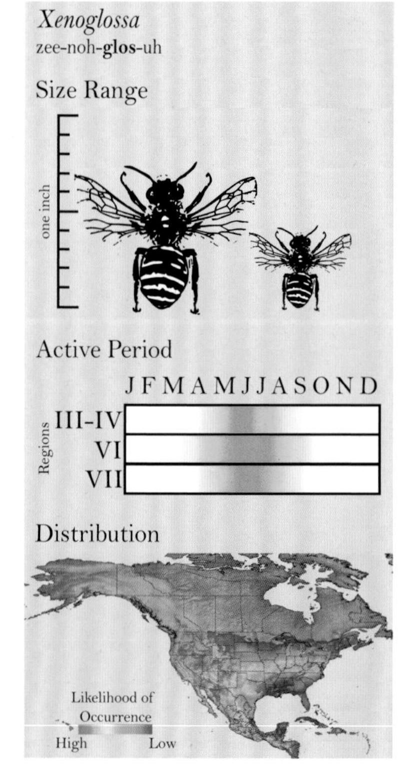

A *Xenoglossa* resting on the petal of a squash flower.

8.7 OTHER EUCERINI

Eucerini is the bee tribe in the Apidae with the most genera found in North America. Four genera were discussed in the previous two sections; the other 10 are described in the following section. *Svastra* and *Tetraloniella* are the most common of these Eucerini, while the others are relatively rare and seldom encountered.

Svastra

The name Svastra is Sanskrit for "sister." It may be that Svastra appeared to be "sister" to Melissoptila, a bee genus with which Holmberg (p.228) compared this bee.

Description

Generally large (even bumble-bee sized), *Svastra* are hulking bees with huge pollen-collecting hairs on the hind legs of females. The antennae of males are long, though not as long as those of other male Eucerini. These bees are commonly seen on sunflowers.

Distribution

Svastra are found only in the Western Hemisphere, from southern Canada south through Central America and amphitropically into South America (see amphitropical distributions, p.212). About 21 species are known in the world. In the United States 16 species occur, but only *Svastra obliqua* is found in Canada.

Diet and Pollination Services

There are both specialist and generalist *Svastra*. A few species specialize on sunflowers (*Helianthus*), but there are also specialists on flowers in the evening primrose family (Oragraceae). One species, *Svastra duplocincta*, specializes on cactus. *Svastra* are also important pollinators of sunflower crops.

Other Biological Notes

Most *Svastra* species are solitary, building and provisioning their own individual nests in the ground. A few species,

A *Svastra obliqua* foraging on a sunflower. This species is a specialist on sunflowers (*Helianthus*); they can be important pollinators of wild and cultivated sunflowers.

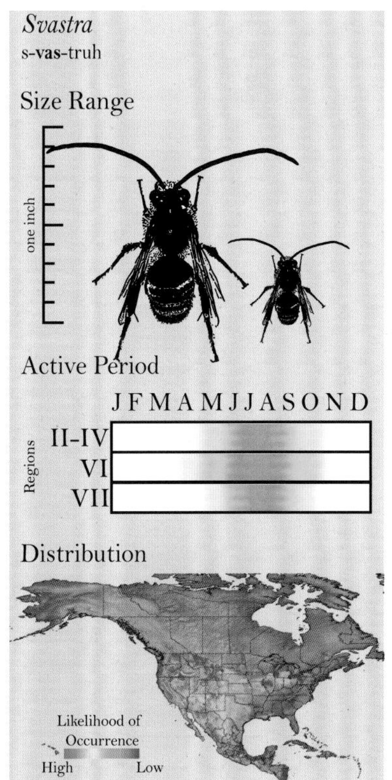

A male *Svastra* on a wildflower. PHOTO BY MARGARETHE BRUMMERMANN.

A female *Svastra* leaving a cactus flower; some Svastra are cactus specialists. PHOTO BY JILLIAN H. COWLES.

however, are opportunistically communal (p.19), and 2 to 12 females may share a nest entrance. Inside the nest, each female builds and fills her own nest cells with pollen.

Solitary and communal species commonly nest in aggregations, with up to 100 females nesting in the same area—their nest holes are often just a few inches apart. Males will hover over these aggregations and intermittently pounce on females traversing between flowers and nests. While nest holes may occasionally have mounds of soil around them, they are more typically bare holes. Nests descend more or less straight down for about 10 inches and then turn so they run horizontally, meandering up and down and branching from time to time. Nest cells are dug vertically and made a bit larger than they need to be. Once the nest cells are excavated, the female fills them back in with a thin layer of soil and a substance she secretes, creating a waterproof cement that lines the entire inside of the cell.

A hot topic

Eucerini are common in the arid southwestern United States, where they dig their nests in exposed hard-packed soils. The desert sun beats down on these barren areas all day. Bees can often be seen chewing at the soil to create a new nest entrance during the heat of the day, when air temperatures are in excess of 100 °F. The ground temperature is even hotter, ranging from 120 to 140 °F! The bees burrow through the scorching soil anyway, apparently oblivious to the flesh-searing temperatures. Even at a depth of a few inches, the soil is still quite warm (more than 80 °F); scientists hypothesize that the rapid development seen in some of these bee larvae might be attributable to the warm temperatures of the incubation chamber.

Tetraloniella
tetra-loh-nee-**el**-uh

Size Range

one inch

Active Period

Region | J F M A M J J A S O N D
VI

Distribution

Likelihood of Occurrence
High Low

Eduardo Ladislao Holmberg (1852–1937)

Holmberg was an Argentine natural historian who first coined the names of several of the Eucerini species discussed here. In addition to expeditions into the mountains of Argentina in search of new specimens to add to the list of known organisms in his country, Holmberg ran the Zoological Gardens of Buenos Aires, gave well-attended lectures at the university, and (as if that weren't enough) published poetry and some of the earliest science fiction novellas to come out of South America.

Tetraloniella

The name Tetraloniella means "diminutive Tetralonia," referring to its similar appearance to the European bee Tetralonia. Tetralonia was described by Maximillian Spinosa in the early 1800s. Spinosa chose this name in reference to the four "areoles" (tetra- means four, halonia refers to the empty cells) making up the majority of the forewing: three submarginal and one marginal cell. The similar-looking Eucera in Europe have just two submarginal cells.

Tetraloniella can be found around the world, mostly north of the equator, though species are found throughout eastern and southern Africa and Madagascar. More than 100 species are known worldwide and 22 species reside in the United States—only one (*Tetraloniella albata*) may make it as far north as Canada. For many species of *Tetraloniella*,

A female *Tetraloniella* foraging on a globe mallow flower (*Sphaeralcea*). PHOTO BY JILLIAN H. COWLES.

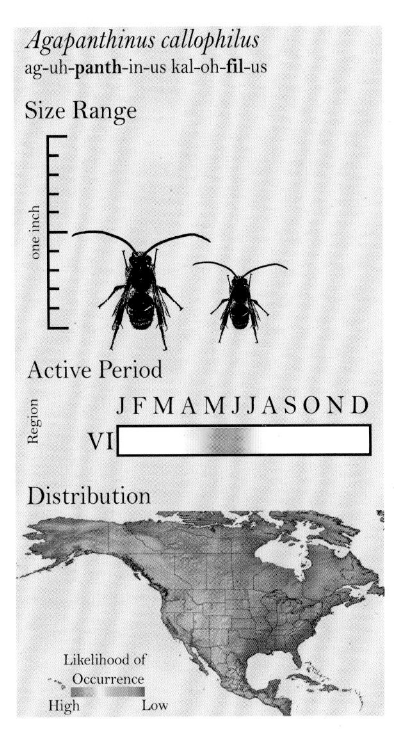

dietary preferences are not known, though both generalists and specialists occur. Several are specialists on flowers in the sunflower family (Asteraceae), including *Helianthus, Coreopsis, Gaillardia,* and *Helenium.* Others frequently visit globe mallow (*Sphaeralcea*), prairie clover (*Dalea*), or sage (*Salvia*). Little is known about the specific nesting habits of *Tetraloniella*. Like all Eucerini that have been studied thus far, *Tetraloniella* nest in the ground. Aggregations of several hundred individuals have been found, and in fact some species may nest communally, even using the same nest for consecutive generations.

Agapanthinus callophilus

Agapanthinus means "lover of flowers."

Only one species of *Agapanthinus* is found anywhere in the world. Restricted to the Baja peninsula of Mexico, and the deserts of California, it is commonly seen in May. Almost nothing is known about the life habits of this bee, since it has only rarely been collected. The eyes appear to be bigger than in other species of bees in the Eucerini tribe, and it has been suggested that this bee flies mostly at dawn or dusk (as is true of other bees with large eyes).

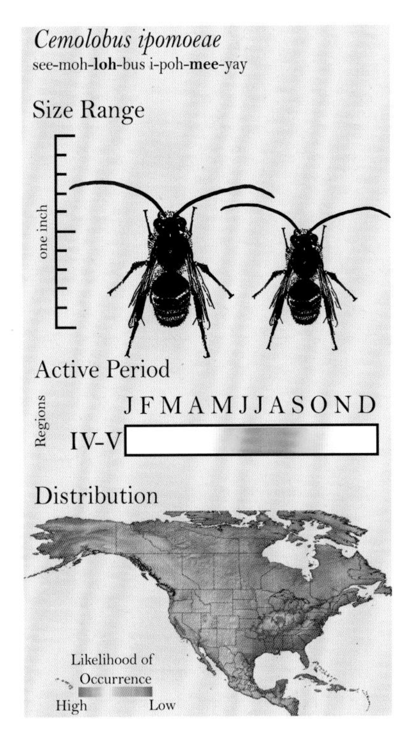

Cemolobus ipomoeae

The name Cemolobus means "lobed snout," a reference to the three-lobed clypeus that distinguishes this genus.

Cemolubus is found only east of the Great Plains, ranging from Pennsylvania south to Georgia, and west across Missouri. It may range south into Mexico as well. Only one species is known. As the species name suggests, this bee is a strict specialist on morning glory flowers (*Ipomoea*), and it is most frequently seen during June and July when those flowers are in full bloom. Generally, *Cemolobus* fly early in the day, before heat and humidity have caused *Ipomoea* flowers to wilt. While a reasonable assumption is that *Cemolobus* nest in the ground, no studies of their nesting habits have been completed.

Gaesischia exul

Gaesischia means "speared hip-joint," a reference to the "coxal thorns" described by the original authors of this genus.

Gaesischia are mainly tropical, being found on the edges of humid areas of Central and South America. There are 36 species, but only one, *Gaesischia exul*, ranges as far north as the United States. There, it reaches only southern Arizona and New Mexico. *Gaesischia* have an amphitropical distribution (p.212), occurring either north of Costa Rica or in southern Brazil. Because of their larger size, *Gaesischia* often fly long distances when foraging and might therefore be important pollinators of subtropical species like the cabbage bark tree (*Andira inermis*). *Gaesischia* nest in the ground, but the specific biology of the North American species has not been researched.

Florilegus condignus

The name Florilegus means "flower gatherer."

Florilegus are smaller Eucerini, found from South America up to the Great Lakes. There are 11 *Florilegus* species in all, but only one is found in the United States. It is relatively widespread east of the Rocky Mountains, ranging to the coast and north to Wisconsin. Interestingly, this same species occurs as far south as Brazil. While *Florilegus condignus* visits a wide variety of flowers, it appears to have some preference for flowers in the pea family (Fabaceae), especially alfalfa. Its preference for alfalfa might make this bee an important secondary pollinator of this crop in the Midwest and eastern United States. Like other Eucerini, *Florilegus* nest in the ground and line their nest cells with a waterproof material.

Martinapis

Martinapis was named by Theodore Cockerell in the early 1900s. In his description of this bee, he explains that he named this genus "after my little son."

Martinapis are uncommonly seen medium-sized bees. Three species are known, one in Argentina and two in the southwestern parts of the United States. Both species are generalists but prefer flowers from the pea family

A *Martinapis* foraging on a legume flower (Fabaceae).

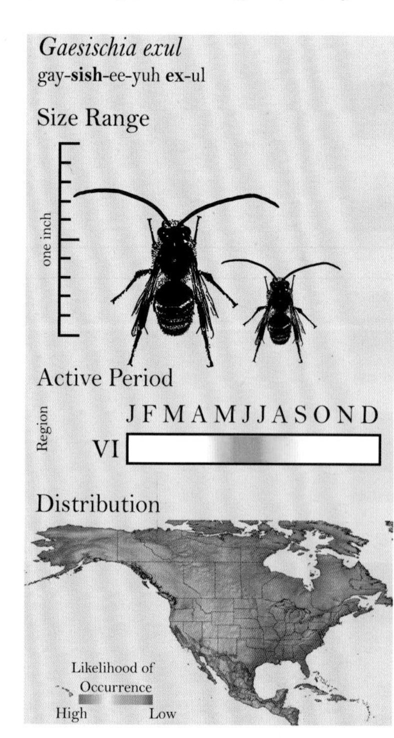

Florilegus condignus
floh-ril-**ee**-jus kon-**dig**-nus
Size Range
one inch
Active Period
Regions
J F M A M J J A S O N D
IV-V
VII
Distribution
Likelihood of Occurrence
High
Low

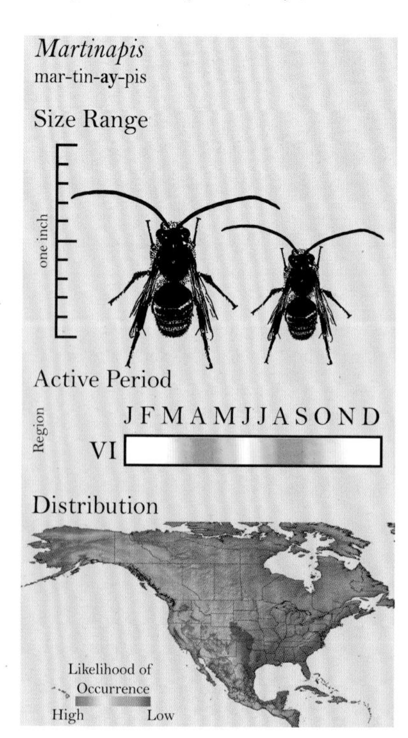

(Fabaceae) and the caltrop family (Zygophyllaceae). Like their relatives, these bees nest in the ground. *Martinapis* is known to dig deeper nests than many bees in the Eucerini tribe, sometimes reaching more than 3 feet in depth. These bees are seen primarily at dusk and dawn.

Melissoptila otomita

Melissoptila means "fuzzy" (-ptilos) "bee" (melissa) in Greek.

Melissoptila is a predominantly tropical genus found mainly in South America and Central America. Roughly 40 species are known, but only one ranges as far north as southern Texas. It is extremely uncommon. The floral preferences of *Melissoptila otomita* are not known, though some other species show a preference for flowers in the family Malvaceae. Like the other long-horned bees, *Melissoptila* are solitary ground nesters.

Syntrichalonia exquisita

The name Syntrichalonia means "with hairy spaces," referring to the light layer of fuzz seen on the wings of many Eucerini.

Syntrichalonia are beautiful large bees with fuzzy golden hair. While the antennae of the males are long, they aren't as long as those of other long-horned bees. Only two species of *Syntrichalonia* are known. One of them is limited to central Mexico, but the other ranges just into the United States, along the border of Arizona, New Mexico, and Texas. *Syntrichalonia exquisita* flies mainly from September through November, but it can be found as early as the end of July. While this bee has been collected most often on flowers from plants in the sunflower family (Asteraceae), it may be because that is the majority of what is in bloom so late in the year.

Simanthedon linsleyi

The name Simanthedon means "pug-nosed bee."

There is one species of *Simanthedon* in the world, found in northern Mexico and southern New Mexico, Arizona, and Texas. *Simanthedon* is a generalist and has been collected on many different flowering desert plants. The nesting biology has not been detailed, but it is likely a ground nester.

A *Syntrichalonia* foraging on a composite flower. PHOTO BY JILLIAN H. COWLES.

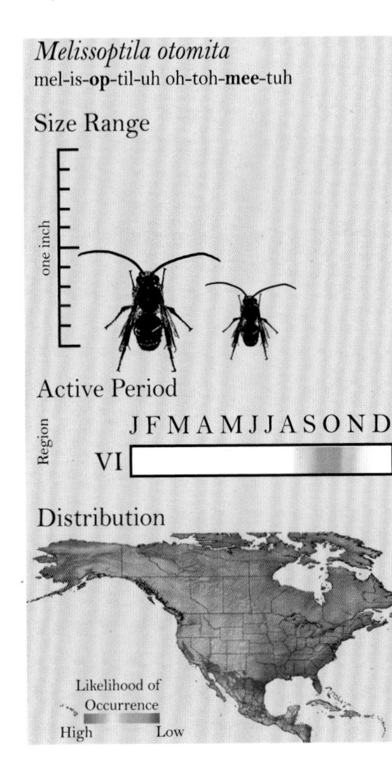

Syntrichalonia exquisita
sin-trik-uh-**loh**-nee-yuh ex-**kwiz**-it-uh

Size Range

one inch

Active Period

Region

J F M A M J J A S O N D

VI

Distribution

Likelihood of Occurrence

High Low

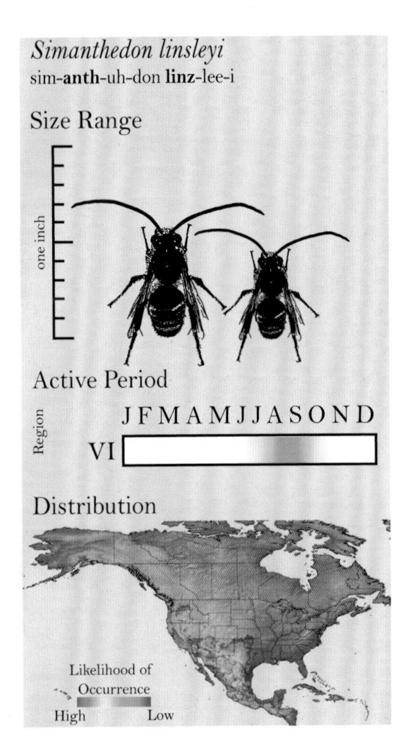

8.8 ANTHOPHORINI

ANTHOPHORA AND *HABROPODA*

The Anthophorini are found worldwide, with more than 700 species in 7 genera. In the United States just 2 genera are found, and they can be very difficult to tell apart (see identification for specifics). Both *Anthophora* and *Habropoda* are known to nest in the ground and so are commonly referred to as digger bees.

Anthophora

The name Anthophora means "flower bearer" (antho- meaning "flower," and -phora meaning "to bear").

DESCRIPTION

These hairy bees are fast flying and extremely common. Ranging from slate grey to rusty red, the hair patterns on these bees are striking.

DISTRIBUTION

Anthophora are found worldwide, with the exception of Australia and other areas of the South Pacific, being especially common in the Mediterranean and South Africa. There are more than 400 species, roughly 50 of which are found in the United States, with 11 extending north into Canada. They are most common in the western United States (fewer than 10 species are found in the east). One species, *Anthophora plumipes,* was introduced into the United States in the 1980s or early 1990s from Europe. Its range is currently Washington, D.C. and Maryland.

DIET AND POLLINATION SERVICES

Anthophora includes both specialists and a huge number of generalist bees. Specialists are associated with a wide array of distantly related flowers, including the evening primrose family (Onagraceae) and creosote (*Larrea tridentata*).

Though not as loyal as specialists, *Anthophora* that are generalists can be helpful for pollination. One species,

A male *Anthophora* drinking nectar from a manzanita flower (*Arctostaphylos*). *Anthophora* and *Habropoda* are both capable of buzz pollinating these kinds of flowers.

Shake it off

Many *Anthophora* and *Habropoda* do well in cool weather, when other bees don't fly. Some species will fly when temperatures are below 60 °F, temperatures common during the early hours of the day, and in the cool months of early spring. *Anthophora* and *Habropoda* are better at "shivering" to warm up when it is cold out than are most other bees; once they're warmed up, flying keeps them warm in the same way that jogging warms a human. However, some scientists think that these bees may get too warm in the middle of a hot day, which is why they are more likely to be in their burrows than on flowers in the early afternoon. *Andrena* (section 3.1) are another bee that does well in cold weather, and it is not uncommon to see them motionless on flowers or rocks first thing in the morning. They are actually shivering inside, warming up their bodies to commence the day's activities.

Anthophora
anth-**ah**-foh-ruh

Size Range

one inch

Active Period

Regions	J F M A M J J A S O N D
II-IV	
VI	
VII	

Distribution

Likelihood of Occurrence
High Low

Anthophora are excellent fliers. They often hover in front of flowers as they sip nectar through their long tongues.

An *Anthophora* hovering near a Russian sage flower (*Perovskia*).

An *Anthophora* on a cat's eye (*Cryptantha*). With their long tongues they can sip nectar from narrow flower tubes.

Anthophora urbana, will visit a smorgasbord of flowers. Scientists have found, however, that fields of tomatoes in California with greater numbers of *Anthophora urbana* living in the vicinity produce more tomatoes than fields with fewer of these bees, likely because they are very good at buzz pollination. Their usefulness is not limited to crop plants, either. The large hairy bodies of *Anthophora* mean that they can carry more pollen than many other bees. Generalist *Anthophora* have been found to be important pollinators of a number of wildflowers, including beardtonuges (*Penstemon*), lupines (*Lupinus*), and evening primroses (*Oenothera*).

An *Anthophora* foraging on a milkvetch (*Astragalus*).

An *Anthophora* surveying a Rocky Mountain bee plant (*Cleome*). Photo by B. Seth Topham.

An *Anthophora* manipulating the complex petals of a milkvetch flower (*Astragalus*).

Other Biological Notes

Anthophora are active for a longer period of the year than *Habropoda*, with some species flying in the spring (February or March in California), and others flying in the summer and into the fall (October).

Almost all *Anthophora* nest in the ground, often in aggregations of several hundred to a thousand nests. A few species are communal, with several females sharing a nest entrance, but with their own individual nest cells inside. Much the same way a dog digs with its front paws, many digger bees burrow in the ground with their front legs to start a nest, then use their mandibles to loosen dirt that is more compacted as they go deeper. Dirt is shoveled out of the nest as the bee backs out, and piled around the nest entrance, like a tiny gopher mound. Only one species, *Anthophora furcata*, nests in twigs instead of in the ground.

Anthophora females are often harassed by males, to the point where the males may attempt copulation once every

Most *Anthophora* are gray colored, but some, like this *Anthophora pacifica*, are almost completely black.

An *Anthophora* resting on a beardtongue (*Penstemon*) flower.

Above: An *Anthophora* hovering near a scorpionweed flower (*Phacelia*). *Anthophora* often keep their legs tucked tightly under their bodies when flying, unlike other genera.

Left: Male *Anthophora*, like the one pictured here, often have a lot of white markings on their faces. Notice the extended labrum, below the clypeus, not often visible in bees.

three seconds. Sometimes many males will fight for one freshly emerged female, forming a 'ball' of bees with as many as 20 or 30 males surrounding one female. Females have to maneuver to avoid males in midair, hide from them on flowers, and physically fight them off when they try to mount.

Anthophora are known as digger bees because they nest in the ground. This female begins digging her nest by removing small rocks with her mandibles.

An *Anthophora* digging her nest in a sloping bank.

As steep banks erode, they often expose old nests of *Anthophora*. Here, several tunnels and old nest cells are visible in an eroded bank.

Like many bees in the family Apidae, *Anthophora* males often sleep by biting onto twigs or stems to keep their place before falling asleep.

Habropoda

Habropoda means "graceful or delicate foot" (habro- meaning "graceful" or "delicate," and poda meaning "foot").

DISTRIBUTION

Habropoda are found most commonly in the Northern Hemisphere (Asia, North America, Europe, and Central America), though they can also be found in northern Africa. There are more than 50 species worldwide, and

A *Habropoda* on a flower of creosote bush (*Larrea tridentata*). Photo by Hartmut Wisch.

21 can be found in the United States and 3 in Canada. As with *Anthophora,* they are more concentrated in the west; nearly a third of them are found only in California. Only one species of *Habropoda* occurs east of the Mississippi River (*H. laboriosa,* the southeastern blueberry bee).

Diet and Pollination Services

Like the genus *Anthophora*, *Habropoda* includes species that are generalists and others that are specialists. As an example, *Habropoda depressa* will visit any number of spring-blooming flowers in California. In contrast, *Habropoda pallida* favors flowers of creosote bush (*Larrea*) over others, though it will occasionally visit other species. In the east, the southeastern blueberry bee (*Habropoda laboriosa*) seems to prefer blueberry flowers (*Vaccinium*) and trumpet flowers (*Gelsemium*) over all others. This bee has proven to be a much more efficient pollinator of blueberries, which it buzz pollinates (p.21), compared with most other bee species, including the commercially used honey bee. In the absence of the southeastern blueberry bee, fewer blueberries are produced.

Bears should thank the bees too...

One scientific study concluded that a single female southeastern blueberry bee visits almost 50,000 rabbiteye blueberries (*Vaccinium ashei*) in its lifetime, contributing to the production of nearly 6000 blueberries.

Other Biological Notes

Like *Anthophora*, *Habropoda* nest in the ground, often in aggregations of several hundred individuals. Some species are very particular about the type of soil in which they nest. *Habropoda miserabilis*, for instance, will nest only in sand dunes along the coast of Washington, Oregon, and California. *Habropoda* make just one cell that encloses just one egg at the end of each nest.

As with most bees, the male *Habropoda* emerge from the nest in which they developed earlier than the females. The newly emerged males of some species will wait near the nests where females haven't yet emerged. They are presumably listening for the sounds (or feeling the vibrations) of females digging their way to the surface. Impatient little bees, they will dig down to a female as soon as they hear her. Often several male bees will fight to be the

A *Habropoda* foraging on a legume flower. Notice how the color and thickness of the first tergal segment are the same as the thorax. This is not uncommon in *Habropoda*. Photo by L. Saul-Gershenz

one that digs down to a female. As with other animals, the bigger bee usually wins the right to mate.

Not all *Habropoda* larvae from the previous year emerge; instead, some individuals will stay in the nest for 2 years and maybe longer; some bee species will delay emergence for 7 or even 10 years. Why all the bees from the previous year don't emerge together is not known, but it may have something to do with avoiding parasites, or with responding to rainfall cues (see bet-hedging, p.17).

Above: A male *Habropoda* resting on a leaf in Illinois. Like *Anthophora*, male *Habropoda* often have white faces.

Below: A female *Habropoda* drinking from a manzanita flower (*Arctostaphylos*). Photo by Alice Abela.

When the time is right...

Bees spend the majority of their lives in the bee equivalent of solitary confinement. The nest is dark, isolated from nearly all sensory input, and sealed off even from brothers and sisters. In this place, how do bees know when it is time to dig their way out of the nest and finally meet the rest of the world? The answer is complicated and not well understood. To some extent, the natural biological clock of the bee guides them, as does their genetic predispositions (i.e., "if mom emerged in April then I will too!"). Amazingly, bees can also sense certain environmental cues about the world beyond from inside their nest cells. Specifically, changes in temperature, as the ground thaws in the spring, can be a clue to a bee that it might be safe to emerge. Bees may even begin the journey from their nest to the open air and then stop if the surface soil seems too chilly—no longer in their nest, but not yet fully exposed either. Changes in soil moisture associated with either spring or monsoonal rains are both indicative of pending floral blooms and seem to inform bees that it is time to move out. Bees that wait for a few years before emerging are more of a mystery, and scientists are still puzzling out how some of a bee's offspring may stay dormant, while others emerge each year.

8.9 CENTRIS

The name Centris means "spined", but the particular bodily spine to which the name refers is unclear, and the original author, Fabricius, does not elaborate in his description.

Description

Centris are big burly bees; they fly fast and can stop on a dime. These bees commonly fly during the hottest times of the day, when other bees are resting and waiting for the temperature to drop. Many *Centris* are adapted to carrying floral oils instead of, or in addition to, nectar. They are often specialists on plants that produce large quantities of such oils. This practice has led to one common name, the oil baron bees.

Distribution

In excess of 200 species of *Centris* are known. All are found in the Western Hemisphere, and 90% of them in South America. Only 23 species occur north of the Mexican border, though they are mostly restricted to the southern half of the United States (none make it as far north as Canada). One species is recorded from Kansas, a few have been found in northern California, three have been found in Florida, and all other species are found in the arid southwest.

One of the species in Florida, *Centris nitida*, is a recent (1997) introduction to the southern half of the state. It is native to Central America and the northern part of South America. In both Florida and Central America this bee is a specialist on *Byrsonima lucida*, an oil-producing plant that is visited by the native Florida *Centris* as well. Though they share a host plant, it does not appear that the non-native is having any

Johann Christian Fabricius (1745–1808)

Centris is but one of the many bees named by Fabricius more than 200 years ago. An avid naturalist from a very young age, Fabricius studied under Linnaeus, the father of all modern taxonomy and the originator of the idea of a genus and species name for every organism (Linnaeus gave the first bee he described its scientific name: *Apis mellifera*). In the less than 25 years he spent studying insects, Fabricius named nearly 10,000 species of insects, including many bees.

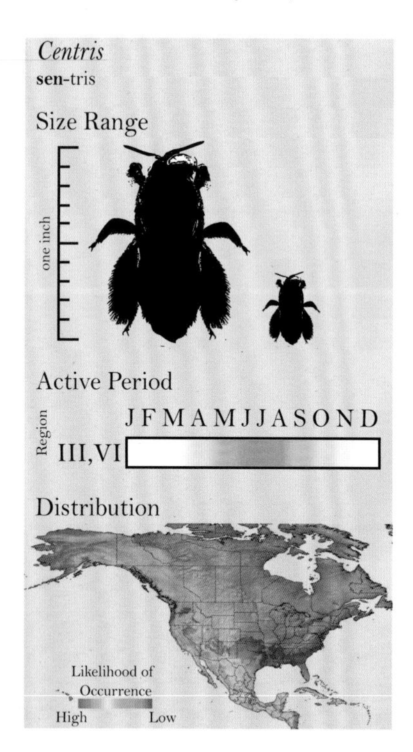

A female *Centris* on an indigo bush flower (*Psorthamnus*). Male *Centris* often have yellow or green-colored eyes, but the females of many species have bright red eyes.

A *Centris* collecting pollen from a palo verde flower (*Parkinsonia*). While some female *Centris* have red eyes, other species do not. Photo by Lon Brehmer and Enriqueta Flores-Guevara.

impact on the success of the native; in fact, it seems that the presence of this additional specialized visitor may help promote fruit set in the plant they both visit. Considering that the plant is endangered, the bee's arrival in southern Florida may be a good thing.

Diet and Pollination Services

Centris are mostly generalists, but a few are pollen specialists. For example, *Centris californica* appears to be a specialist on *Wislezenia*, and perhaps other plants in the family Cleomaceae. Some bees that are generalists do appear to prefer certain desert flowers, though. Desert willow (*Chilopsis*) and palo verde (*Parkinsonia*) seem to be favorites of many *Centris* species.

Centris includes a number of bees that, in addition to collecting pollen, collect oil. The oils are brought back to the nest where they are used either in nest construction (i.e., for lining the walls or creating plugs to seal nest cells once

A *Centris* foraging on a flower of creosote bush (*Larrea*). Photo by Margarethe Brummermann.

A *Centris* on an ironwood flower (*Olneya*). Photo by Margarethe Brummermann.

A male *Centris* perching on a leaf, guarding his territory from any perceived intruders. Photo by Hartmut Wisch.

they are completed), or mixed with pollen for their larvae to eat. They may use these oils instead of nectar, or they may collect both nectar and oils.

Why these bees collect oils is not entirely known. The oils are often very fatty and probably a good source of energy for growing larvae. They also (like all oils) repel water. It is thought that the first species of *Centris* to collect oils were those living in rain forests. Ground-nesting bees in rain forest environments have a hard time preventing detrimental molds from growing in the nest cells. Using the oils to line the nest cell walls may help keep the cells drier in these humid environments. At least six species in the arid southwest have lost the ability to collect oils—they have no, or very reduced, oil-collecting hairs on their front and mid legs. The dry soils of the desert mean a smaller chance of mold growing in nest cells. In these cases the time-consuming process of collecting oils may not be offset by its benefits in these dry habitats. However, *Centris atripes*, found from Texas west to California, does retain the ability to collect oils, specifically from white ratany (*Krameria grayi*).

In the tropics, *Centris* are important pollinators of a number of trees, including Brazil nut (*Bertholettia*), cashew (*Anacardium*), and West Indian cherry (*Malphigia*). And, because they can buzz pollinate (p.21), they are hypothesized to be important native pollinators of eggplants.

Other Biological Notes

Centris are all solitary bees and do not share the duties of gathering pollen or laying eggs. A great diversity of nesting habits can be found among species of *Centris*. While most North American species nest in the ground, a few species nest in dead wood. *Centris nitida* is one such wood nester, and it was likely introduced into Florida by being transported as larvae in a piece of wood.

Among the ground-nesting species, a few nest in large aggregations, but others nest alone. *Centris pallida*, for example, is known to nest in aggregations of several thousand individuals. Males emerge first in the spring, digging their way from the nest cell straight to the surface. They then hover close by, waiting for females to emerge.

Built-in butter knives

Bees that collect oils from flowers have very special adaptations on their front and mid tibiae (segments of the legs); long hairs, shaped like butter knives and spatulas. Oil-producing flowers produce oil in one of two ways. Either (1) the oil is secreted just below the leaf's surface, forming blisters known as elaiophores, or (2) the oil is secreted through the ends of special hairs called trichomes. When a *Centris* lands on an oil-producing flower, she uses her legs to rub these hairs, or to push on the elaiophores, breaking the tissue's surface and releasing the oils. She scrapes up the oil with the special hairs on her legs, and then, in midair, wipes her front and mid legs in the joint of her back legs, transferring the oil to her scopae (tufts of pollen-collecting hairs on the hind legs). The scopae of *Centris* are also modified—very coarse, and widely spaced for storing oils.

A male *Centris* on an indigo bush flower (*Psorthamnus*).

As a female gets close to the surface, a male will assist from above, digging his way down to her. It appears they can smell the females as they burrow up from beneath the surface; to be sure, one scientist killed a handful of female bees and buried them. With no sound cues, the males were still able to find the females, digging down to the still corpses, presumably following their lingering aroma.

Centris pallida males range widely in size, with large ones nearly half as big again as the smallest ones. Very large males are bullies, pushing smaller ones out of the way to get to emerging females. One would think that over time, all males would be large, since they mate more frequently; scientists have found no change in the size of males over many years, however. It is thought that the females vary the amount of pollen they give their male offspring—those given more pollen to eat become the large males. Why the females don't provide the same amount of pollen to all their male offspring, however, remains a mystery.

Males also look for mates around flowering plants. It is not uncommon to see large numbers (more than 100 individuals) of *Centris* flying around palo verde (*Cercidium*). These bees stake out territories near the plant, releasing pheromones as they hover. If you watch these bees near their territories, you can see the males extend their back legs as they hover, dropping compounds onto plant leaves. These territories are tiny—only a cubic yard—but a male *Centris* will guard that space ferociously, chasing away anything that gets too close (dragonflies, other bees, flies, falling petals or leaves, and even pebbles tossed in the vicinity of the bee).

Some species of *Centris* dig deep burrows in the ground, just to put one nest cell at the bottom; they then fill in the entire burrow and dig another one for the next nest cell. Other species put many nest cells in each burrow. Finally, some *Centris* do not dig a nest at all, but reuse nests made by other organisms. For example, in South America, *Centris* are known to build nests inside termite mounds, in the nests of some leafcutter ants, inside fissures in caves, in the burrows of larger animals, or even in abandoned mines.

Some like it hot

Unlike other bees in hot desert environments, *Centris* can be seen flying in the middle of the day (for comparison see squash bees, section 8.6). When bees fly, their flight muscles generate enormous amounts of heat. That fact combined with the extreme temperatures of the early summer desert means that *Centris* are often flying very close to the upper limits of the temperature their bodies can handle—nearly 120 °F. There is some evidence that they may slow down their wing beat in order to maintain a (barely) tolerable temperature.

Many Apidae can look very similar to each other, but *Centris* are distinct with their bright-colored eyes and bushy pollen-collecting hairs (scopa). Interestingly, after a *Centris* dies, its colorful eyes turn brown.

A male *Centris* resting on a twig. Photo by Alice Abela.

A *Centris* visiting a salt cedar flower (*Tamarix*). Photo by Jillian H. Cowles.

8.10 *BOMBUS*

The name Bombus comes from the Greek word bombos, which means "a buzzing sound", referring to the low hum these bees make as they fly gracefully around flowers. The common name "bumble bee" can be traced back to the word bombelen in Middle English (AD 1200–1500), which means "to hum." In fact, prior to the 1920s, bumble bees were more often called "humble bees," also a reference to the soft droning inherent in their foraging activities. The term "humble bees" was used both by William Shakespeare in A Midsummer Night's Dream *and by Charles Darwin in* On the Origin of Species. *A few popular articles in the 1920s about* Bombus *referred to them as "bumble bees" and the new name took.*

Description

Bumble bees are medium to large hairy bees that all belong to the genus *Bombus*. Bumble bees are among the few bees native to North America that are truly social, with a queen and workers.

Distribution

Around the world over 250 bumble bee species are known, but fewer than 50 of them occur in North America. Bumble bees are found primarily in the Northern Hemisphere (North America, Europe, and Asia) but some native populations also exist in parts of South America. Also, a few species have been introduced to New Zealand and Tasmania. Although bumble bees can be found in most habitats in the Northern Hemisphere, ranging from the southwestern deserts to the tundra, these bees are most diverse in mountainous regions. Worldwide, the areas with highest bumble bee diversity are in Europe and Asia, particularly in the Himalayas and the Alps.

Bumble bee or bumblebee? Honey bee or honeybee?

The term *bumble bee* is just as often written *bumblebee* in the United States and Canada, and knowing which style is correct can be tricky. Most dictionaries stick with the concatenated version (honeybee), but most scientists separate the two. Their logic is that the "bee" part of the word is an accurate reflection of the insect's identity, and so it can stand on its own. There are many kinds of bees, they argue: sweat bees, carpenter bees, honey bees, and bumble bees among them. In cases where the second part of the word does *not* accurately reflect the kind of insect, it should be concatenated. So while *house fly* is correct (because house flies are indeed flies), *butter fly* is not (because butterflies are not flies).

A bumble bee foraging on a dandelion (*Taraxacum*).

A bumble bee foraging on a beardtongue flower (*Penstemon*).

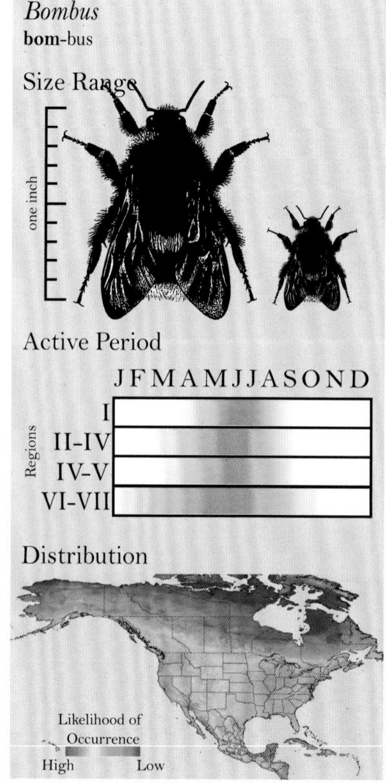

Why don't we have bumble bee honey?

Like honey bees, bumble bee workers collect copious amounts of nectar, which they bring back to the hive for storage. Unlike honey bees, however, the bumble bee workers do not dehydrate the stored nectar, turning it into honey (section 8.11). Instead this nectar is used by bumble bees, along with pollen, to feed the developing young. Because bumble bee hives begin anew each year, there is no need to store large amounts of nectar as honey to sustain the workers through the winter the way that honey bee colonies must. This is why we don't see bumble bee honey in grocery stores.

Bumble bees are most diverse in mountainous areas; unlike most bees they are rare in North America's deserts. One species, however, *Bombus sonorus*, is common in the deserts of the southwest. PHOTO BY LON BREHMER AND ENRIQUETA FLORES-GUEVARA.

DIET AND POLLINATION SERVICES

Bumble bees are generalists and can be found foraging on a wide variety of plants. Some species have extraordinarily long tongues, enabling them to feed on nectar from plants with deep flower tubes like beardtongues (*Penstemon*), foxgloves (*Digitalis*), monkshoods (*Aconitum*), and columbines (*Aquilegia*).

Studies have shown that for many crops, pollination by bumble bees produces bigger fruit, faster fruit set, and larger yields than other pollination methods, most specifically honey bee pollination. First, bumble bees have a distinct advantage over European honey bees when it comes to retrieving pollen from some plants: they can buzz pollinate. They are therefore much more effective pollinators of some important crops, specifically those with flowers requiring buzz pollination (p.21). These plants include tomatoes, peppers, eggplants, potatoes, and even some berries like blueberries. Second, bumble bees have been shown to be faster workers than honey bees, often visiting twice as many flowers per minute. Finally, researchers have estimated that bumble bees will do at least eight times more work than a honey bee because bumble bees can remain active in colder temperatures, and they can carry more pollen.

OTHER BIOLOGICAL NOTES

Often, bumble bees are the first bees seen in spring and the last bees out in the fall. Bumble bees have special adaptations that allow them to be active in colder weather and colder climates than most other bees. In addition to their thick and insulating coat of hair, bumble bees often

ABOVE: A bumble bee flying by a hollyhock flower (*Alcea*).

RIGHT: The most common bumble bee in the eastern United States is known as the common eastern bumble bee (*Bombus impatiens*), pictured here on a blanket flower (*Gaillardia*). PHOTO BY DONNA K. RACE.

A bumble bee maneuvering among cat's eye (*Cryptantha*) flowers. PHOTO BY B. SETH TOPHAM.

bask in the sun to warm themselves before they head out to forage. When sun and fuzz aren't enough, bumble bees can actually generate heat internally by shivering their flight muscles. These bees can uncouple their wings from their flight muscles, allowing them to contract the muscles without flapping their wings. Those muscle contractions can raise the internal temperature of the bee, making them significantly warmer than their surrounding environment. In fact, bumble bees can't take off and fly until their flight muscles are above 80 °F; by shivering their flight muscles to warm up, they can actively forage in temperatures much too cold for other bees. Recently, researchers have noticed some bumble bees (*Bombus terrestris*) in urban Britain remain active during the winter season, foraging when

Up, up, and away!

Recent studies have found that not only can bumble bees fly at cold temperatures, they can also fly at extremely high elevations. Scientists put bumble bees into a pressure chamber in which they had altered the air density to mimic high-elevation air and found that bumble bees could theoretically fly at elevations up to 500 feet higher than Mount Everest. Most helicopters can't fly that high.

A large queen bumble bee foraging on a hollyhock flower (*Alcea*).

Bumble bees, like their cousins the honey bees, have a pollen basket (corbicula) on their hind legs rather than scopa. Here you can see the moist mixture of pollen and nectar stuck to the bumble bee's back leg.

temperatures are as low as 37 °F. The fact that both queens and workers have been observed in the winter suggests that a second generation of bee colonies may be active during these months.

Bumble bees are eusocial (meaning they are truly social). Typically, in the spring a new queen bumble bee will emerge from her winter hibernation and begin searching for a nest site. Once a suitable site is established, the queen will collect pollen and nectar. When the nest is sufficiently stocked with pollen and nectar, the queen will lay eggs on the pollen mass and incubate them by covering the eggs with her abdomen (like a bird on her brood). The queen stays in the nest from this point on, attending to her growing larvae, and leaving the nest only if more food is needed. About a month after they are laid, this first batch of eggs will have developed into adult workers. These workers then take over both the foraging duties and care of the next round of the queen's offspring. The nest continues to grow through the summer, with the queen laying more eggs, which develop into more workers. At the end of the season, the queen will produce a set of male offspring, followed closely

A mating pair of bumble bees. Often, because male bumble bees are much smaller than females, the female continues to fly around with the male still attached to her abdomen.

by a set of new queens. These males and new queens leave the nest to find mates. After mating, the males die and the new queens find a place to hibernate through the winter. The old queen and all remaining workers die at the end of the year, and the cycle begins again the next spring.

Most bumble bee species make their nests in the ground, often in preexisting cavities like abandoned rodent burrows, in piles of wood, or in leaf litter. Occasionally, some bumble bees will nest in building foundations.

Bumble bees are one of the only bees native to the United States and Canada that are truly social, meaning they live in a hive with a single reproductive queen and multiple sterile workers. Unlike honey bee queens, a bumble bee queen lives for only a single year. This annual cycle generally keeps bumble bee hives much smaller than the hives of honey bees. Most mature bumble bee colonies consist of fewer than 200 bees, although some can have as many as 1000 individuals. For comparison, honey bees may have around 60,000 bees in a single colony. All bumble bee workers are female. They perform a variety of duties including foraging for nectar and pollen, feeding the growing young, feeding the queen, cleaning the nest, and defending the hive when needed. Worker bumble bees generally do not lay eggs, either because they have underdeveloped ovaries due to the suppression of a hormone needed for ovary development, or because of physical aggression from the queen. Either way, the queen is the primary reproducer and the workers simply maintain a healthy colony. Male or drone bumble bees do not participate in the activities of the workers.

ABOVE: A bumble bee foraging on a thistle flower (*Cirsium*).

ABOVE MIDDLE: A male bumble bee resting on a leaf. PHOTO BY ALICE ABELA.

ABOVE RIGHT: Bumble bees are often found high in the mountains. Here a bumble bee visits a flower blooming around 10,000 feet above sea level in the Uintah Mountains of Utah.

RIGHT: A bumble bee hovering in front of a flower as she drinks nectar with her long tongue.

8.11 *APIS MELLIFERA*

The name Apis is a very old word whose original roots may be Egyptian, though it is also related to the Greek work for "swarm." Mellifera means "honey-bearing" in Latin.

Apis mellifera are not native to North America, but were brought here by European colonists in the 1600s. Honey bees differ from most other bees in North America in two key ways: (1) they are social (with a queen and workers living in a hive) and, (2) related to their sociality, they produce large amounts of honey.

Description

Often, when people think of a bee they picture a honey bee. Honey bees are common, medium-sized bees in the genus *Apis*. In North America, all honey bees are members of the same species, *Apis mellifera,* also known as the European honey bee.

Distribution

Apis mellifera is found worldwide with the exception of Antarctica, the Arctic Circle, and the Sahara Desert (although a wild colony was discovered at an isolated oasis in Libya). While the European honey bee, because of its domestication by humans, is extremely widespread, six other *Apis* species also exist. They are native to Asia and the Middle East and are not as far ranging.

Unlike many of the similar-looking native North American bees, honey bees often dangle their legs down as they fly from flower to flower.

Diet and Pollination Services

Honey bees are true generalists and will collect pollen or nectar from nearly every type of flower. They have even been seen gathering pollen from plants that are typically wind pollinated like corn, pine trees, or sagebrush. In addition to collecting pollen to feed developing larvae, honey bees spend a fair amount of time collecting nectar, which is stored as honey. Honey bees collect nectar from flowers and transport it back to the hive. Here, it is partially digested, then regurgitated into honeycomb, and finally thickened into honey when the bees buzz their wings over the honeycombs and excess water is evaporated. In wild colonies, the honey is used as a food source during colder months, when floral resources are scarce. In domestic colonies, the honey is harvested and bees are often fed a diet of sugar water during cold months.

Honey bees have become the most widely used pollinator in agricultural systems, largely because of their easily manageable, transportable, and extremely large

Reinventing the wheel

Even though Europeans brought honey bees to the New World in order to harvest their honey, native peoples of Central America have been harvesting honey from other native bees for thousands of years. Some stingless bees in the genus *Melipona*, a distant relative of the honey bee, were cultivated by the Maya and used for their honey production. Interestingly, a 14-million-year-old fossil of a close relative of modern honey bees was discovered in Nevada. This discovery suggests that honey bees, or at least their relatives, were once more widespread than they currently are.

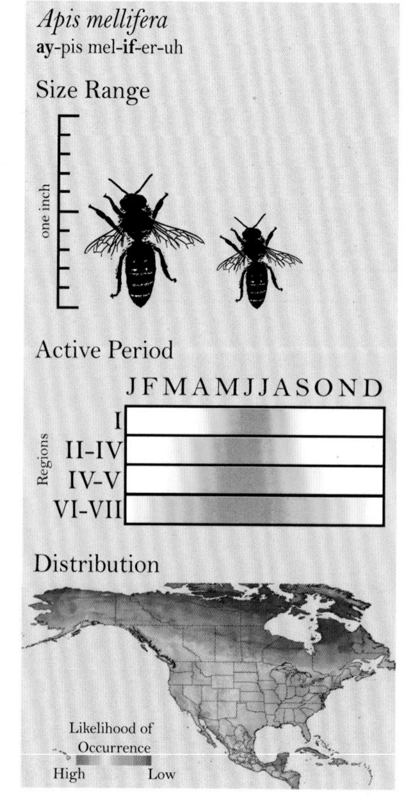

Even though only one species of honey bee is found in North America, this bee is highly variable in coloration, even within the same colony. These two images illustrate some of the variability in honey bee color.

colonies. In fact, honey bees account for nearly 80% of all crop pollination in the United States. Because of their wide use as pollinators of commercial crops, the value of honey bees is estimated to be around $15 billion annually. Most beekeepers use modern box-like hives to house their colonies, which can be easily transported from one crop to another. This allows beekeepers to rent out their colonies for crop pollination across the country. Some are shipped from Michigan to California to Florida in the span of a few months in order to provide pollinators for the crops grown in each of these states at different times of year.

Other Biological Notes

One aspect of honey bee biology that makes them particularly valuable for agriculture is that they are active primarily on the basis of temperature rather than seasonality the way other bees are. Honey bees are most active between 60 and 105 °F, though they can forage at temperatures as low as 55 °F. This characteristic makes them valuable pollinators of crops that bloom before many native bees are active. For example, almond trees in California will flower in February, earlier than most native bees typically emerge, but honey bees will effectively pollinate the almond trees as long as the temperature is warm enough. The ability of honey bees to forage at such a large range of temperatures means that they are active somewhere in North America during every month of the year.

A honey bee hovering near a century plant flower (*Agave*).

People often think of honey bee hives as gray balls dangling from tree branches. In reality, honey bees do not build enclosed hives; instead, they build in cavities, like hollow trees or crevices in rocks. Pictured here is an abandoned hive of wild (feral) honey bees built in a small rock cavity. Notice the empty combs made from wax. Photo by Lindsey E. Wilson.

While many styles of hives are used by beekeepers, the most common are boxes. This type of hive is used by commercial beekeepers because they are easy to transport and can be stacked together for efficient use of space. Photo by Mike Wells of Harvest Lane Honey.

Despite the common depiction of honey bee hives as oval-shaped structures, covered in a paper-like substance and hanging from a tree branch as a tempting snack for a cartoon bear (think of Yogi Bear and Winnie the Pooh), honey bees actually nest in cavities like hollow trees or rock cavities. The paper nests we see hanging from tree branches belong to wasps.

Predators seem to inordinately target honey bees and their colonies (to feed on the honey and the larval bees). As a result these bees have evolved some unique defenses. First, inherent to their social structure (see below for a description of honey bee sociality), they are able to defend their hive as a coordinated group, with several hundred or thousand bees attacking a predator. Second, honey bee workers have a unique sting with barbs on it, like a harpoon. When a honey bee stings a predator, the barbs on the sting cause it and the attached venom sack to lodge in the skin of the attacker. The venom sack continues to pump venom through the sting into the skin of the predator, potentially inflicting more damage than barbless stings inflicted by other bees. One result of the honey bee's sting and venom sack becoming lodged in the attacker is that when the bee pulls away from her victim, she suffers a mortal wound; her sting and venom sack are ripped out of her abdomen. A honey bee worker can thus sting only once, dying soon after stinging something. Despite this deadly defensive strategy, a colony consisting of tens of thousands of worker bees will be better off losing a few hundred members in defense of the hive than having the entire hive destroyed.

Dozens of books have been written about honey bees, specifically about their social structure. Because of this, we will give just a brief overview of honey bee sociality. Honey bees are known to be eusocial (truly social), meaning they live in a colony that participates in cooperative care of the young and there is a division of labor with reproductive and nonreproductive groups. A honey bee colony typically consists of a single reproductive queen, several thousand drones (fertile male bees), and tens of thousands of sterile workers, which are all female.

QUEENS. Queen honey bees are adult females that have mated. Typically a queen will mate with 12–15 drones. She then stores the sperm from these drones in a special organ called a spermatheca. In this organ she will keep the sperm alive for the remainder of her life, which can

Custom bee houses

In addition to the natural cavities used by honey bees for nests, they will also nest in a variety of man-made structures. Large clay jars were traditionally used in parts of the Mediterranean and Middle East. Bee skeps, resembling a basket made of straw turned upside down, were customarily used in Europe. Hollow logs, often called bee gums, were also periodically used by early beekeepers. Modern beekeepers often use box-like structures with removable frames, making the harvesting of honey much more efficient than other hive structures. These modern beehives are also useful in that they can be neatly stacked together for easy transport from one location to another. Thus beekeepers can move their colonies from one field to another, enhancing the bees' role as pollinators of agricultural crops.

Honey bees are the only bees that lose their stings when they sting you

No other bee in North America has barbs on its sting, so other bee stings don't catch in the skin the same way; therefore, all other bees can sting you more than once. They won't, however, unless thoroughly provoked.

be from two to seven years. A queen bee can choose whether she wants to lay an unfertilized egg, which will develop into a male bee, or to use some sperm from her spermatheca to fertilize an egg, which will develop into a worker or new queen (see haplodiploid sex determination, p.15). At the peak of egg laying, a queen can lay up to 2000 eggs a day—more than her own body weight. This feat is made possible by the hoard of worker bees meticulously attending to the queen's every need, including feeding and cleaning her.

WORKERS. Worker honey bees are females that lack full reproductive capabilities. Worker bees have a much shorter life span than the queens, typically living only a few months, even fewer during the summer season. Over the entirety of their lives, worker bees' duties include cleaning brood cells, feeding larvae, feeding drones, feeding the queen, producing wax, building brood cells and honeycomb, sealing full honeycombs, attending the queen, storing pollen, removing dead larvae and workers, fanning the hive to maintain a constant temperature, carrying water, foraging for pollen and nectar, and defending the colony. These tasks are taken on sequentially, with worker bees playing different roles as they mature. A young worker bee is responsible for cleaning the brood cells, the areas of the hive where larvae grow and develop. Older worker bees function as nurse bees, responsible for feeding the colony's larvae. These larvae will develop into either more workers or future queens, depending on the diet they are fed by the nurse bees. After being a nurse bee, a worker honey bee focuses on wax production, then on repairing the wax cells used to house larvae and honey. Wax is produced by the worker through special glands on the abdomen. These roles: cleaning, nurse bee, and wax production, last about half a worker's life; the remainder is spent foraging.

Honey bee workers exchange information about good foraging resources by dancing inside the hive. This process, known as the waggle dance, can portray information about the distance and direction of a particular resource from the hive, so that other workers will be more efficient in their foraging (see waggle dance, p.20).

Why a honeycomb?

The characteristic hexagonal shape of the individual cells in a honeycomb are so-shaped for a reason: it turns out that six-sided walls are the most efficient use of wax, allowing bees to produce the greatest number of individual cells with the least amount of wax.

Often honey bees will construct brood cells (where they raise the next generation of bees) and honey cells (where they store honey) in close proximity to each other. Here, worker bees are building brood cells near the middle of the comb and honey cells near the corners. The completed brood cells are capped with a darker orange cap, while the honey cells are capped with a lighter wax cap. PHOTO BY MIKE WELLS OF HARVEST LANE HONEY.

DRONES. Drones are fertile male bees. Drones do not perform tasks around the colony the way workers do; their main role is to mate with a receptive queen (from a colony other than their own). The life expectancy of a drone is typically about three months.

KILLER BEES

Killer bees, or more appropriately, Africanized honey bees, are a hybrid of two subspecies of honey bee. Subspecies are separate populations that are still capable of interbreeding—the domestic dog and the Mexican wolf, for example, are both *Canis lupus*, but they represent separate subspecies. The African honey bee variety (*Apis mellifera scutellata*) was crossbred with a European honey bee variety (*A. mellifera iberica* or *A. mellifera ligustica*) in hopes of developing a honey bee capable of thriving in the tropics,

Bomb-sniffing bees

Honey bees have been trained to act as bomb detectors. While this may conjure images of a swarm of bees flying around an airport, buzzing to alert authorities to bombs in people's luggage, the reality is much less exciting. Like all bees, honey bees excel at detecting small amounts of airborne chemicals (this is how they find flowers as they fly around a landscape). Researchers have used this ability to train honey bees to react to minute amounts of chemicals, specifically those found in explosive compounds. Similar to the way one would train a dog to identify the odors of drugs, trainers reward honey bees with sugar water when the bees correctly sense a particular explosive compound; using their extended tongues, the honey bees suck up the sugar water. Eventually, the bees stick their tongues out in expectation of the reward every time the chemical compound is detected. Trained bees can then be exposed to different air samples. If a particular compound is present, a bee will stick out its tongue, alerting the trainer that explosives are present in the area.

where honey bees typically do not fare well. While these Africanized hybrids are nearly impossible to distinguish from typical European honey bees based on physical characteristics, Africanized bees exhibit more defensive behaviors than any of the European varieties of honey bee. The sting of the Africanized bees is no more potent than that of a typical European honey bee, but the Africanized bee is more easily provoked and will attack in greater numbers than typical European honey bees. This hyperdefensive behavior makes the Africanized bees more dangerous than European honey bees and has led to the deaths of pets, livestock, and, in some cases, people. In addition to being overly aggressive, Africanized bees are less choosy about where they build their hives than are European bees. The newer hybrid will colonize smaller areas and is more likely to nest in ground cavities than its European cousin, potentially allowing greater contact with humans as they are more likely to establish in urban or suburban areas. Africanized varieties cannot survive long periods of time without foraging, and they do not fare as well when the weather is cold. This has prevented them from establishing in places with harsh winters or very dry summers.

Disappearing Honey Bees

An important topic seen frequently in the news is the "disappearance" of honey bees. The cause of honey bee declines is currently a focus of numerous scientific studies, but the reason remains elusive. Bees have always suffered from a number of issues related to nesting close together (mites, viruses, fungi, etc.). The current epidemic, wherein all the workers in a hive disappear (leaving the queen behind), is called colony collapse disorder (CCD). CCD is concerning because our agriculture system relies heavily on honey bees for pollination, and the shortage of healthy hives to transport to flowering crops means much-reduced fruit set. It is important to realize that this phenomenon is not one in which all the bees die (with dead workers filling the hive), but rather a case where something causes all the worker bees to leave the hive and not return. Similar things have happened in the past; for example, symptoms similar to those of CCD were documented as far back as 1869. In the early 1900s, a similar phenomenon was called disappearing disease. While an explanation for CCD remains uncertain, scientists think that other factors may be exacerbating the problem (pesticides, parasitic mites, viruses, and climate change).

Are bee populations declining?

There is mounting evidence that at least some bees other than the honey bee are also experiencing population declines. The culprit appears to be, in large part, the compounding effects of modern human activities. As an example, consider the bumble bee. Substantial evidence shows precipitous declines in the population sizes of numerous European bumble bee species over the last 50 years, with several now extinct and a number on the brink. Declines among North American species have been harder to verify (there is less historical data, and many areas of the west remain largely unsampled). One species (*Bombus franklini*) has not been seen in nearly 10 years. The most commonly seen bumble bee in the west was once the western bumble bee (*Bombus occidentalis*). It is now extremely rare in most of its range. Similar declines have been documented for several eastern species. A direct cause-and-effect relationship that universally explains bumble bee declines has yet to be discovered, but several factors may be intensifying the problem. Habitat loss due to agricultural practices and urbanization can both negatively impact bumble bees, more so than other bees, because their foraging distances are so much bigger and because they are social (i.e. they need more resources than an empty lot can provide). Pesticides seldom target one particular insect, and beneficial insects are an unfortunate by-product. Lastly, bumble bees are often kept in close quarters inside green houses for use in pollinating green house tomatoes and other crops. In such close proximity, they are often prone to outbreaks of various pathogens. Though the bees live inside greenhouses, it is apparent that they often escape, potentially spreading with them large loads of mites, fungi, and other harmful parasites. The specifics are still being pieced together, but it has been found that the bumble bees in decline have greater levels of infection than species that are not.

8.12 *EUGLOSSA*

The name Euglossa means "well-tongued," referring to the extraordinarily long tongue of these bees.

Description

Euglossa is a group of tropical bees that include some of the most striking of all bees, often with bright metallic green-, blue-, or even copper-colored integument (skin). The male *Euglossa* have an interesting habit of collecting scent compounds, and because many species collect the floral scents from orchid flowers, these bees are often called orchid bees.

Distribution

Euglossa are native to the tropics of Central and South America. Just a few species are found outside tropical environments, with a handful ranging north through the mountains of northern Mexico. Though mostly tropical, *Euglossa* have been collected in habitats ranging from rainforests at sea level to cloud forests above 6000 feet in elevation. Recently some populations of *Euglossa dilemma* have established in the United States, specifically in southern Florida. A few *Euglossa* specimens have also been collected in southern Arizona and Texas, but no evidence has been found that breeding populations have settled in these western areas. Instead, these western observations are thought to be of male bees that have strayed north of the breeding populations found to the south in Mexico.

Diet and Pollination Services

Across their range, *Euglossa* use a variety of plants and seem fairly opportunistic in terms of the plants they will visit for both pollen and nectar. In Florida, *Euglossa dilemma* is often seen visiting blue and violet flowers, especially morning glory, pickerel weed, and porterweed. In addition to visiting flowers for pollen and nectar, female *Euglossa* collect resin they use in nesting. *Euglossa dilemma* is known to collect resin from the flowers of a plant called *Dalechampia* and may also visit cashew and conifer trees

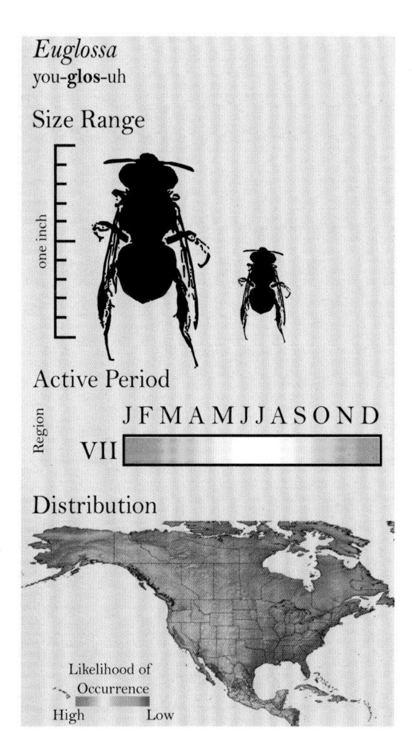

A *Euglossa* foraging on porterweed (*Stachytarpheta*). Photo by Alan Chin Lee.

for resin. Male *Euglossa*, like all male bees, visit a wide variety of plants for nectar. Unlike other bees, though, *Euglossa* males also collect scent compounds and actually store these chemicals on their specially modified hind legs. Male bees will collect these perfumes from a variety of sources, including fungi, rotting vegetation, and a range of flowers, most notably orchids. It is thought that *Euglossa* use these scent compounds in species recognition and that they might also be used in attracting females. Perhaps the male with the strongest scent is more attractive to a female. Because many *Euglossa* males visit orchids to collect their scents, they play a crucial role in the pollination of many orchid species.

Like all Apidae, *Euglossa* have very long tongues. Photo by Alan Chin Lee.

Other Biological Notes

Euglossa are cavity nesters, and females use plant resins to construct nests in a variety of natural and artificial cavities. Nests are often used by successive generations. Most *Euglossa* are solitary nesters, but some share nests with other females, often their sisters. Sometimes a mother will establish a nest with one or more of her daughters. This behavior of sharing nests among females representing multiple generations indicates that some primitive sociality exists among these bees. Many bees in North America mass-provision their nests, meaning they leave all the pollen needed by their young before they seal the nest and leave. In contrast, *Euglossa* provide pollen and nectar for their young as the larvae grow. This is called progressive provisioning and is not found in most bees in North America.

A *Euglossa* drinking nectar from a firebush flower (*Hamelia*). Photo by Alan Chin Lee.

Part-time Residents

A second genus of orchid bee occasionally strays into the United States: *Eulaema*. These iridescent beauties collect oils, and visit orchids, just as *Euglossa* do, and they are likely important pollinators of orchid flowers in the areas of the Central American tropics where they are abundant. In the United States, these bees are very rarely seen along the Arizona and New Mexico borders.

A *Euglossa* sleeping, legs curled close to the body. Male *Euglossa*, like many other bees, sleep by biting onto a twig or leaf and dangling there, supported only by their mandibles. Photo by Alan Chin Lee.

9

POLLEN THIEVES

Cleptoparasitic bees, like those pictured here, look more like wasps than bees. Their color and hair patterns are striking and as varied as the bees on which they rely. While many of these bees are seldom seen, others are as common as their hosts. Those pictured here are presented roughly to scale.

Though most bees fill their days visiting flowers and collecting pollen, some genera take advantage of the hard work of others. These thieving bees sneak into the nest of an unsuspecting "normal" bee (known as the host), lay an egg near the pollen mass being gathered by the host bee for her own offspring, and then sneak back out. When the egg of the thief hatches, it kills the host's offspring and then eats the pollen meant for its victim. Sometimes called brood parasites, these bees are also referred to as cuckoo bees, because they are similar to cuckoo birds, which lay an egg in the nest of another bird and leave it for that bird to raise. They are more technically called cleptoparasites. *Clepto* means "thief" in Greek, and the term *cleptoparasite* refers specifically to an organism that lives off another by stealing its food. In this case the cleptoparasite feeds on the host's hard-earned pollen stores.

Many species of cuckoo bees are known, and they are not all in the same bee family; the lifestyle has in fact evolved at least 27 times, each in a different bee lineage. Cleptoparasitic bees parasitize only other bees. While some genera are very common and as frequently seen as their hosts, others are rare or seldom observed or collected. The normal methods of learning about bees don't always help when studying cuckoo bees. First, because they don't visit flowers for pollen, build nests, or (for males) establish

Size varies

Interestingly, any one cleptoparasite species can vary greatly in size, though individuals are never bigger than the host. The cause for this size variation is unknown.

Emry's rule

Cuckoo bees often parasitize bees that are closely related to them, usually in the same bee family. This strange pattern is known by the name of the scientist who first noted it.

A species of *Triepeolus* roots for nectar on a sunflower.

A *Sphecodes* resting on a flower before she heads out to look for the nests of her host, bees in the family Halictidae.

A *Nomada* bee drinking nectar from an aster (*Aster*). *Nomada* look more wasplike than many other cleptoparasitic bees.

A *Nomada* sunning itself on a leaf. Cleptoparasitic bees can often be seen resting on leaves, twigs, and flowers as they wait for a chance to infiltrate their host's nest.

territories, it is difficult to observe their day-to-day behaviors; they have no central location from which they work. Second, they are less common than their hosts (or their hosts would go extinct). And, finally, observing their behavior, growth, and development inside a nest is difficult. As a result, researchers do not know a lot about these bees. Rather than elaborate for each genus, as we have done for all other bees in this book, we will summarize what little is known about each of the cleptoparastic bee genera, focusing on those that are common and most likely to be seen.

While the behaviors of cleptoparasitic bees differ from genus to genus, the general plan of attack is the same. An adult female crawls into the nest of her host and lays an egg in a cell that is in the process of being filled with pollen. The egg is very tiny, often laid in a little crack or plastered on the side of the nest so the host bee is less likely to see it. The adult cleptoparasite may lurk outside, near the nest entrance, laying eggs in successive nest cells as the host bee prepares them. When the egg hatches, the larva often has gigantic pincer-like mandibles (jaws), and it uses these to kill the host larva or egg. It then eats the pollen left by the host mother for her (now-dead) offspring.

In North America, there are cleptoparasitic bees that belong to three of the bee families discussed previously: Apidae, Halictidae, and Megachilidae. Apidae has the most genera of cleptoparasites (it also has the most genera of other bees, too), whereas Megachilidae is represented by just three cleptoparasite genera, and Halictidae by two.

One-hit wonders

A handful of cleptoparasites have been found only rarely in the United States, usually along the Mexican border, and never in Canada. These bees are usually known from one location, from one particular collecting event, and by fewer than a dozen specimens. While they are technically bees of the United States, they are probably not long-term residents. More likely, they were blown off course during a storm or hopped over the border during a particularly wet and flower-rich year. Alternatively, they could be residents that are so rare that, despite relatively intensive collecting along the border, they have managed to avoid capture. These rare bees are *Temnosoma smaragdinum*, *Mesoplia dugesi*, *Brachymelecta mucida* (known from one specimen collected in Nevada around 1878), *Odyneropsis apache*, and *Rhopalolemma* (known from two specimens representing two species: *Rhopalolemma robertsi*, collected in California once, 70 years ago, and *Rhopalolemma rotundiceps*, collected once in Arizona in 1997). Other than this mention, we do not include them in our book.

Stelis is a parasitic bee that visits species in the Megachilidae bee family. The yellow markings on the abdomen of this otherwise dark bee are a good way to identify it.

IDENTIFICATION TIPS

A female lacking pollen-collecting hairs is either *Hylaeus* (section 4.2) or a cleptoparasite. With cleptoparasites, the general body shape is often the most diagnostic feature—spines and thickened integument (skin) are common (they often have to defend themselves against the females whose nests they invade). Look for flattened white hairs, pointy body parts, and a hovering behavior over the ground (as they search for bee nests). The identifying features of all the parasitic bees are presented here, rather than for each of the three sections. Many are difficult to tell apart; without need of pollen-collecting structures and nest building apparati, they lack many of the features commonly used to tell other bees apart. We provide guidelines for telltale features, but without a microscope and considerable knowledge of bee genitalia, settling for the family or tribe to which a parasite belongs may have to be satisfactory.

A ring under the collar

Because, like wasps, cleptoparasites do not collect pollen, they often look more wasplike than beelike. Their bodies are rather hairless, with spindly legs that lack scopa, and they are often seen hovering over the ground in a manner similar to a wasp hunting for prey. Even the biologists have had trouble distinguishing between the two. Look for the silvery hairs on the face, and the close-together antennae, which are common to many wasps. With a microscope you may also be able to see the pronotal lobe. On all bees there are branched hairs on this lobe. Even the bees with very few hairs exhibit a few feathered hairs near the pronotal lobe. Wasps have hairs on their pronotal lobes too, but the hairs are simple, and without any splits.

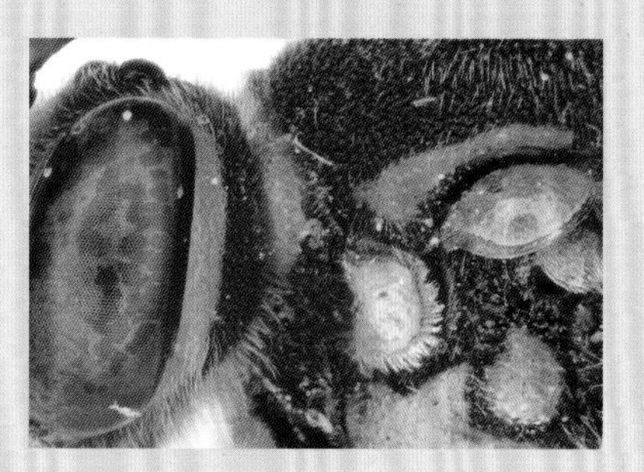

Bee

Wasp

CUCKOO BEES: APIDAE (SECTION 9.1)

TRIBE NOMADINI (*NOMADA*)

Nomada are one of the easiest cleptoparasites to identify. With striking yellow, black, white, and even red markings, they are easily seen on flowers and as they hover close to the ground, looking for empty bee nests. They frequently parasitize *Andrena* nests; if you've determined that *Andrena* are in the area, *Nomada* are likely too.

Identifying Features of Tribe Nomadini (*Nomada*)

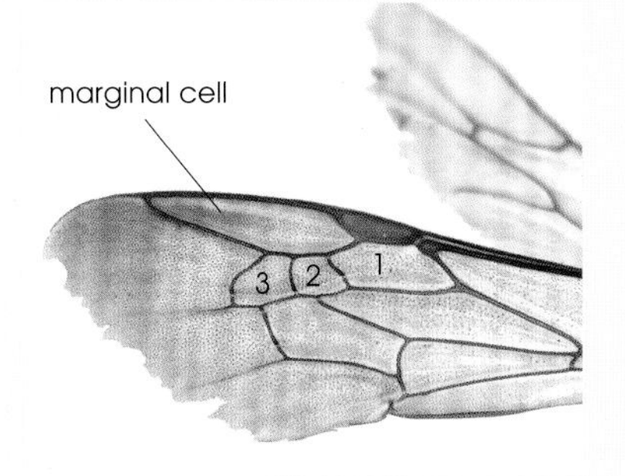

LEFT: The apex of the marginal cell with *Nomada* is sharply pointed and ends right on the wing margin. There are usually three submarginal cells, with the third one being relatively short. A few species have just two submarginal cells.

RIGHT: In *Nomada* the mandible and the compound eye are not perfectly aligned, with the compound eye shifted slightly back from being centered over the mandible base—compare this with *Brachynomada*.

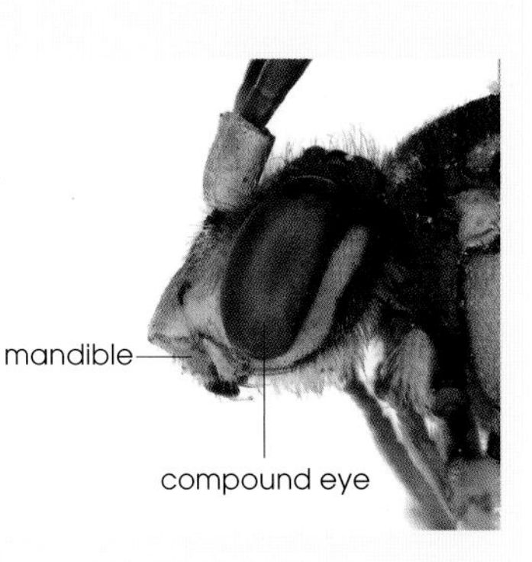

Similar Species to *Nomada*

Brachynomada, later in this section

Wasps can also look similar to *Nomada*. Watch for silver hairs on the face of wasps, and branched hairs around the edge of the pronotal lobe—wasps have no branched hairs.

TRIBE AMMOBATINI (*OREOPASITES*), Identifying Features

LEFT: *Oreopasites* are minute, with a red abdomen and rust red to soft gray thorax. The hairs may be short, but they are not scale-like.

RIGHT: Unlike those of many other parasitic bees, the antennae on *Oreopasites* are not particularly short or stubby.

Similar Species to *Oreopasites*

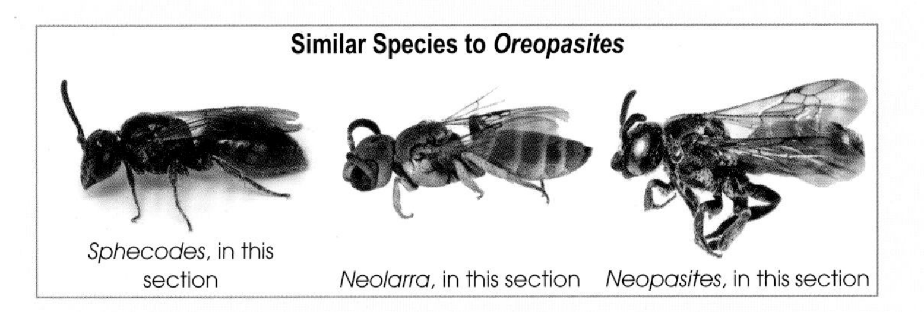
Sphecodes, in this section | *Neolarra*, in this section | *Neopasites*, in this section

TRIBE NEOLARRINI (*NEOLARRA*), Identifying Features

LEFT: Miniscule, with either a black or red abdomen, these tiny bees are often collected along with *Perdita* (section 3.3). The hairs on the body are short and flattened to resemble scales.

RIGHT: The wing of a *Neolarra* is very different from that of all other bees. The marginal cell is tiny—smaller, almost, than the stigma it abuts. It may have one or two submarginal cells, but the second is often quite small. The majority of the wing is veinless.

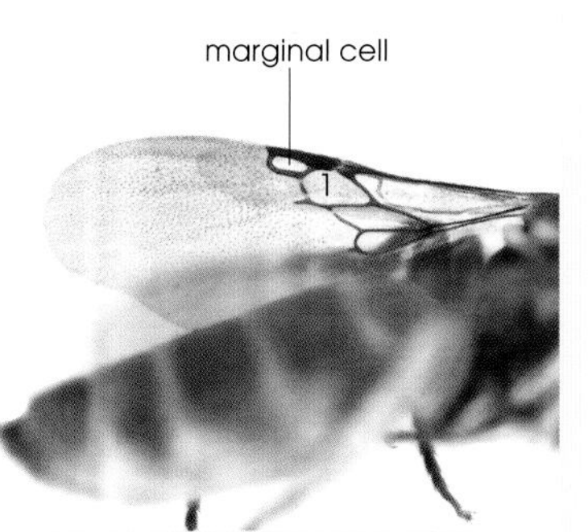

continued overleaf

TRIBE BIASTINI (*NEOPASITES*), Identifying Features

LEFT: Another group of very small bees, *Neopasites* have either a red or a black abdomen (usually red) and short hair that stands erect. These hairs are not flattened. The antennae are fairly short and stubby. If you look at the end of the abdomen there is no pygidial plate.

RIGHT: On the wing of *Neopasites*, there are two submarginal cells. The marginal cell is relatively long and pointed at its tip. The veins and accompanying cells take up a good portion of the wing. These features can distinguish it from the similar-looking *Neolarra*.

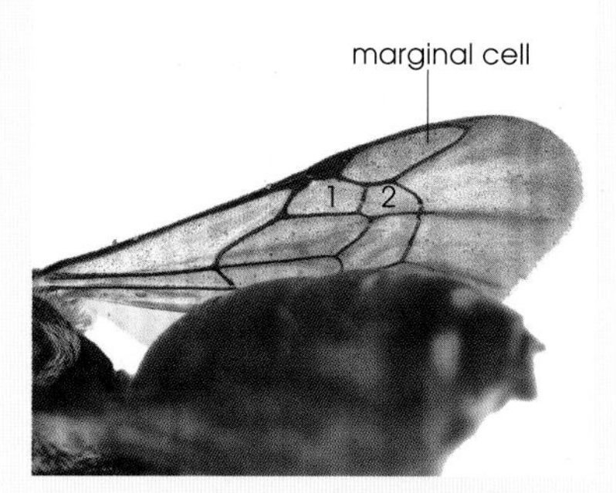

TRIBE OSIRINI (*EPEOLOIDES*), Identifying Features

These small to medium-sized bees have a hunchbacked look in profile. They are black, often shiny, but covered in very short, feathery hairs. The eyes of the males come together at the top, giving the face an almost cross-eyed appearance. The clypeus sticks out from the face more than on most cleptoparasites. The legs are long, and the tibia and femur of the hind leg are robust. The wing has three submarginal cells.

TRIBE EPEOLINI (*EPEOLUS* AND *TRIEPEOLUS*), Identifying Features

Triepeolus

Epeolus

Of the parasitic Apidae, the Epeolini are among the most recognizable. With a characteristic "smiley face" marked in short woolly hair across the thorax, the look is unmistakable (see the text for images of the smiling face). These bees tend to be larger than many parasites, and a close inspection of the thorax may reveal spines on the back end of the scutum.

TRIBE BRACHYNOMADINI (*BRACHYNOMADA*, *PARANOMADA*, *TRIOPASITES*)

Paranomada *Brachynomada* *Triopasites*

Brachynomadini is a tribe of small bees with the same overall look as *Nomada*. Generally they are black, but some have the thorax and abdomen red. They differ from most *Nomada* in that they never have yellow or white markings on their bodies, and they are smaller. *PARANOMADA* PHOTO BY USGS BEE INVENTORY AND MONITORING LAB.

Identifying Features of Tribe Brachynomadini (*Brachynomada, Paranomada, Triopasites*)

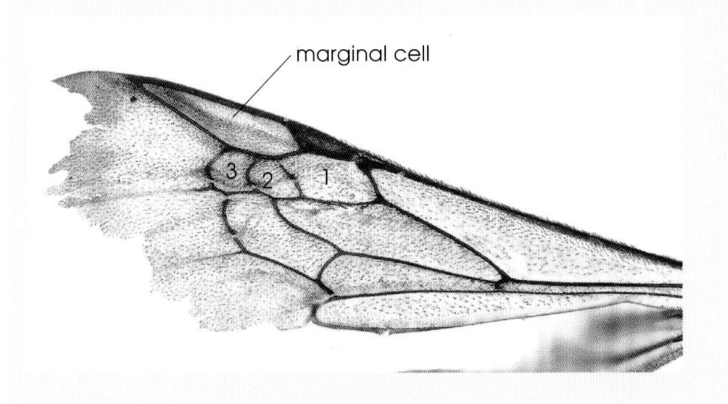

ABOVE: There are two or three submarginal cells on the wing. The marginal cell ends just off the wing margin. It is rather sharply pointed (while it is more rounded at the tip in *Nomada*, the bee genus that most closely resembles the Brachynomadini).

RIGHT: The mandible is more centered under the eye in the Brachynomadini than it is in the similar-looking *Nomada*, because the eye is tilted forward.

compound eye
mandible

TRIBE HEXEPEOLINI (*HEXEPEOLUS*), Identifying Features

Hexepeolus are striking bees with white stripes on the often red abdomen that break right in the middle of each tergal segment. The axillae (little projections on each side of the scutellum) are not sharply pointed.

TRIBE AMMOBATOIDINI (*HOLCOPASITES*), Identifying Features

These striking bees are petite, often with a bright red abdomen, and patches of white scale-like hair on the abdomen, the thorax, and even the back of the head. The eyes of these bees are hairy. The antennae are relatively long, not stubby as in many small parasitic bees. Note on the wing that the marginal cell is rounded at its tip, which is not on the wing margin. Also, the second submarginal cell is much smaller than first.

TRIBE TOWNSENDIELLA (*TOWNSENDIELLA*), Identifying Features

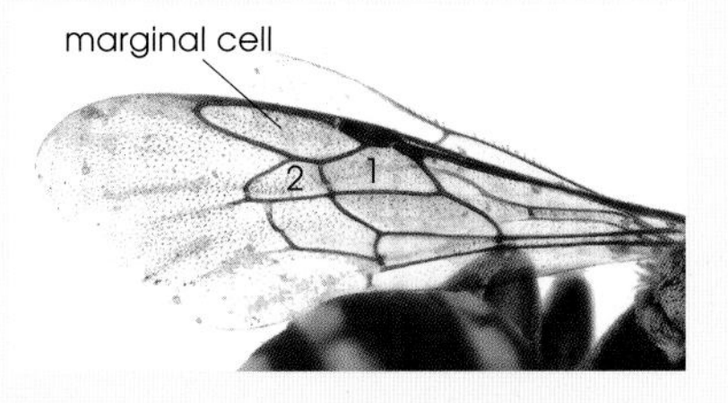

The wing on a *Townsendiella* bee has two submarginal cells, though the second is relatively short and angled noticeably away from the marginal cell. The marginal cell is rounded at the tip and bent gently away from the wing margin, and it is much bigger than the stigma.

ABOVE LEFT: *Townsendiella* are beautiful tiny bees, often with a red abdomen. They have short hair, pressed flat against the body, and the hairs themselves are flattened, so they resemble scales or shingles. A look at the tip of the abdomen of females reveals a pygidial plate.

continued overleaf

TRIBE PROTEPEOLINI (*LEIOPODUS SINGULARIS*), Identifying Features

LEFT: *Leiopodus* are fairly large cleptoparasites with a chunky, knobby thorax. They have appressed white hairs on the thorax and abdomen in distinct patches. The legs often have a red hue, as do the two little "knobs" on either side of the scutellum.

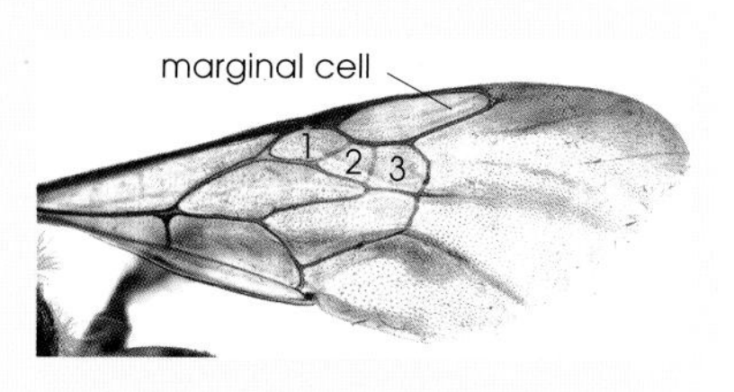

With *Leiopodus* the wing has three submarginal cells of roughly equal size; the marginal cell is rounded at its tip and ends off the wing margin.

TRIBE ERICROCIDINI (*ERICROCIS*), Identifying Features

RIGHT: *Ericrocis* are thick bees, with a large thorax and bright white-and-black markings on the thorax and the abdomen; on the abdomen these markings are often wavy, and the middle area contains no white hairs, giving the appearance of a black line down the middle of the abdomen. Females have a yellow clypeus, whereas that of males is black.

TRIBE MELECTINI (*BRACHYMELECTA*, *MELECTA*, *XEROMELECTA*, *ZACOSMIA*)

Xeromelecta *Zacosmia* *Melecta*

Perhaps the most striking of all the cleptoparasitic bees, the Melectini tend to be larger, and they often sport brightly colored hairs on the thorax. The thorax often has an abrupt hump in it, about even with the tegula, giving the bees a bulky appearance.

Identifying Features of Tribe Melectini (*Brachymelecta*, *Melecta*, *Xeromelecta*, *Zacosmia*)

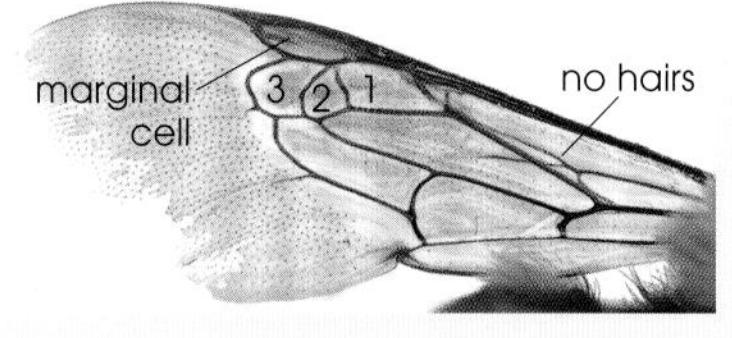

Like those of the Anthophorini they often parasitize, *Melecta* wings are bare in the middle area, with no little hairs. The marginal cell is rounded or oval-shaped with no evident point. It is short, barely longer than the end of the last submarginal cell.

Similar Species to the Melectini

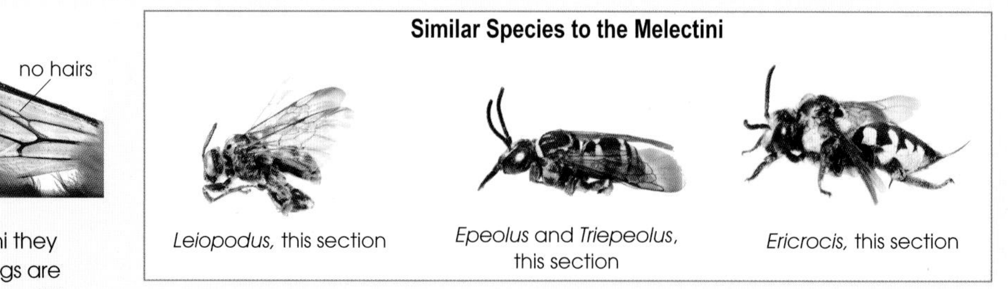

Leiopodus, this section

Epeolus and *Triepeolus,* this section

Ericrocis, this section

TRIBE BOMBINI (*BOMBUS (PSITHYRUS)*), Identifying Features

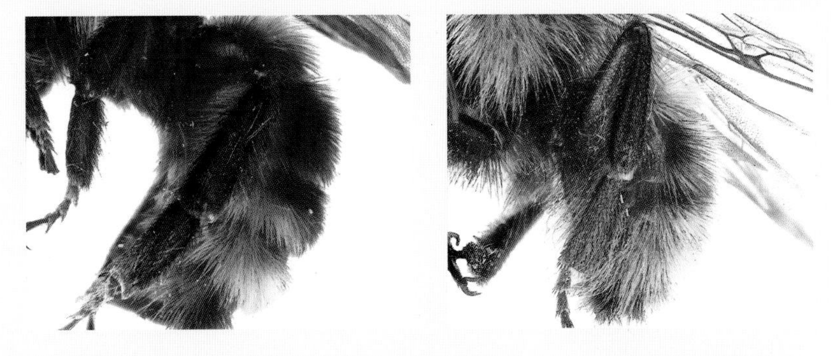

Psithyrus and their hosts, the nonparasitic bumble bees, look remarkably similar (perhaps this helps the parasites sneak into the bee nests in the first place). Both are hairy, with yellow, black, and even orange stripes across the abdomen, and they are both large; a close look reveals important differences, however.

Because *Psithyrus* don't collect pollen, females have lost the corbicula (pollen basket) on their hind legs, which are instead lightly hairy (compare the left image with the image of a bumble bee corbicula in the bumble bee identification section). Even in the males (right), which also lack pollen-collecting hairs, the center of the hind tibia of nonparasites is rather hairless, whereas in *Psithyrus* it is uniformly hairy across the whole leg segment.

Similar Species to *Psithyrus*

Bombus (section 8.10)

CLEPTOPARASITIC BEES: HALICTIDAE (SECTION 9.2)

Sphecodes are the only common parasitic bee in the family Halictidae. They are in the Halictini tribe along with *Halictus* and *Lasioglossum* and share several tribal characters with these bees. Specifically, *Sphecodes* have an arcuate basal vein (see chapter 6 for images). They differ most obviously from the nonparasitic bees in that the females lack pollen-collecting hairs. Additionally the body tends to be strongly punctate (covered in tiny pock marks), and the abdomen is almost always red, though in a few species it is only partially (or rarely not at all) red.

CLEPTOPARASITIC BEES: MEGACHILIDAE (SECTION 9.3)

Dioxys

Coelioxys

Coelioxys and *Dioxys*, in the tribe Dioxyini, are distinct among parasites. The extremely pointed abdomen is an easy clue to their identity. They are generally black-and-white bees with strong white stripes across the abdomen.

Identifying Features of Megachilidae

LEFT: The sides of the thorax of *Coelioxys* have long pointy spines that hang over the propodeum. The spines are rounded or entirely missing in *Dioxys*, making it easy to distinguish them. RIGHT: *Stelis* have the body shape of a Megachilidae: rounded and stout. The abdomen generally has yellow or ivory markings, almost stripes, but often not complete across the width of the abdomen. On the middle tibia there are two spines at the tip, which are distinct and separate it from other Megachilidae (as does the lack of pollen-collecting hairs on the females).

9.1 CUCKOO BEES: APIDAE

The family with the most slackers

The Apidae bee family contains more cleptoparasitic bees than any other family of bees.

Cleptoparasitic bees in the Apidae bee family fall into two subfamilies: the Nomadinae and the Apinae. While all the Nomadinae likely arose from the same pollen-stealing ancestor a very long time ago, the cleptoparasitism observed among species in the Apinae likely resulted from several unique evolutionary events. Therefore, while all the Nomadinae share certain behavior patterns (with subtle differences), the Apinae parasites are more different from each other.

The Nomadinae are known for parasitizing nest cells that are open and in the process of being filled with pollen. They place their eggs inside the walls of the nest cells of the preferred host bees. Some Nomadinae put their eggs only part way into the wall, while others dig a small divot out of the wall, place their egg in this nook, and then seal it off with a bit of mud. Many Nomadinae probably return to the same nest over several days, laying eggs in each successive cell built by the host female. The larvae that hatch from the intruder's egg look like little monsters. They have enormous pincer mandibles and thickened integument (skin) on their heads. Rather than downward-facing mandibles, as seen in adult bees (even adult parasites), the mandibles are turned so they face directly out in front of the larva. Recall that bee larvae are generally legless grubs barely capable of forward motion; in contrast, the larvae of these cleptoparasites are extremely mobile. They fiercely attack the host's egg, or larvae if the egg has already hatched, and kill it using their scythe-like mandibles. The parasite's larvae then eats the host pollen, continuing to grow as any other bee, emerging the following year in sync with its hosts. Nomadinae parasitize a wide array of bees in many families (all of them ground nesters). All bees in the Nomadinae subfamily are cleptoparasites. These bees are split into 11 tribes, 9 of which include genera found in the United States.

In contrast, the Apinae subfamily is a large group that includes, in addition to some cleptoparasites, almost all the large fuzzy bees in the Apidae bee family. Four tribes that include cleptoparasitic bees from the Apinae are found in the United States and Canada. All the Apinae cleptoparasites parasitize closely related bees (i.e., bees in the Apinae subfamily, and often in the same tribes).

A *Nomada* searching for the nest entrance of her host, an *Andrena*.

SUBFAMILY NOMADINAE

Tribe Nomadini

NOMADA

Nomada is the Latinized version of the word "nomad," which, as in English, means wanderer. One can guess that early observers of Nomada watched them wander haphazardly inches above the ground in search of unoccupied nests.

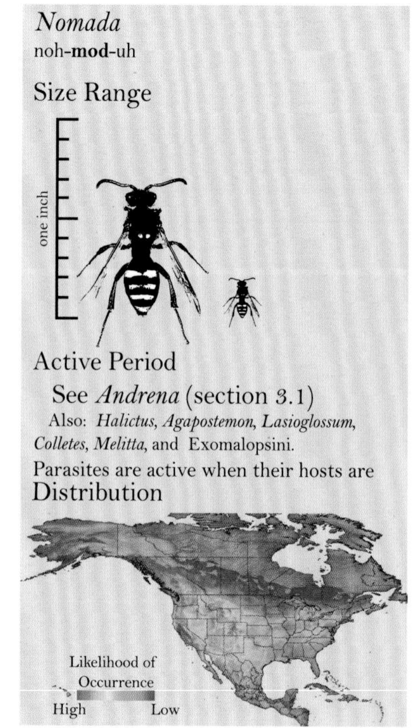

Nomada are among the most commonly encountered parasitic bees. Their extensive yellow and red markings and hairless bodies give them an appearance more similar to wasps than to other bees. They range in size from a miniscule 0.1 inch all the way to 0.6 inch. Around the world, *Nomada* are found all across Europe, south as far as South Africa, east to Japan, and even into India, the Philippines, and Australia. In the United States, *Nomada* are found as far north as Alaska and from coast to coast. There are more than 700 species worldwide. Nearly 300 of them occur in the United States but less than 40 are in Canada. Most species of *Nomada* are cleptoparasites of *Andrena*, and quite often when large aggregations of *Andrena* are nesting close together, many *Nomada* can also be seen perusing

Nomada come in a variety of colors. In North America they are the only bees that are entirely red.

A *Nomada* nectaring from a fleabane (*Erigeron*); they never forage for pollen on flowers.

A *Nomada* on a desert composite flower. Photo by Hartmut Wisch.

A *Nomada* drinking nectar from a yarrow flower (*Achillea*).

Nomada are often colored various shades of yellow, red, and black. Many species have red legs.

A colorful *Nomada* perches on a flower. Photo by Jillian H. Cowles.

the area, looking for a nest where the female is not at home. *Nomada* tend to fly more slowly than *Andrena*; this, along with their bright colorations, makes them easy to spot. If they enter a nest and find that the female *Andrena* is already inside, they may perch on a blade of grass or some other high spot and watch the nest entrance, waiting for the *Andrena* to leave before sneaking back in. Some species of *Nomada* in the United States and Canada are also parasitic on *Melitta*, *Colletes*, *Agapostemon*, *Lasioglossum*, *Halictus*, *Exomalopsis*, and *Eucera*; however, as far as researchers know, each individual *Nomada* species visits just one kind of bee nest and does not drop an egg in, say, a *Melitta* nest and also a *Colletes* nest. These bees will lay up to four eggs in one host nest cell, which also includes the host's egg. The female inserts the egg(s) halfway into the cell wall, so that half sticks out into the cell and the other half is embedded. The first *Nomada* larva to hatch quickly destroys all the other eggs (including its siblings), using its scissors-like mandibles before feeding on the pollen mass left for the host's larva.

Nomada, like other cuckoo bees, do not have a nest to sleep in at night, so many of them bite onto a twig and sleep dangling in the air. PHOTO BY DANA ATKINSON.

The scent of a woman

Oddly, male *Nomada* secrete odors that smell remarkably similar to the smells secreted by female host bees. Why a male *Nomada* would need to smell similar to a female host is entirely unknown.

Tribe Ammobatini

OREOPASITES

The name Oreopasites means "mountain Pasites"; oreo (or oro) is Greek for "mountain," and Pasites is a European cleptoparasite.

Oreopasites are rarely seen bees with white matted-down hair on their thoraxes (backs), and red abdomens. These bees are miniscule, from only 0.05 to 0.27 inch long. Though uncommon, their range is expansive, and they can be seen from southern Canada all the way through Mexico. There are 11 species in the world, all in the Western Hemisphere, and 10 of them occur in the United States and Canada. They use the pollen provisions gathered by bees in the Andrenidae bee family, mostly *Calliopsis* and *Perdita*.

A small *Oreopasites* on a cat's eye (*Cryptantha*) flower. PHOTO BY HARTMUT WISCH.

Tribe Neolarrini

NEOLARRA

Neolarra means "new Larra." Larra is a genus of wasp that this tiny bee closely resembles. The first scientist to examine this dead specimen thought it was a new kind of wasp, not a new bee.

Like *Oreopasites*, *Neolarra* are also microscopically small (0.05–0.25 inch), with slim bodies and red or black abdomens. They have a coating of short white hairs over most of their bodies, giving them a ghostly appearance. Found only in North America, they sneak into bee nests of

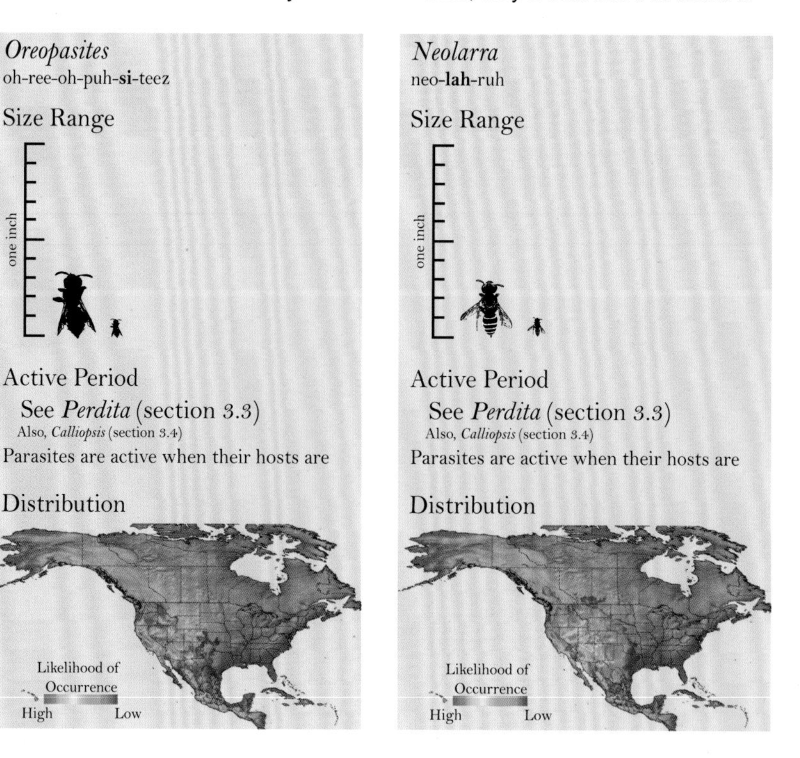

One of the smallest cuckoo bees is *Neolarra*. They match the size of their tiny hosts: *Perdita*. PHOTO BY HARTMUT WISCH.

the equally tiny *Perdita*. *Neolarra* are concentrated mostly in the southwest, but a few species can be seen as far north as Alberta and Saskatchewan in Canada, and as far east as Tennessee and Georgia. There are 14 described species, all of which can be found in the United States, and at least 2 of which occur in Canada.

Tribe Biastini

NEOPASITES

Neopasites means "new Pasites," referring to the similarity between this bee and the European genus Pasites. The genus Pasites was described by Louis Jurine in 1807; he didn't say why he named the bee Pasites, but the name is Greek for "all the same." In his tome on classifying bees and flies, though, he describes it as having characteristics similar to several other cleptoparasitic bees he was identifying (thus they are all alike). One can imagine that with a turn-of-the-seventeenth-century microscope, most things did look all alike!

These tiny bees fly aimlessly over the ground: small, nondescript gray gnat look-alikes that are nearly impossible to recognize. Their antennae are stout and thick, but so small that they are hard to see without a microscope. *Neopasites* are cleptoparasites of *Dufourea* and concentrated in California, though some range as far east as southern New Mexico. There are five species; all can be found in the United States, but none make it as far north as Canada.

Tribe Osirini

EPEOLOIDES PILOSULA

Epeoloides means "Epeolus look-alike," a reference to the vague similarities between this bee and the bee genus Epeolus, a much more common cleptoparasite.

For a cleptoparasite, this species is fairy unremarkable, with no bright hair bands or skin colorations; however, in profile it appears hunchbacked, as the thorax is distinctly hump-shaped. Only two species of *Epeoloides* are known in the world, one in Europe, and one in the United States and Canada (*Epeoloides pilosula*). It is found only in the east, ranging from Quebec and Nova Scotia, Canada, south to Georgia and west as far as the Dakotas. This bee is extremely rare and is recognized as endangered in Canada. Prior to 2002, it had not been collected in North America since the 1960s. It is a parasite of the also rare oil-collecting bee *Macropis*.

Tribe Epeolini

EPEOLUS

What the original identifier (Pierre Latreille) had in mind when he named this bee isn't known. Epeolus likely means "without a tail" and is perhaps a reference to the modifications, associated with the parasitic lifestyle, to the abdomens and genitalia of both males and females.

Epeolus are pretty bees with a chunky thorax and mesmerizing white-and-black stripes of close-cropped hair on their abdomens. A few species also have red markings

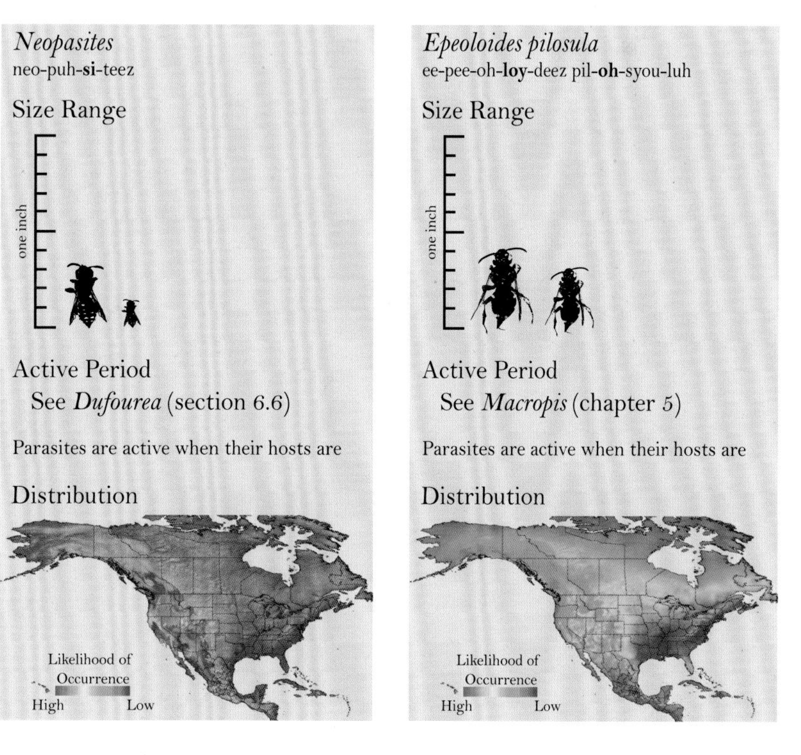

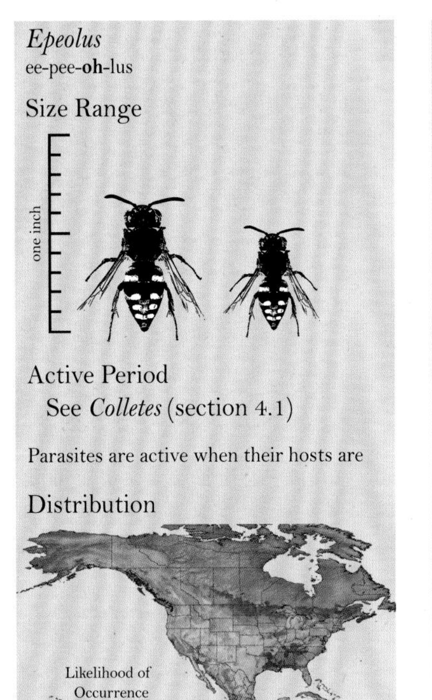

Triepeolus
tri-ee-pee-**oh**-lus

Size Range

one inch

Active Period
See Eucerini (sections 8.5–8.7)

Parasites are active when their hosts are

Distribution

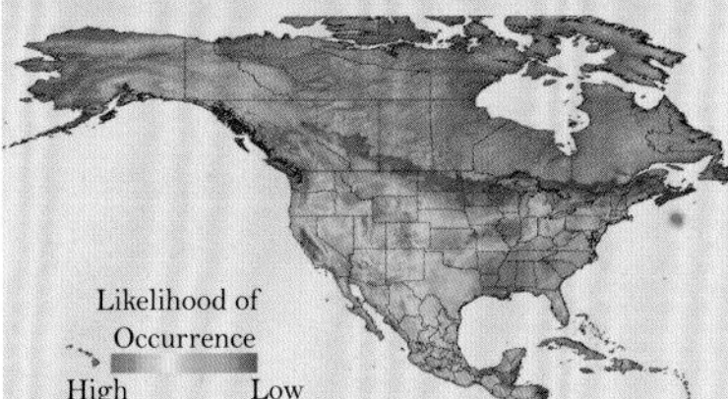

A small *Epeolus* bee. Photo by Jillian H. Cowles.

An *Epeolus* bee drinking nectar from a composite flower (Asteraceae). Photo by Jillian H. Cowles.

on the abdomen (e.g. *Epeolus zonatus*, which is found only in South Carolina, Georgia, and Florida). *Epeolus* are common around the world, ranging from Finland south through Africa and east all the way to Japan. They are rare south of the Himalayas. In the Western Hemisphere they can be found from subarctic Alaska through Mexico, and from coast to coast. They range in size from 0.2 to 0.6 inch. There are 57 species north of Mexico, and 12 in Canada. All species of *Epeolus* are cleptoparasites of *Colletes*. *Colletes* have interesting nest cells in that they line them with cellophane-like layers to protect the eggs inside from moisture and bacteria in the soil. A female *Epeolus* has spines on the end of her abdomen, and she uses these to pierce a U-shaped hole in the lining, laying an egg between two of the layers. She secretes a bit of a "glue" at the same time she lays the egg, so it adheres to the lining with the tip just flush with the inside of the cell.

An *Epeolus* bee on a composite flower (Asteraceae). Though there is a passing resemblance to *Nomada*, *Nomada* are mostly hairless, with integumental hair patterns. *Epeolus* in contrast have hair patterns created by appressed hairs. Photo by Margarethe Brummermann.

Triepeolus

The name Triepeolus refers to the fact that this bee looks like Epeolus except that it has three-segmented maxillary palps (mouth parts) instead of the two-segmented maxillary palps of Epeolus.

The striking black-and-white appearance of these bees makes them unmistakable as they cruise low over the ground, searching for host bee nests. In fact, the thorax often looks as though it is smiling at you, with two white eye dots, and a grin denoting the bottom of the scutum. To round off the dramatic look, they often have red legs. Some are quite large, well over half an inch, while others are only

Above: *Triepeolus* are striking bees, often with contrasting black and pale yellow stripes.

A *Triepeolus* resting on a mallow flower (*Malva*).

This Triepeolus looks as though he is foraging on the flower. In fact he is fast asleep, antennae relaxed out front, legs tucked tightly underneath.

Many *Triepeolus* appear to have a smiley face on the thorax. PHOTO BY JILLIAN H. COWLES.

a quarter of an inch long. They are an eclectic group of cleptoparasites and can be found associated with nests of Eucerini, *Anthophora*, *Centris*, *Melitoma*, *Ptiloglossa*, *Xenoglossa, Protoxaea*, and *Dieunomia*. *Triepeolus* will find a host as she is gathering pollen and follow her back to the nest. With so many potential hosts, it is no surprise that *Triepeolus* can be found coast to coast and all the way north to the lower provinces of Canada. These bees are also found in other countries, occurring in South America, south and central Europe, and as far east as Russia and Japan. There are more than 150 species in the world, and about 100 occur in the United States; 6 are in Canada. A female *Triepeolus* bee will follow a nest tunnel as it wends its way a foot or more down to an open cell. There, she lays an egg into the cell wall, so that just one end sticks out a little. The egg chorion (shell) is covered by a sort of hinged lid (called an operculum).

While most *Triepeolus* have a striped abdomen, some species have spots. PHOTO BY MARGARETHE BRUMMERMANN.

Tribe Brachynomadini

BRACHYNOMADA

Brachynomada means "little Nomada," the larger genus they closely resemble.

Several *Brachynomada* species have red abdomens. These bees are found from the northern Great Plains south through Central America. There are 6 species in North America, with probably 16 species worldwide, all in the

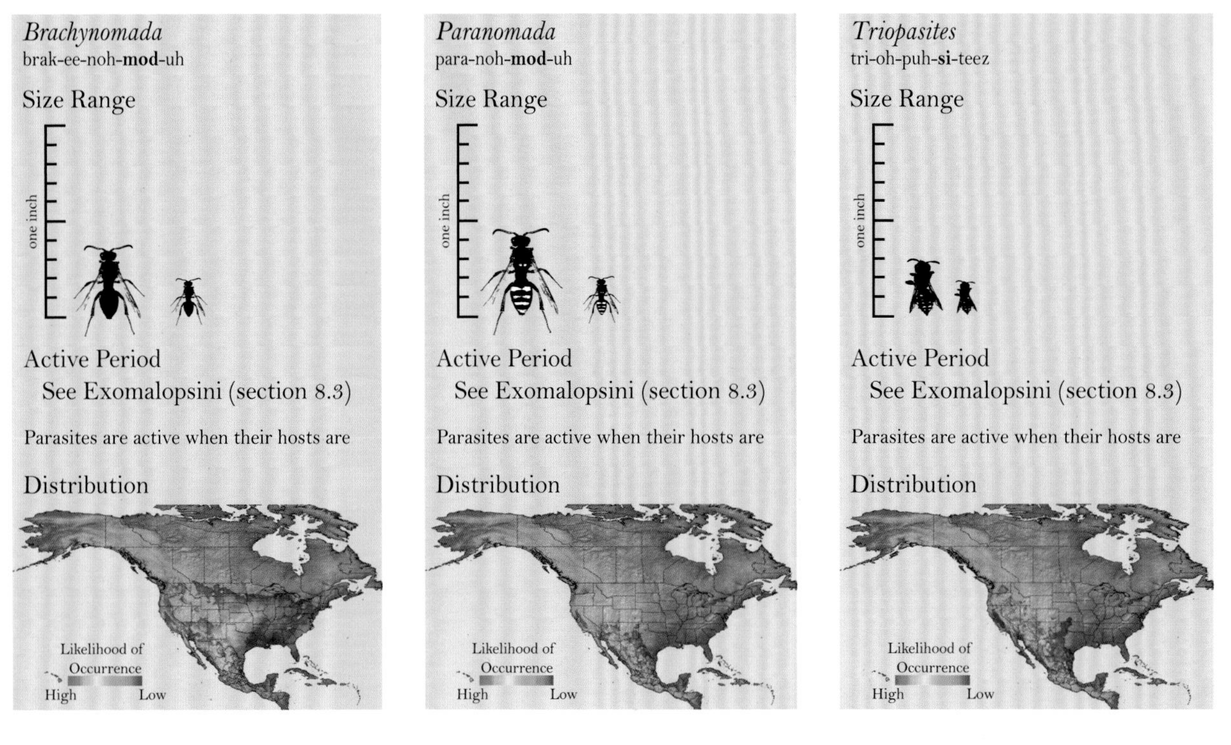

Western Hemisphere. They are tiny bees (0.15–0.35 inch) and specialize on the nests of the small bees known as *Anthophorula*.

PARANOMADA

Paranomada means "alongside Nomada," referring to the similarity in appearance between this bee and the more common genus Nomada.

Though they look superficially similar to the *Brachynomada*, *Paranomada* are flatter and smoother. They parasitize bees in the genus *Exomalopsis*. *Exomalopsis* often nest communally (p.21) with several females sharing the same nest entrance. This makes it more difficult for *Paranomada* to find the nest unoccupied (any one of several females could always be home). It is thought that their flat and smooth bodies enable them to press up against the side wall of the nest tunnel when a female passes and thus remain undetected. Three species occur, found only in southern Arizona and southern California, although they range as far as southern Mexico.

TRIOPASITES

Triopasites means "third Pasites," likely referring to the similarity of this genus to the European genus Pasites.

Like many cleptoparasites, these bees are gnat-sized (0.15–0.25 inch). Only two species of *Triopasites* are known, both of which occur occasionally in the United States. They are sometimes seen in the southwest, from southern California across to western Texas. Like *Brachynomada*, they parasitize *Anthophorula*.

Tribe Hexepeolini

HEXEPEOLUS RHODOGYNE

The name Hexepeolus means "six Epeolus," referring to the six-segmented palps (mouthparts) characteristic of these bees.

These are medium-sized white-striped bees, often with a red abdomen underneath the white bands of hair. Only one species of *Hexepeolus* is known in the world. It feeds on pollen gathered by species of the equally uncommon *Ancylandrena;* it is therefore found only in the southwestern United States, where its host occurs.

This *Hexepeolus* perches on a rock near its host's nest. *Hexepeolus* are known to attack the nests only of *Ancylandrena*. PHOTO BY MARGARETHE BRUMMERMANN.

Sleeping under the stars

Because cleptoparasitic bees don't build their own nests, they have no "home" to sleep in at night. Instead, many cleptoparasites sleep by biting onto the edge of a leaf or twig, locking their mandibles (jaws) and hanging on all night.

A specialist by default

Many of the bees robbed by *Holcopasites* are specialists. This means that the *Holcopasites* larvae eat just one kind of pollen, and it is not of their choosing. Interestingly, adult *Holcopasites* are often seen visiting the same flowers on which the host bee specializes, suggesting that they are inadvertently raised to be "specialists" of sorts.

This *Holcopasites* rests on a flower after drinking some nectar. These striking bees parasitize bees in the bee family Andrenidae. PHOTO BY JILLIAN H. COWLES.

Hexepeolus "pace" back and forth over areas where they have found bee nests, laying eggs in the same nest tunnels again and again with each new cell built and prepared by the host. With their larger size and white markings, it is easy to see this species as they are out on patrol. Eggs are deposited in cracks in the host's cell wall, pressed into it so that the egg is entirely flush with its surface. Often several *Hexepeolus* eggs will be found in the same nest cell; the strongest survives as the hatched larvae battle it out for ownership of the pollen mass. Usually the first to emerge wins, because it is larger, but this is not always the case.

Tribe Ammobatoidini

HOLCOPASITES

Holcopasites means "furrowed Pasites" in Greek, and the name is probably a reference to the very bumpy exoskeleton of this little bee.

Holcopasites have a red abdomen with darkened patches at the ends, and a dark black roughened thorax. With the antennae arising near the middle of the face, white spots, and a ripply-looking thorax, they are distinctive under a microscope. Though they are striking, they are so small that they are often overlooked (0.1–0.3 inch). They can be found from central Canada south to southern Mexico, and from coast to coast, though they are seldom seen in the Pacific Northwest. There are 16 species, 15 of which can be found in the United States and Canada. *Holcopasites* parasitize bees in the subfamily Panurginae (Andrenidae), including *Protandrena* and *Pseudopanurgus*. Like most cuckoo bees, *Holcopasites* usually occupy their host's nest for only a few minutes; however, some individuals have been known to spend over 30 minutes inside.

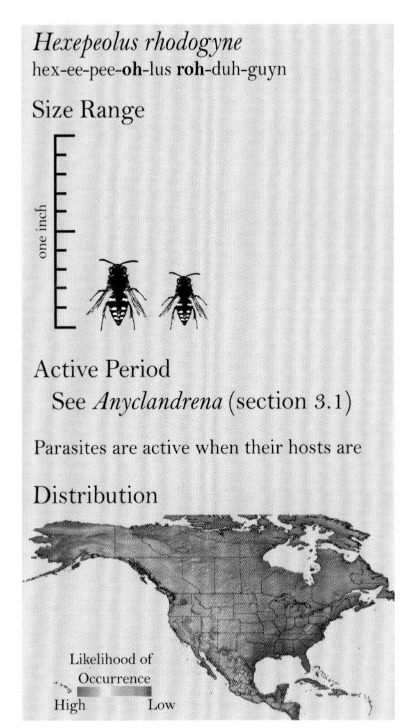

Holcopasites
hohl-koh-puh-**si**-teez
Size Range

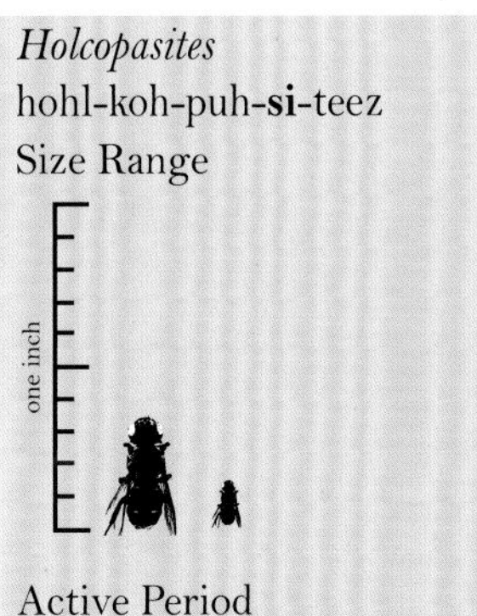

Active Period
See Panurginae (sections 3.2–3.4)

Parasites are active when their hosts are

Distribution

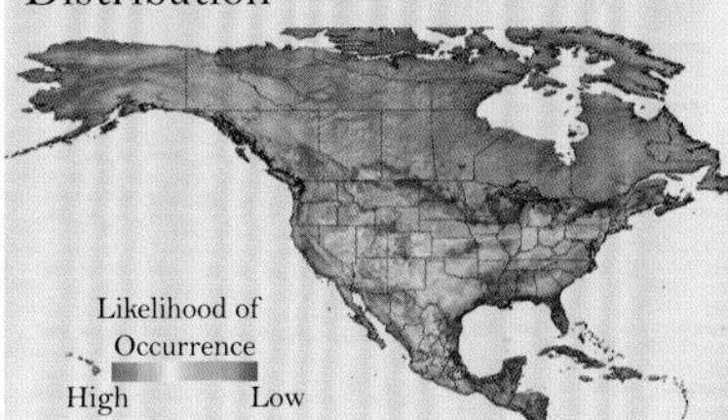

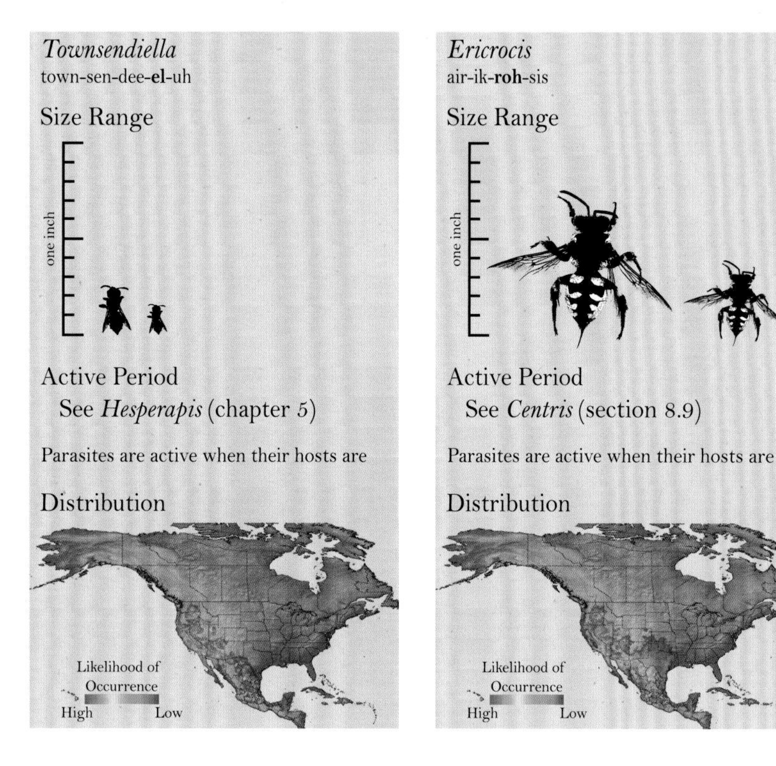

Tribe Townsendiellini

TOWNSENDIELLA

The genus Townsendiella is named after Charles Henry Tyler Townsend, an entomologist who was a professor alongside Cockerell in Las Cruces, New Mexico, before becoming director of the entomological stations of the United States and an entomological assistant in the Bureau of Entomology of the USDA. He eventually became an important entomologist in Sao Paulo, Brazil, where he died in 1944.

Townsendiella are small bees, never more than a quarter of an inch long. Three species are known, and they are all eye-catching, with rust-red abdomens and even, on some species, red antennae. They are found in the southwest, from central California east to New Mexico, and south into the dry regions of Mexico. They appear to be cleptoparasites of *Hesperapis,* with each species of *Townsendiella* attacking a different species of *Hesperapis.*

APINAE

Tribe Protepeolini

LEIOPODUS SINGULARIS

Leiopodus was named by Frederick Smith, who unfortunately did not expand on his rationale for the name he chose for this bee. The name breaks down literally to "smooth stalk or foot." The feet are not smooth, however, and one can only imagine the part of the bee to which he was referring.

Only one species of *Leiopodus* is known north of the Mexican border. This striking bee (0.25–0.5 inch) does not make it very far north, being found mainly from southern California across to western Texas. In the world, only four others have been identified, occurring from central Mexico south all the way to Argentina. *Leiopodus* is a cleptoparasite in nests of Emphorini bees (section 8.4): *Diadasia*, *Melitoma*, and *Ptilothrix*. Rather than flying randomly above the ground, *Leiopodus* will lurk near nesting sites (many Emphorini nest in aggregations), crouching on rocks or plants with wings pressed flat, abdomens low, and antennae forward and straight. One will wait until a female is seen leaving a nest and then sneak in. If this bee is seen, the host bee will often chase it down, knocking it out of the air as the small cuckoo bee attempts to flee. *Leiopodus* visit the same nests over and over, as successive nest cells are built by the host. Like other cleptoparasites, the mother inserts an egg into the wall of a nest cell that is in the process of being filled with pollen by the host mother. The parasite mother caps the tiny egg cavity with a bit of mud, and when the parasite hatches, it digs out from the wall. Using its needle-thin mandibles, it then destroys the host's egg or larva.

Tribe Ericrocidini

ERICROCIS

Ericrocis means "very woolly" in Greek.

Ericrocis are rare bees, jet black, but with striking white hair, so short it looks like a military buzz cut. These bees

Ericrosis are large, beautiful bees covered with short, velvety hair.

Ericrosis, like all cuckoo bees do not specialize on pollen. They can therefore be found on a variety of flowers, always visiting for nectar. Photo by Jillian H. Cowles.

An *Ericrosis* resting on a flower as it searches for the nests of its host, *Centris*.

are relatively large, ranging from 0.33 to 0.67 inch. These bees parasitize *Centris* and are restricted to the range of their host (the southwestern United States, the Great Plains, and occasionally the Gulf coast). *Ericrocis* break into already closed cells to lay an egg. They dig a small hole through the mud wall constructed by the host to cap a finished nest cell, then turn around and lay an egg through the hole onto the food mass inside.

Tribe Melectini

Melecta

Melecta means "honey collector" when translated from its ancient Greek roots.

Melecta are adorable bees with big, fuzzy thoraxes and sometimes with beautifully white-spotted abdomens. They look more like a typical bee than do any of the other cleptoparasites. These furry bees parasitize members

While many cuckoo bees look like wasps, *Melecta* look more like other bees. Photo by Rick Avis.

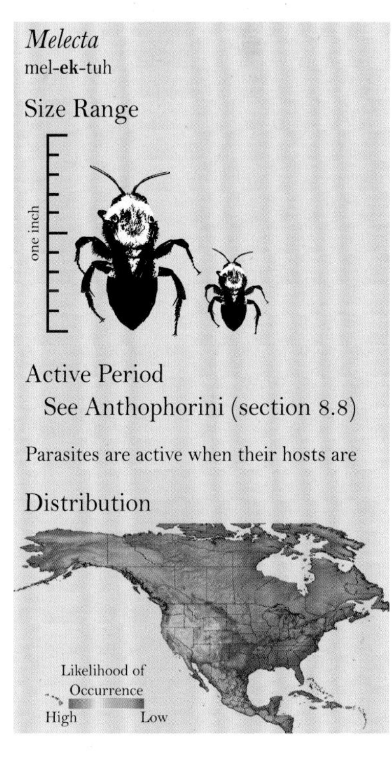

of the equally hairy Anthophorini. Anthophorini are widespread, and *Melecta* are only slightly less so, being found from British Columbia south past the Mexican border and east to the Great Plains (a few species are rarely collected along the east coast). Five species are known in the United States and Canada. Females break into closed cells and lay an egg on the cell wall inside, then reclose the cell with mud made with soil and spit.

Xeromelecta

Xeromelecta means "dry Melecta," referring to the fact that most species are found in desert environments of the United States.

Worldwide, six species of *Xeromelecta* are known; three are found only in the Caribbean, and the other three occur in the United States and Canada, making it as far east as the Great Plains. All three species parasitize Anthophorini. Many are striking, with stark white-and-black stripes on the abdomen, and noticeably thickened antennae. A few species lack the white markings, though, and are jet black with rusty brown hair on the thorax. In all cases, the thorax has many knobs and bumps.

A *Xeromelecta* drinking nectar before it searches for the nests of its hosts, *Anthophora* and *Habropoda*. Photo by Hartmut Wisch.

Zacosmia maculata

Zacosmia means "exceptional Osmia." Za- is a Greek prefix intended to emphasize the words following it.

Zacosmia are the smallest bees in the tribe Melectini at only 0.2-0.3 inch. Covered in very short white hairs, with touches of brown, these bees have an extremely domed thorax, giving them the appearance of football linebackers with substantial shoulder pads. Little but burly, these bees are found from eastern Washington and southern Alberta south through central Mexico. They parasitize one subgenus of *Anthophora*, the relatively small *Heliophila*. Only one species of *Zacosmia* is known.

Xeromelecta
zero-mel-**ek**-tuh

Size Range

Active Period
See *Anthophora* (section 8.8)

Parasites are active when their hosts are

Distribution

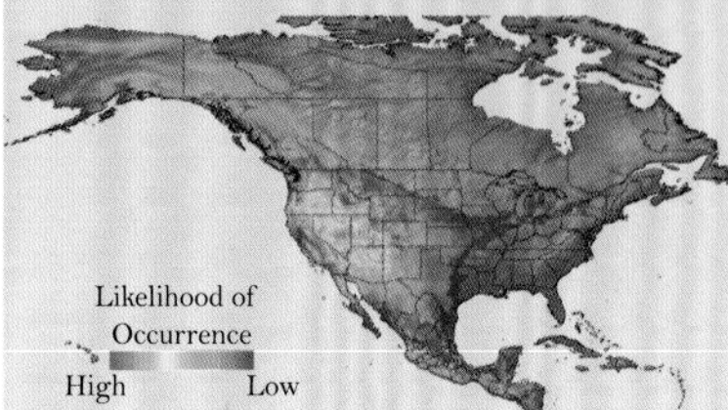

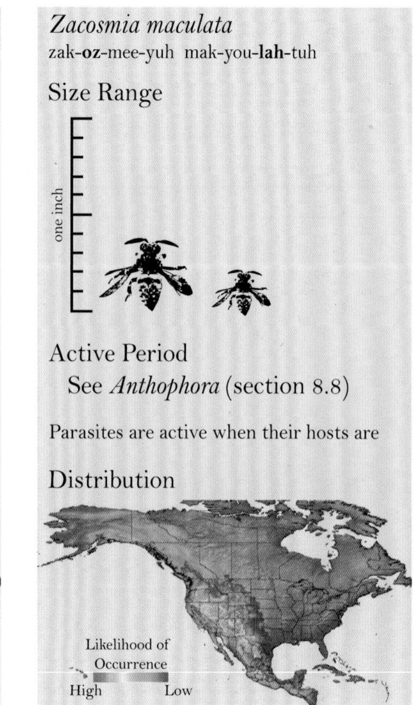

A Pioneer

While G. A. Sandhouse (1896–1940) was by no means the most prolific bee researcher of the early twentieth century, this scientist was important for another reason: she was a woman. Women scientists were exceedingly rare in the early 1900s, especially entomologists. Grace studied bees under the tutelage of Dr. Cockerell publishing the long-standing key to *Osmia* for her PhD, and identifying bees for the USDA for many years.

A *Zacosmia* drinking nectar from a composite flower. PHOTO BY HARTMUT WISCH.

Tribe Bombini

BOMBUS (PSITHYRUS)

Psithyrus means "whispering" in Greek, whereas Bombus means "humming." The name Psithyrus is perhaps a reference to the subtle and sneaky nature of these usurping bees—whispering from flower to flower instead of boldly humming.

One very unique cleptoparasite is actually a bumble bee: the cuckoo bumble bee. Belonging to the same genus as other bumble bees (*Bombus*), and looking remarkably similar to them, it is in its own subgenus: *Psithyrus.* There are 29 species of *Bombus* (*Psithyrus*), 6 of which occur in the United States and Canada. These bees do not form colonies like other bumble bees, but instead take over the colonies of nonparasitic bumble bees. A *Psithyrus* female locates a bumble bee nest, creeps into it, and either kills or subdues the current queen. She then takes her place as the egg producer for the colony. The workers protest, of course, but she quells the uprising with a combination of physical attacks and pheromones. Thus subdued, the host workers will continue to collect pollen and nectar for the growing offspring, raising parasitic bees instead of regular bumble bees. When these parasites reach adulthood, they leave the nest to find other bumble bee nests to usurp.

While the cuckoo bumble bees look much like other bumble bees, they do not collect pollen. Instead, they take over the nest of a different *Bombus* species and cause the workers to raise their parasitic offspring. PHOTO BY JAMES P. STRANGE.

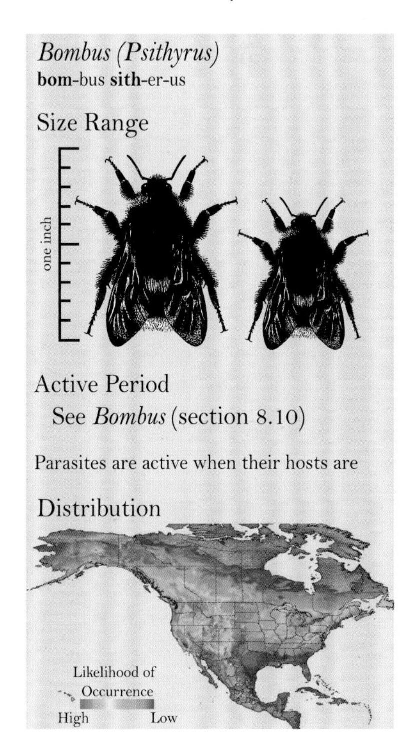

9.2 CUCKOO BEES: HALICTIDAE

Only two cleptoparasitic bee groups are placed in the Halictidae bee family. One (*Sphecodes*) is common—more common, even, than some nonparasitic bees. The other (*Temnosoma*) is rare, seldom seen except in southern Arizona (p.255).

SPHECODES

Sphecodes is Greek for “wasp-like” a reminder that distinguishing these bees from wasps sometimes takes a careful eye.

Sphecodes range in size from 0.15 to 0.6 inch. With a blood-red abdomen and shiny black thorax, this bee is easy to recognize. These cuckoo bees are found around the world, with the exception of Australia. A total of 325 species have been named around the world; about 75 of them occur in the United States and 25 in Canada. In North America they occur as far north as northern Saskatchewan and Alberta provinces, and they are even found in Alaska. They range from coast to coast. While most species parasitize other bees in the Halictidae bee family (especially *Lasioglossum*, *Halictus*, *Augochlora*, *Augochlorella*, and *Augochloropsis*), they have also been observed usurping pollen loaves in nests of *Colletes*, *Andrena*, *Calliopsis*, *Perdita*, and *Dasypoda*. Interestingly, *Sphecodes* females enter the nest of their host and kill the host eggs themselves, replacing it with one of their own. This behavior differs from that of other cleptoparasitic bees that leave their larvae to destroy the host egg. It also appears that females of some species of *Sphecodes* reside in the nests of their hosts along with their hosts. Many bees in the family Halictidae exhibit some form of sociality, where bees cooperate to build and otherwise prepare nest cells for their offspring. Finding *Sphecodes* bees in the midst of a social nest suggests that these species may be social parasites that take over the role of “queen bee” and continuously lay eggs in place of those of the host species.

LEFT: *Sphecodes* are generally fairly small and can often be seen flying low over the ground as they search for the nests of their hosts, other bees in the family Halictidae.

BELOW: A *Sphecodes* sipping nectar from a fleabane (*Erigeron*).

A *Sphecodes* taking off after visiting a flower.

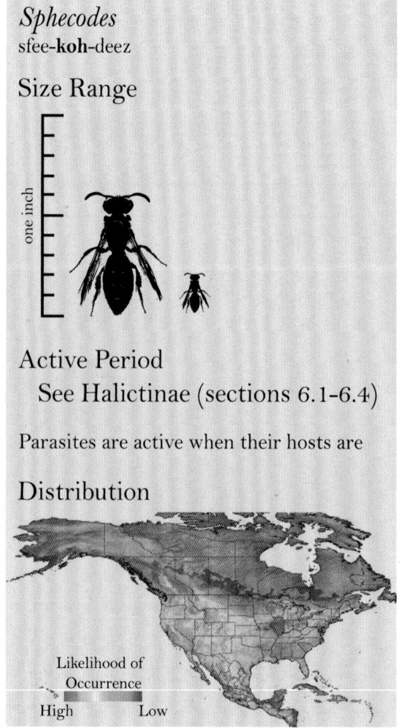

9.3 CUCKOO BEES: MEGACHILIDAE

A female *Coelioxys* on a daisy. Note the extremely pointed abdomen, seen only in *Coelioxys* and its close relative *Dioxys*.

Three genera of cleptoparasitic bees are found in the Megachilidae bee family. Unlike all the cleptoparasites mentioned in the other chapters (among the Apidae and Halictidae), the Megachilidae include bees that are cleptoparasitic on twig-nesting and wood-nesting bees.

Coelioxys

Coelioxys is Greek for "sharp belly," a reference to the characteristically pointy abdomen seen in all species.

Coelioxys is one of the most diverse of the cleptoparasitic bee genera, with over 500 species found around the world; they are especially abundant in South America. In the United States and Canada, *Coelioxys* are found all the way to the Arctic Circle and from California to Florida. North of Mexico, 45 species are known, with 10 in Canada. The majority parasitize bees in the genus *Megachile*, but some species in Europe parasitize *Anthophora*, and records also exist of *Coelioxys* in *Trachusa* and *Hoplitis* nests. Not only are they diverse, these bees are easy to identify—they have an extremely pointed abdomen, black thorax, and often red legs or abdomen. They can be large for a cleptoparasite; up to 0.75 inch long.

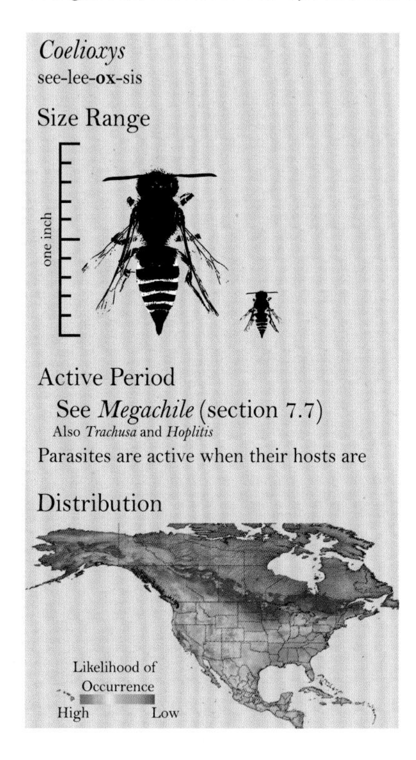

A male *Coelioxys* forages for nectar on a Russian sage flower (*Perovskia*).

A male *Coelioxys;* note the spines on the thorax and at the end of the abdomen.

A colorful *Coelioxys* on a mesquite flower (*Prosopis*). PHOTO BY JILLIAN H. COWLES.

Coelioxys use their pointy abdomen to break a hole into closed leafy nest cell walls of their hosts. They lay an egg inside, and when it hatches (which is almost immediately), it uses tweezer-sharp mandibles to snip the egg or young larva of the host bee in half.

DIOXYS

Dioxys means thoroughly sharp in Greek, a reference to the pointy abdomen.

These bees are less common than *Coelioxys*. Around 19 species have been identified in the world, including 5 north of the Mexican border, where they range from eastern Texas north to Colorado, Wyoming, and eastern Oregon. They parasitize other Megachilidae, mostly *Anthidium*, but also *Megachile* and *Osmia*. Species in Europe are not very specific in their host preferences, with one species being found in nests of at least three different host species.

STELIS

The name Stelis means "to stand," or "stand alone." In 1806, Dr. Georg Wolfgang Franz Panzer renamed this bee from Apis, which it was previously called. In his description

A *Dioxys* warming up on a rock on a cold day.

Dioxys
di-**ox**-sis

Size Range

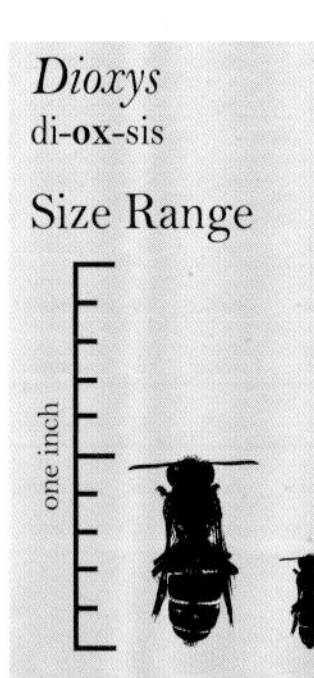

Active Period
See Megachilidae (chapter 7)

Parasites are active when their hosts are

Distribution

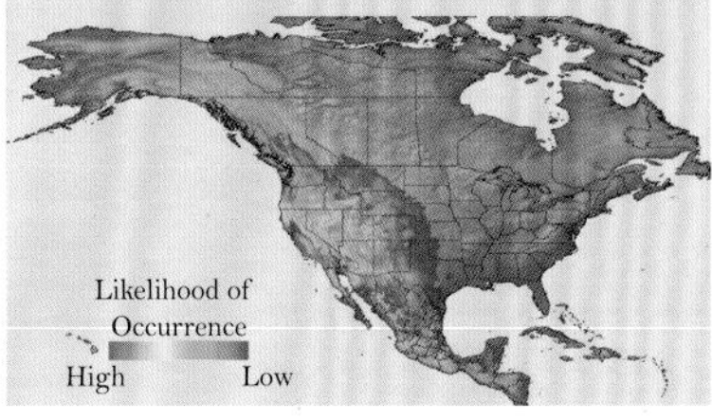

Stelis
stee-lis

Size Range

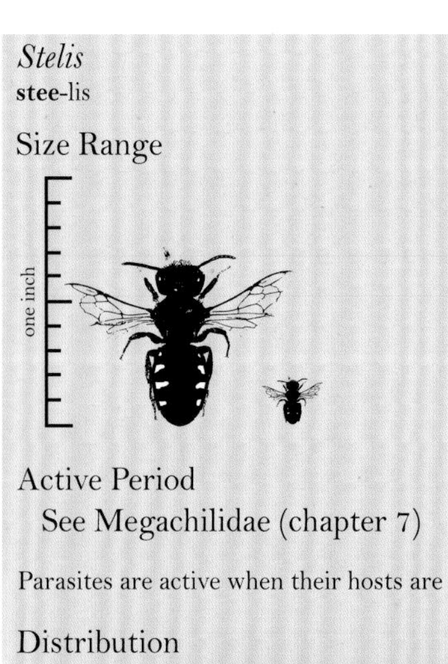

Active Period
See Megachilidae (chapter 7)

Parasites are active when their hosts are

Distribution

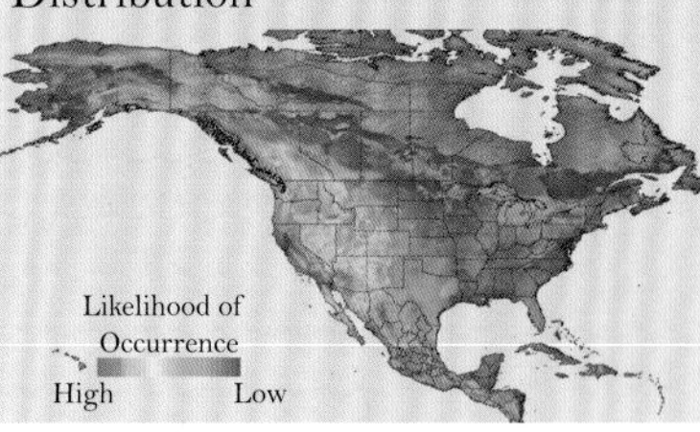

he points out that this bee can hardly be grouped with other Apis, intimating that its vestiture makes it stand out (stand alone) as a unique genus.

Stelis are beautiful bees, often metallic bluish-green, though many species are jet black. They have beautiful ivory, yellow, or white dots or stripes on their abdomens. Slightly more than 100 species can be found around the world, concentrated in the Northern Hemisphere, including 51 species in the United States and Canada. They occur well above the Arctic Circle (most bees don't), and from Maine and Florida west to British Columbia and southern California. *Stelis* parasitize a wide array of bees, mostly in the family Megachilidae. *Stelis* sneak into nests still being filled with pollen and lay an egg that, when it hatches, kills the host larva or egg.

Stelis are often metallic blue or green with white or yellow stripes on their abdomen.

A *Stelis* searching for the nearby nest entrance of a wool-carder bee (*Anthidium*; see section 7.5).

A *Stelis* resting on a desert wildflower.

Appendix

GUIDE TO THE PRONUNCIATION OF BEE NAMES

VOWELS

Short vowels are indicated by being combined with the consonant that follows

a as in cat: (**an, at, ap,** etc.)

e as in pet: (**et, ep, el,** etc.)

EXCEPTION: for **er** in the middle of a word, which is pronounced like "urge"

i as in dip: (**ip, it, id,** etc.)

NOTE: i is often followed by two consonants in a word (iss, ill), indicating a short sound. We have eliminated the second consonant to be consistent, but the sound is the same. Thus **is** is pronounced as in "kiss," **il** as in "mill"

o as in ox (**op, ot, od,** etc.)

NOTE: In scientific words what should be a short o vowel sound is sometimes pronounced as a long vowel (oh vs. ah). It seems to go either way. When the sound can be confused, we clarify with **ah**.

u as in cup (**un, up, ut**, etc.)

When preceded by a consonant or at the end of a word, followed by an h (e.g., **luh, puh, suh**)

Long vowels—these are generally not followed by a consonant, but there are exceptions: **eet**, **eez.**

a as in pane: **ay**

e as in meet: **ee**

i as in sigh: **i**

When preceded by a consonant, i maintains its long sound. Thus **si** is pronounced "sigh," **ti** is pronounced "tie," **gi** is pronounced "guy"

o as in goat: **oh**

u as in suit: **oo**

Odd Vowels

eu: at the beginning of a word, pronounced and written like **you**. Occasionally has a consonant in front; thus **syou** is said as in "I miss you"

ia: at the end of a word, written as **ee-yuh**

oi: in the middle of a word, written **oy**

er: at the beginning of a word, written as **air**

CONSONANTS

Most are as you would expect. We list exceptions below:

zh = as in "vision"

j = soft j as in "jay"

This applies to words where a g may indicate the same sound (thus *Panurgus* is pronounced "pan-er-jus"

g = a hard g, as in "guy," "gay," or "gash"

th at the end of the syllable = soft sound as in "anther" or "panther" Occasionally at the beginning of a word, as in "thrix"

k = hard k as in "mack"

z = z as in "zoo"

f = f as in "find"

INDEX

Page entries in **bold** refer to pages with illustrations.

PHOTOGRAPHIC ACKNOWLEDGMENTS

Africa Gomez 19.

Alan Chin Lee 251, 252, 253.

Alice Abela 23, 80, 91, 103, 106, 114, 237, 241, 245.

B. Seth Topham 8, 51, 54, 87, 192, 203, 205, 222, 233, 244.

Bob Peterson 21.

Dana Atkinson (www.abundantnature.com) 97, 104, 119, 137, 264.

Dick Belgers 116.

Donna K. Race 243.

Hartmut Wisch 80, 90, 114, 115, 119, 131, 136, 138, 140, 145, 146, 169, 175, 205, 210, 220, 236, 240, 263, 264, 265, 272, 273.

Jaco Visser 24.

James H. Cane 16, 18.

James P. Strange 273.

Jillian H. Cowles 23, 24, 69, 79, 84, 85, 86, 88, 89, 93, 94, 107, 109, 119, 127, 139, 141, 142, 143, 144, 150, 160, 163, 179, 181, 182, 183, 187, 188, 209, 211, 217, 225, 228, 229, 231, 241, 263, 266, 267, 269, 271, 276.

Joel Gardner 117.

Laurelin Evanhoe and Matthew Haug 18, 171.

Leslie Saul-Gershenz 15, 16, 141, 142, 237.

Lindsey E. Wilson 10, 51, 52, 56, 68, 69, 247.

Lon Brehmer and Enriqueta Flores-Guevara 92, 205, 239, 243.

Margarethe Brummermann 81, 108, 136, 228, 240, 266, 267, 268.

Michael Orr 81, 144.

Mike Wells of Harvest Lane Honey 20, 248, 249.

Rick Avis 25, 271.

Robert Klips 216.

Sonny Mencher 217.

Steve Scott 138.

Susanna M. Messinger 78, 79, 128.

Thom Wilson 216.

USDA Bee Biology and Systematics Laboratory 21, 27, 28, 29, 30, 31, 33, 34, 36, 37, 44, 45, 46, 47, 74, 99, 112, 125, 150, 152, 154, 155, 156, 157, 158, 187.

USDA W. P. Kemp 15.

USGS Bee Inventory and Monitoring Lab 45, 47, 98, 258.

All other photos are by the authors.